Methods of Experimental Physics

VOLUME 12

ASTROPHYSICS

PART C: Radio Observations

METHODS OF EXPERIMENTAL PHYSICS:

L. Marton, *Editor-in-Chief*

Clair Marton, *Assistant Editor*

Volume 12

Astrophysics

PART C: Radio Observations

Edited by

M. L. MEEKS

Massachusetts Institute of Technology
Cambridge, Massachusetts
and
Northeast Radio Observatory Corporation
Haystack Observatory
Westford, Massachusetts

1976

ACADEMIC PRESS · **New York** **San Francisco** **London**
A Subsidiary of Harcourt Brace Jovanovich, Publishers

ACADEMIC PRESS, INC.
111 Fifth Avenue, New York, New York 10003

United Kingdom Edition published by
ACADEMIC PRESS, INC. (LONDON) LTD.
24/28 Oval Road, London NW1

Library of Congress Cataloging in Publication Data   (Revised)
Main entry under title:

Astrophysics.

    (Methods of experimental physics, v. 12)
    Includes bibliographical references.
    CONTENTS: pt. A.   Optical and infrared, edited by
N. Carleton; pt. B. Radio telescopes, pt. C. Radio observation
edited by L. Marton and M. L. Meeks.
    1.   Astronomical spectroscopy.   2.   Photometry,
Astronomical.   3.   Spectrum, Infra-red.   I.   Carleton,
Nathaniel, (date)   ed.   II.   Marton, Ladislaus
Laszlo, (date)   ed.   III.   Meeks, Marion Littlejohn,
(date)   ed.   IV.   Series.
QB465.A8   522'.6   73-17150
ISBN 0–12–475953–X (v. 12, pt. C)

CONTENTS

4. Single-Antenna Observations

CONTRIBUTORS

Numbers in parentheses indicate the pages on which the authors' contributions begin.

J. A. BALL, *Center for Astrophysics, Harvard College Observatory and Smithsonian Astrophysical Observatory, Cambridge, Massachusetts* (46)

NORMAN BRENNER, *Department of Earth and Planetary Sciences, Massachusetts Institute of Technology, Cambridge, Massachusetts* (284, 296, 299, 303)

JOHN R. DICKEL, *University of Illinois Observatory, Urbana, Illinois* (1)

M. A. GORDON, *National Radio Astronomy Observatory, Green Bank, West Virginia* (277)

C. HAZARD, *Institute of Astronomy, University of Cambridge, Cambridge, England* (92)

CARL HEILES, *Astronomy Department, University of California, Berkeley, California* (58)

G. RICHARD HUGUENIN, *Department of Physics and Astronomy, University of Massachusetts, Amherst, Massachusetts* (78)

L. T. LITTLE, *Electronics Laboratory, University of Kent, Canterbury, Kent, England* (118)

J. M. MORAN, *Center for Astrophysics, Harvard College Observatory and Smithsonian Astrophysical Observatory, Cambridge, Massachusetts* (174, 228)

GUY POOLEY, *Mullard Radio Astronomy Observatory, University of Cambridge, Cambridge, England* (158)

A. E. E. ROGERS, *Haystack Observatory, Northeast Radio Observatory Corporation, Westford, Massachusetts* (139)

IRWIN I. SHAPIRO, *Department of Earth and Planetary Sciences and Department of Physics, Massachusetts Institute of Technology, Cambridge, Massachusetts* (261)

ROBERT F. C. VESSOT, *Center for Astrophysics, Harvard College Observatory and Smithsonian Astrophysical Observatory, Cambridge, Massachusetts* (198)

D. R. W. WILLIAMS, *Radio Astronomy Laboratory, University of California, Berkeley, California* (19)

G. T. WRIXON*, *Bell Telephone Laboratories, Crawford Hill Laboratory, Holmdel, New Jersey* (58)

STANLEY H. ZISK, *Haystack Observatory, Northeast Radio Observatory Corporation, Westford, Massachusetts* (296)

* Present address: Department of Electric Engineering, University College, Cork, Ireland.

FOREWORD

I have already had the pleasure to welcome to our treatise Part A of the Methods of Astrophysics, edited by Professor N. Carleton; Parts B and C, edited by Dr. M. L. Meeks, are introduced here. They contain a presentation of the methods used in radio astrophysics, a branch of physics which in a short time developed to a degree that in some respects it surpasses the present methods used in optical astrophysics.

In earlier volumes I used the opportunity offered by the Foreword to announce expected additions to this series. Professor D. Williams, who edited both the first and second editions of our Molecular Physics volume, kindly consented to edit a volume on spectroscopy. Professor G. Weissler is organizing a volume on vacuum technology methods, and discussions are underway to cover several other areas of physics.

My warmest thanks to Dr. Meeks and his authors for a remarkable and great effort in putting together these volumes. I hope the scientific community will agree with my judgment on its outstanding quality.

L. MARTON

PREFACE

Many different fields have contributed to the methods of radio astronomy, for example, antenna theory, the engineering of radiometers and digital systems, and computer programming for data analysis and display. We intend that these volumes provide a guide to the various methods currently used in radio astonomy whether the reader be an astonomer, a graduate student, a physicist working in a related field, or an engineer developing equipment for a radio telescope.

Although most of the thirty-one authors represented are radio astronomers, the others are specialists who have contributed directly to the advancement of radio astronomy through work in their own field. Interdisciplinary collaboration, particularly between electrical engineers and astronomers, has been fruitful in the development of new techniques and devices in this field. We hope that these volumes will encourage further collaboration by bringing together descriptions of the essential methods from many different points of view. The dissimilar backgrounds of the contributors have, however, presented some editorial problems. It has not proven feasible to standardize the notation throughout the volumes; for example, frequency is designated by v in some chapters and f in others.

Volumes 12B and 12C were conceived as a single-volume companion to Professor N. Carleton's Volume 12A dealing with the optical and infrared regions of the spectrum. When all chapters for the radio volume were completed and edited, the decision was made to publish the material in two volumes rather than one.

Let me take this opportunity to thank all the contributors for their efforts, cooperation, and patience in the preparation of this volume.

M. L. MEEKS

CONTENTS OF VOLUME 12, PARTS A AND B

CONTRIBUTORS TO VOLUME 12, PARTS A AND B

PART A

G. G. FAZIO, *Smithsonian Astrophysical Observatory, Cambridge, Massachusetts*

DONALD M. HUNTEN, *Kitt Peak National Observatory, Tucson, Arizona*

GERALD E. KRON, *US Naval Observatory, Flagstaff Station, Flagstaff, Arizona*

D. W. LATHAM, *Smithsonian Astrophysical Observatory, Cambridge, Massachusetts*

F. J. LOW, *Lunar and Planetary Laboratory, University of Arizona, Tucson, Arizona*

JOHN L. LOWRANCE, *Princeton University Observatory, Princeton University, Princeton, New Jersey*

G. H. RIEKE, *Lunar and Planetary Observatory, University of Arizona, Tucson, Arizona*

F. L. ROESLER, *Department of Physics, University of Wisconsin, Madison, Wisconsin*

HERBERT W. SCHNOPPER, *Department of Physics, Massachusetts Institute of Technology, Cambridge, Massachusetts*

DANIEL J. SCHROEDER, *Thompson Observatory, Beloit College, Beloit, Wisconsin*

K. SERKOWSKI, *Lunar and Planetary Laboratory, University of Arizona, Tucson, Arizona*

RODGER I. THOMPSON, *Steward Observatory, University of Arizona, Tucson, Arizona*

E. J. WAMPLER, *Lick Observatory, Board of Studies in Astronomy and Astrophysics, University of California, Santa Cruz, California*

ANDREW T. YOUNG, *Department of Physics, Texas A & M University, College Station, Texas*

PAUL ZUCCHINO, *Princeton University Observatory, Princeton University, Princeton, New Jersey*

PART B

B. F. C. COOPER, *Division of Radiophysics, Commonwealth Scientific and Industrial Research Organization, Epping, NSW, Australia*

R. K. CRANE, *Lincoln Laboratory, Massachusetts Institute of Technology, Lexington, Massachusetts*

JOCHEN EDRICH, *Department of Electrical Engineering and Denver Research Institute, University of Denver, Denver, Colorado*

TOR HAGFORS,* *Lincoln Laboratory, Massachusetts Institute of Technology, Lexington, Massachusetts*

J. C. JAMES, *Teledyne Brown Engineering, Research Park, Huntsville, Alabama*

M. L. MEEKS, *Haystack Observatory, Northeast Radio Observatory Corporation, Westford, Massachusetts*

HAYS PENFIELD, *Center for Astrophysics, Harvard College Observatory, Cambridge, Massachusetts*

R. M. PRICE, *Physics Department and Research Laboratory of Electronics, Massachusetts Institute of Technology, Cambridge, Massachusetts*

W. V. T. RUSCH, *Department of Electrical Engineering, University of Southern California, Los Angeles, California*

JOHN RUZE, *Lincoln Laboratory, Massachusetts Institute of Technology, Lexington, Massachusetts*

J. W. WATERS,† *Research Laboratory of Electronics, Massachusetts Institute of Technology, Cambridge, Massachusetts*

W. J. WELCH, *Radio Astronomy Laboratory, University of California, Berkeley, California*

R. WIELEBINSKI, *Max-Planck Institut für Radioastronomie, Bonn, German Federal Republic*

K. SIGFRID YNGVESSON, *Electrical and Computer Engineer Department, University of Massachusetts, Amherst, Massachusetts*

---

* Present address: Norges Tekniske Hogskole, Trondheim, Norway.
† Present address: Jet Propulsion Laboratory, California Institute of Technology, Pasadena, California.

4. SINGLE-ANTENNA OBSERVATIONS

4.1. Observations of Small-Diameter Sources*

4.1.1. Introduction

We define a small-diameter source as one which has an angular extent smaller than the half-power beamwidth of the telescope, and we wish to determine as many properties of such a source as possible from continuum measurements. As our starting point we have the antenna temperature T_a measured with the telescope directed at a given position in the sky with coordinates θ_0, ϕ_0. This antenna temperature is the convolution of the response of the antenna with the true distribution of brightness in the sky. This relation can be expressed by the convolution integral

$$T_a(\theta_0, \phi_0) = \frac{1}{2k} \int_{-\infty}^{\infty} \int_{-\infty}^{\infty} A_e(\theta_0 - \theta, \phi_0 - \phi) b(\theta, \phi) \, d\theta \, d\phi, \quad (4.1.1)$$

where A_e is the effective area of the antenna in the given direction, b is the unknown brightness distribution of the source, and k is Boltzmann's constant. In this equation, and throughout this section, we have assumed that b is nonzero over only a small part of the sky so that the limits of the integral can be taken as infinity.

The factor $\frac{1}{2}$ in Eq. (4.1.1) applies only to unpolarized sources and is due to the fact that an antenna can respond to only one sense of polarization at any given time. For a polarized source, the resultant antenna temperature would be different for different configurations of the feed. In general, the total flux density of a source is the sum of the signals received in two opposite senses of polarization. In practice, these are usually either two orthogonal linear planes or the two circular senses of polarization. The principles of polarimetric observations are discussed in Sections 4.2.6. and 5.1.3. We shall discuss later (Section 4.1.3) some of the practical problems in determining the polarization, but for now let us consider the properties of an unpolarized source as measured in a single polarization. The evaluation of A_e has been discussed elsewhere (Chapters 1.3 and 1.5), so we shall assume that it is known. We shall also assume in our discussion that appropriate corrections

---

* Chapter 4.1 is by John R. Dickel.

1

have been made to T_a for atmospheric absorption, etc. (see Chapter 2.3). Then measuring T_a, we can solve Eq. (4.1.1) for the unknown brightness distribution of the source b. The function obtained by direct inversion of the integral is termed the "principal solution,"[1] and, in general, this solution will not perfectly represent b because of an incomplete response of the antenna to the detailed brightness fluctuations of the source as described below.

4.1.2. Solution of the Convolution Integral

4.1.2.1. General Procedure. The brightness distribution of a radio source can be represented as a summation of Fourier components each of which is a sinusoidal temperature distribution of some amplitude and spatial frequency. This, of course, is an angular distribution, and thus the spatial frequency is the number of cycles per radian across the sky rather than the more usual time varying function. Similarly, the antenna's response pattern and the observed antenna temperature can each be represented as a Fourier series. The spatial-frequency spectra of these functions can be found from their Fourier transforms which give the amplitudes of each spatial frequency contained in the distributions. The Fourier transform of $T_a(\phi, \theta)$ is written as

$$\overline{T_a(S_\theta, S_\phi)} = \int_{-\infty}^{\infty} \int_{-\infty}^{\infty} T_a(\theta, \phi) e^{-2\pi i(\theta S_\theta + \phi S_\phi)} \, d\theta \, d\phi, \qquad (4.1.2)$$

where S_θ and S_ϕ are the spatial frequencies in each coordinate and the other symbols have the same meaning as before; similar expressions exist for the Fourier transforms of the other two functions.[2] Also, let is recall that the original function can be obtained from its Fourier transform by an additional Fourier transformation.

These Fourier transforms can be used to solve the convolution integral by application of the convolution theorem which states that the Fourier transform of the convolution of two functions is equal to the product of their Fourier transforms, or, if Eq. (4.1.1) holds, then

$$\overline{T_a(S_\theta, S_\phi)} = \frac{1}{2k} \overline{b(S_\theta, S_\phi)} \, \overline{A_e(S_\theta, S_\phi)}. \qquad (4.1.3)$$

Therefore, after measuring T_a and A_e, we can take their Fourier transforms and substitute into Eq. (4.1.3) to obtain the Fourier transform of b from which the desired function b itself can be obtained by a final transformation.

[1] R. N. Bracewell and J. A. Roberts, *Aust. J. Phys.* **7**, 615 (1954).

[2] The Fourier transform and Fourier analysis in general are discussed in many mathematical textbooks. One good reference is R. N. Bracewell, "The Fourier Transform and Its Applications." McGraw-Hill, New York. 1965.

Note that Eq. (4.1.3) is an algebraic product so that any spatial frequency for which the response of the antenna is zero will not be found in T_a and will thus be missing in the derived b. This becomes serious in practice because the response patterns of all antennas are diffraction patterns which have zero response to some spatial frequencies. In particular, the finite size of an antenna creates a cutoff above which no higher spatial frequencies can be recorded. The value of this cutoff frequency depends upon the shape of the antenna aperture but is approximately equal to 1/half-power beamwidth.[3] It is not possible to improve this limit by taking running means of closely spaced samples. Bracewell and Roberts have shown that "an observed distribution is completely determined by measurements spaced at equal discrete intervals which are at least as narrow as $\frac{1}{2}\phi_c = \frac{1}{2}S_c^{-1}$" where S_c is the cutoff spatial frequency and ϕ_c is an angular interval. In other words, we obtain no new information by sampling the data at intervals closer than $\frac{1}{2}$ beamwidth, and thus for a source which is smaller than the beamwidth we can obtain only three unique data in any one coordinate or a total of five data for the source. (The sample at the peak position is a duplicate in both coordinates.) The information generally obtained is the position and extent of the source in each coordinate and the antenna temperature at the position of maximum emission.

An important quantity describing a radio source is the flux density given by

$$S_f = \iint_{\text{source}} b(\theta, \phi) \, d\theta \, d\phi. \tag{4.1.4}$$

To relate this to the observable quantity, T_a, let us integrate the observed distribution over the sky:

$$\int_{-\infty}^{\infty} \int_{-\infty}^{\infty} T_a(\theta_0, \phi_0) \, d\theta_0 \, d\phi_0$$

$$= \frac{1}{2k} \int_{-\infty}^{\infty} \int_{-\infty}^{\infty} d\theta_0 \, d\phi_0 \int_{-\infty}^{\infty} \int_{-\infty}^{\infty} A_e(\theta_0 - \theta, \phi_0 - \phi)b(\theta, \phi) \, d\theta \, d\phi. \tag{4.1.5}$$

The integration over θ_0, ϕ_0 on the right-hand side of the equation is a function only of the antenna's response and so, in principle, can be evaluated. The process is difficult in detail, but the result can be represented by a constant C for any given antenna. Thus we obtain

$$\int_{-\infty}^{\infty} \int_{-\infty}^{\infty} T_a(\theta_0, \phi_0) \, d\theta_0 \, d\phi_0 = \frac{1}{2k} \int_{-\infty}^{\infty} Cb(\theta, \phi) \, d\theta \, d\phi, \tag{4.1.6}$$

[3] E. G. S. Silver (ed.) *in* "Microwave Antenna Theory and Design." McGraw-Hill, New York, 1949.

and then substitution of Eq. (4.1.4) into Eq. (4.1.6) gives

$$S_f = \text{const} \int_{-\infty}^{\infty} \int_{-\infty}^{\infty} T_a(\theta_0, \phi_0) \, d\theta_0 \, d\phi_0 \,. \qquad (4.1.7)$$

In practice, the integration is usually terminated a few beamwidths from the center of the source and the value of const is determined empirically by integration of the observed distribution for sources of known flux density (see Chapter 1.5). This gives us a normalization factor with which to determine the flux density of any source by integrating the volume under the observed T_a-curve.

This is the strictly correct method to measure the flux density and certainly the most accurate one. But often in practice it is simpler and faster to measure only the peak antenna temperature of the source and then apply a correction for the size of the source. This correction factor can be defined by combination of Eqs. (4.1.1) and (4.1.4):

$$S_f = 2kT_a(\theta_0, \phi_0) \, \frac{\iint_{\text{source}} b(\theta, \phi) \, d\theta \, d\phi}{\iint_{\text{source}} A_e(\theta_0 = \theta, \phi_0 - \phi) b(\theta, \phi) \, d\theta \, d\phi}$$

$$= \frac{2kT_a(\theta_0, \phi_0)}{A_e(\theta_0, \phi_0)} \, F, \qquad (4.1.8)$$

where F is the desired factor. Thus, in order to determine the flux density we need additional information about the distribution of the brightness of the source. Also, to evaluate the actual surface brightness of a small-diameter source we must know its size and shape as well as its flux density. Therefore, with the limit of only five data available from the observations as discussed above, we must rely on additional independent information.

4.1.2.2. Representations for Source Shape. If higher resolution studies are available to provide the true shape of a source, then its parameters can be found by direct solution of Eq. (4.1.8) either analytically or numerically depending upon the nature of the function. In many cases, however, little information is available and various approximations must be made.

The response pattern of most antennas can be well approximated by a Gaussian distribution of the form

$$A_e(\theta, \phi) = A_e(\theta_0, \phi_0) \exp\left\{-\left[\frac{(\theta - \theta_0)^2}{0.36W_\theta^2} + \frac{(\phi - \phi_0)^2}{0.36W_\phi^2}\right]\right\}, \qquad (4.1.9)$$

where the W's are the half-power beamwidths and the other symbols have the same meaning as before. As a simplest approximation, we assume that the brightness distribution of the source can be represented by a similar function, and recalling that the Fourier transform of a Gaussian is another Gaussian, we see that solution of the convolution integral is very straightforward—it can readily be shown that the relation for the half-power width of the source in a given coordinate becomes

$$W_S = (W_R{}^2 - W_A{}^2)^{1/2}, \qquad (4.1.10)$$

where the subscripts S, R, and A stand for the source, observed response, and antenna, respectively. The correction for the size of the source (Eq. (4.1.8)) becomes

$$F = [1 + (W_{S\theta}^2/W_{A\theta}^2)]^{1/2}[1 + (W_{S\phi}^2/W_{A\phi}^2)]^{1/2} \qquad (4.1.11a)$$

or

$$F = \frac{W_{R\theta}}{W_{A\theta}} \frac{W_{R\phi}}{W_{A\phi}}. \qquad (4.1.11b)$$

The accuracy of the correction depends critically upon the true shape of the radio source, but the Gaussian approximation appears reasonable for many galaxies and other sources with a central condensation. Another possible representation for such a source is a straight exponential function similar to that used by Menon[4] in an analysis of the Orion Nebula.

Many sources such as the planets and some elliptical galaxies appear as uniformly bright disks or ellipses in the sky, and if a function of this form is convolved with the Gaussian beam pattern, the correction factor F has the form

$$F = a/(1 - e^{-a}), \qquad (4.1.12)$$

where $a = 0.693(W_{S\theta}W_{S\phi}/W_{A\theta}W_{A\phi})$.[5] A comparison of Eqs. (4.1.11) and (4.1.12) shows that the correction is larger for a Gaussian source than for a uniform one, with the magnitude of the difference depending upon the size of the source. Also, the diameter of a disk-shaped source is larger than the half-power width of a Gaussian source with the same observed response width between half-power points. For example, a source with an observed response width in each coordinate of $W_R = 1.41 W_A$ has an apparent width $W_S = W_A$ for a Gaussian representation but $1.06 W_A$ for a disk. The corresponding correction factors F are 2.00 and 1.42, respectively.

In some cases it is possible to approximate a compound source by a combination of two or more simple components. Some "core-halo" galaxies, for example, can be considered as two concentric Gaussians with different widths and central intensities. The emission from Jupiter is a combination of thermal emission from the planetary disk plus a more extended elliptically shaped component from the planet's radiation belts. For all these symmetri-

[4] T. K. Menon, *Publ. Nat. Radio Astron. Observ.* **1**, 1 (1961).
[5] C. M. Wade (unpublished), quoted by D. S. Heeschen, *Astrophys. J.* **133**, 322 (1961).

cal sources the position can be given unambiguously as that of the peak, which is, of course, also the position of the centroid of the observed distribution.

In the case of nonsymmetrical and other complex sources, the factors must be evaluated numerically and the parameters of any source of unknown structure must be treated with extreme caution. For example, Fig. 1a shows

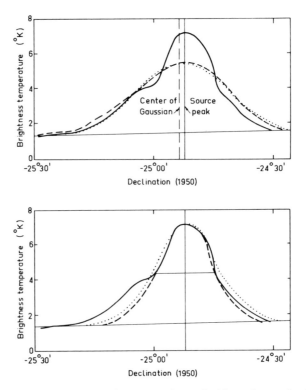

FIG. 1. (a) The record of a complex source observed with a telescope having the same half-power width as the source, where (—) represents the true distribution, (– –) the convolved distribution, and (·) the equivalent Gaussian. (b) Two different Gaussian representations of the complex source, where (—) represents the true distribution, (– –) the Gaussian with the same width as the source, and (·) the Gaussian with the same area under the curve as the source.

the result of convolving a complex source with a Gaussian beam having the same half-power width as the source. (The "true distribution" in this case is an actual scan across the supernova remnant A4 made by D. K. Milne and the author using the 210-ft telescope at Parkes with a beamwidth of 8 arcmin at a wavelength of 11 cm.) Also shown is the equivalent Gaussian having the same width as the observed response. The integrated flux density for the Gaussian representation is 0.97 times the true value, which is probably acceptable but for the fact that the indicated position of the peak of the source is shifted by about 1/10 beamwidth. Thus, without prior knowledge of the structure of the source, it would be difficult to make an optical identification or similar analysis of the data. There is no standard definition of the position of a complex source, and so whenever a position is given it should be clearly stated whether it is that of the centroid or the observed peak.

Let us also use Fig. 1 to demonstrate the danger in using mean values for a source rather than the actual distribution of brightness. Using the Gaussian fit of Fig. 1a, and assuming that the source is also a Gaussian, we can solve Eq. (4.1.10) for the apparent width of the source. This is shown in Fig. 1b as the distribution with the same area under the curve as the source. Such a representation has a width between half-power points which is too large by a factor of 1.15. Alternatively, the Gaussian distribution with the correct width has an integrated flux density which is only 0.87 times the true value.

The above illustrations clearly show the difficulty in comparing measurements made with different resolutions or analyzed with different approximations. Whenever data are collected to determine the spectrum of a source, care must be taken to ensure that comparable beamwidths and the same assumptions about the structure of the source were used at all frequencies. The absolute flux densities may still be wrong but the shape of the spectrum should be correct.

4.1.3. Observational Techniques

4.1.3.1. Total Intensity Measurements. In practice, any set of observations of a source must include at least six measurements. As well as obtaining the five parameters discussed above, we must measure the level of the background emission at the position of the source. This background level can change across the sky because of spurious antenna effects, such as spillover, and variations in the general emission from our galaxy (note the sloped baseline in Fig. 1 from this cause). Often, measurements at one or two points near the source are sufficient, but in some complicated regions it is necessary to establish the background level by interpolation between a number of points surrounding the source.

There are two general techniques for obtaining the desired six data: the slow scan or drift and the on–off procedure. In the scanning technique, the data are usually sampled rapidly as the telescope is scanned across the source. With steerable telescopes, scans are made in each coordinate; with transit instruments, a series of drift scans spaced at $\frac{1}{2}$ beamwidth intervals provide

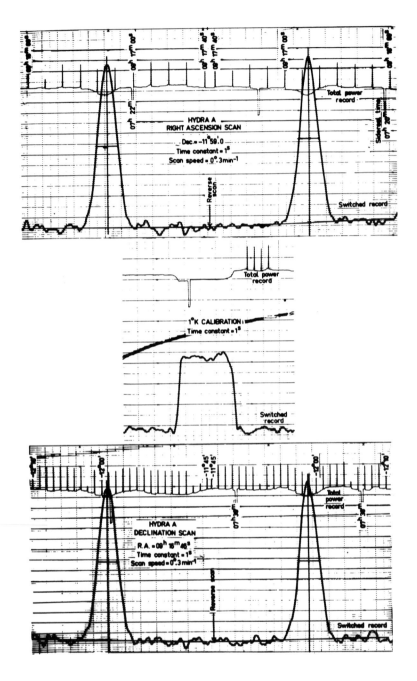

Fig. 2. Scans across the source Hydra A made at a wavelength of 3.4 cm with the 64-m (210-ft) telescope at Parkes. The half-power response width and the position of the source are also sketched.

8

equivalent information. Figure 2 shows an example of a scan through the source Hydra A made at a wavelength of 3.4 cm with the 210-ft telescope at Parkes by D. K. Milne, A. R. Kerr, and the author. The forward and reverse scans provide an easy way to correct for any systematic variation of telescope pointing with the direction of the scan.

The scan should be long enough to reach the background on each end so that a baseline can be interpolated, as shown in Fig. 2, and the intensities of the source measured above it. For a Gaussian distribution, for example, the intensity never falls to zero but reaches 1 percent of the central value at a distance of 1.3 beamwidths from the peak. By then, the signal is generally lost in the noise anyway so that a scan of total length 4–5 beamwidths is usually adequate.

Corrections must also be made when scanning for the finite time constant of the recording device which will not only delay the apparent response to a source but also skew and smooth the observed distribution. As an example of these corrections, Howard,[6] assuming a Gaussian distribution and an RC time constant, has constructed graphs of the corrections as a function of *response width/scan rate · time constant* to be applied to the observed source position, intensity, and width. With the parameters in Fig. 2, the response is not significantly distorted, but it can be seen that the indicated peak of the source is delayed by nearly a time constant in agreement with theoretical analysis.

To reduce the data, the desired response pattern is fit to the data by standard least squares techniques. An example of a computer program to perform such a fit for a Gaussian has been described by von Hoerner.[7] The general procedure is to guess approximate values for the parameters and then minimize the mean square differences between the true and calculated distributions using an iterative technique. The errors in the parameters of sources determined from such fits are usually tedious (but straightforward)˙ to derive from the normal equations. The details, of course, depend upon the distributions assumed, but as a general rule of thumb, the uncertainties in position and size of a source (relative to the beamwidth) are about equal to the fractional uncertainty in the flux density as given by the ratio of the noise level to the intensity of the source. Thus, a source which had a peak flux density of five times the rms noise level would have an uncertainty in position and size of about 0.2 beamwidth.

If the position of the source is known from previous observations and the telescope can be pointed accurately, the on–off technique provides a somewhat more efficient means of obtaining data on a compact source. The telescope is

[6] W. E. Howard, III, *Astron. J.* **66**, 521 (1961).
[7] S. von Hoerner, *Astrophys. J.* **147**, 467 (1967).

set to track successively, for a specified integration time, each of five positions on the source plus baseline points. The advantage of this scheme is that time is not spent observing the source near the edge of the beam pattern where the signal-to-noise ratio is low.

For a Gaussian, the steepest gradient in the brightness distribution occurs at the half-power points, and so these are the most critical positions for the determination of the source size and position. In this region, a change in position by 0.0725 beamwidth will result in a 10% change in intensity so that, for even a moderately weak source, the accuracy quickly becomes limited by telescope parameters and fluctuations in the background. To measure the source intensity it is, of course, best to center the antenna on the source. Here, the pointing accuracy of the telescope becomes of prime importance because a displacement by one-tenth of a beamwidth will result in a decrease in the source intensity to 0.973 of the peak value. The problem becomes particularly acute at high frequencies where the beamwidths are small and the corrections for atmospheric refraction, etc. are relatively large.

For special applications, where some of the parameters of the source are known, it is possible to considerably shorten the observational program. For example, to monitor known quasars for possible variability it is sufficient merely to switch back and forth between the source and a single position nearby. Because the polarization of these sources is known to vary, two measurements at orthogonal positions of the feed must actually be made to record the total flux density This process is repeated at time intervals shorter than the possible period of variability. The only other information needed is an accurate calibration. Observations of the planets also entail special considerations. We not only know their positions and angular sizes, but the planets move across the sky background. It is therefore possible to measure the flux density of the planet plus background relative to the flux density at a reference position. Then later, after the planet has moved away, the background alone can be compared with the reference, and the difference is the flux density of the planet alone. Because the background is eliminated, the result is limited entirely by the system noise and instrumental calibration. The on–off observations of Venus[8] in Fig. 3 demonstrate the variation in the background. This was subsequently removed with the aid of later observations of the same area of the sky.

4.1.3.2. Polarization Measurements.   Part of a polarization observation of Venus (which is unpolarized) taken during the same period as the observations shown in Fig. 3, and using the same on–off procedure, is illustrated in Fig. 4. In this situation, the receiver was Dicke switched between two orthogonal concentric feeds in order to obtain directly one of the polarized intensities (the Stokes parameter S_1, sometimes referred to as Q). The

[8] J. R. Dickel, *Icarus* **6**, 417 (1967).

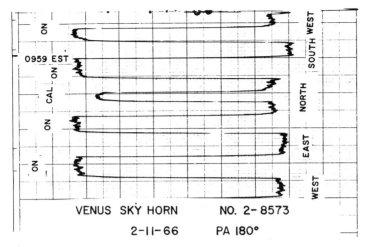

FIG. 3. An on–off measurement of the planet Venus made at a wavelength of 6 cm with the 42.7-m (140-ft) telescope of the NRAO [from J. R. Dickel, *Icarus* **6**, 417 (1967)].

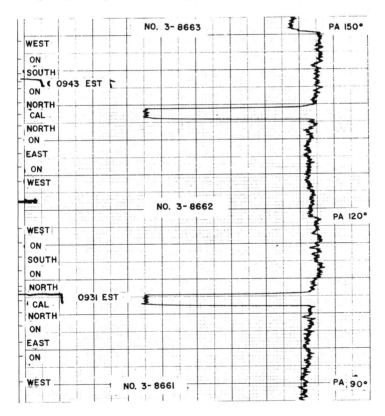

FIG. 4. An on–off polarization observation of Venus made at a wavelength of 6 cm with the 42.7-m (140-ft) telescope of the NRAO. The radiometer was switched between two orthogonal feed elements [J. R. Dickel, *Icarus* **6**, 417 (1967)].

11

record shown is the difference output from the switched radiometer and represents the signal received in the feed at the indicated position angle PA minus that in the feed at $PA + 90°$. As well as a drift in the background, an instrumental polarization of 0.5% at a position angle of $180°$ is detectable on the records. The instrumental effect is apparently caused by distortions of the dish surface which slightly alter the amount of power reflected at different position angles. That there is not a different response in the two feed systems can be verified by making switched-polarization observations at two indicated position angles separated by 90%. In the absence of any different response of the feeds the two resultant outputs should have the same magnitude (note that they will have opposite sign, however, because we always subtract beam 2 from beam 1), and any difference can be attributed to one feed element having a stronger response than the other. Thus, the feed effects can be measured and compensated for in the observation of a single source, but the dish effects must be found from repeated observations of various standard sources whose polarization is known.

Variations in the background polarization from both the sky and spurious ground pickup will seriously affect any polarization measurements. Very often the latter effect is the most significant, and so it would be extremely valuable to measure completely the polarization of a source with the telescope at a fixed position with respect to the ground. This cannot be done by a series of on–off measurements at fixed position angles, and so another method, illustrated in Fig. 5, has been developed to closely approach the desired condition. This is to continuously rotate the polarization feed while tracking a given point in the sky west of the source, then returning to the initial hour angle and repeating the rotation while tracking the source, and finally, repeating the process at a third point equidistant to the east of the source. The polarized signal will have a sinusoidal variation with a period which is $\frac{1}{2}$ the rotation period of the polarizer, and subtraction of the average off-source data from the on-source data will give the polarized signal from the source. The rotation rate of the polarizer should be rapid enough so that the background variation is negligible during one period. In the example, the large spurious polarization of the background is clearly seen and, in this case, is much stronger than the source polarization.

4.1.4. Limits to the Accuracy of an Observation

As radiometer systems continue to improve, it is becoming possible to reach a given sensitivity in shorter and shorter time so that the final accuracy of a measurement is generally limited by factors other than receiver noise. One of these is the increase in system noise caused by the cosmic background itself. At low frequencies the temperature of the sky can reach several

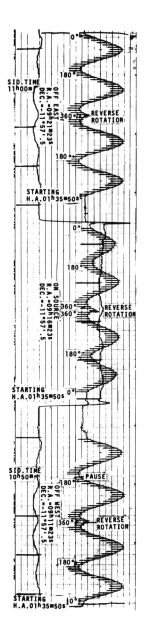

FIG. 5. A polarization observation of Hydra A at a wavelength of 11 cm made with the 64-m (210-ft) telescope at Parkes. Both the switched record (between two orthogonal beams) and the total power record of the signal in one beam are shown.

thousand degrees, and thus the noise fluctuations in the receiver will be increased by this amount. There is no way to remove this effect, but because it is identical to the receiver noise the fluctuations can always be reduced to the required level by sufficiently long integration times.

Inaccuracies in telescope pointing are often the greatest cause of uncertainty in both the position and size of a source. These have been discussed in Chapter 1.5 and will not be considered further. Another serious source of uncertainty with large antennas at high frequencies is temperature gradients across the reflector which distort the paraboloidal shape causing changes in the gain. Use of thermal paints and the enclosing of the antenna in a radome or other controlled environment can help to reduce these effects.

At frequencies greater than about 3000 MHz, an important source of error is the variation in the recorded signal caused by atmospheric fluctuations. This can be largely compensated for by using a dual-beam technique.[9] Two feed elements are placed close together on opposite sides of the focal point of the telescope, and the system switched between these two feeds and synchronously detected. The separation is such that the two resultant beams are only a few beamwidths apart. In the near field of the telescope where the radiation patterns are nearly uniform cylinders, the two beams will overlap almost completely; where they start to diverge significantly because of diffraction depends upon the illumination patterns of the antenna, but it is approximately the Rayleigh distance $D^2/2\lambda$ where D is the diameter of the telescope and λ is the wavelength. Most of the turbulent eddies which cause the fluctuations occur within a few miles of the earth's surface (see Chapter 2.2), well within the near field of most telescopes, and thus generally appear in both beams. The fluctuations caused by these common elements will be cancelled by the switching process, leaving only the contribution from the very small or distant cells which do not lie within both beams. An example of the improvement obtained by this technique is shown in Fig. 6 taken from Baars.[9] Note that even in clear weather there is an improvement in the noise by a factor of two using the dual beam (DB) technique over the single beam (SB) switched versus a load.

An additional refinement of the dual-beam technique involves positioning the two beams alternately on the source. First, the positive-responding beam is positioned on the source while the negative-responding one points to the sky background on one side (for example, let us say east of the source). Next, the negative beam is directed at the source while the positive one points to the sky background on the opposite side of the source (west, in our example). Then, if the switched radiometer outputs at the two positions are subtracted, we can represent the net signal by the equations

[9] J. W. M. Baars, Ph.D. Dissertation, Delft Technol. Univ. (1970).

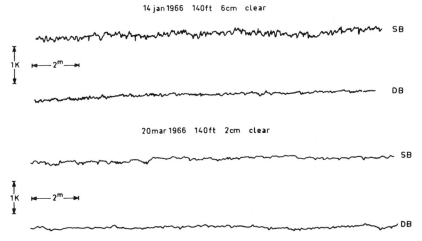

FIG. 6. Sample radiometer outputs at two wavelengths using the 42.7-m (140-ft) telescope at Green Bank to compare the dual beam (DB) technique with that of switching a single beam versus a comparison load (SB) [from J. W. M. Baars, Ph.D. Dissertation, Delft. Technol. Univ. (1970)]. The time and intensity scales are shown for both records.

$$\text{signal} = \begin{bmatrix} \text{first position} \\ (\text{source} + \text{background}) \\ - \text{background east} \end{bmatrix} - \begin{bmatrix} \text{second position} \\ \text{background west} \\ - (\text{source} + \text{background}) \end{bmatrix}$$

$$\text{(4.1.13a)}$$

$$\text{signal} = 2 \cdot \text{source} + 2 \cdot \text{background} - \begin{bmatrix} \text{background east} \\ + \text{background west} \end{bmatrix}$$

$$\text{(4.1.13b)}$$

Then, if the variation in the background is linear, we can set

$$\text{background} = \frac{\text{background east} + \text{background west}}{2}$$

and the net result is

$$\text{signal} = 2 \cdot \text{source}. \qquad \text{(4.1.13c)}$$

Thus, because one beam is always on-source, the signal-to-noise ratio is twice that of an ordinary on–off measurement. This technique can also be used with the scanning procedure by directing the scan along the axis between the two beams. The recorded signal will first deflect one direction and then the other, with the maximum difference between twice the single-beam antenna temperature as shown in Fig. 7 from Batchelor et al.[10]

[10] R. A. Batchelor, J. W. Brooks, and B. F. C. Cooper, IEEE Trans. Antennas Propag. 16, 228 (1968).

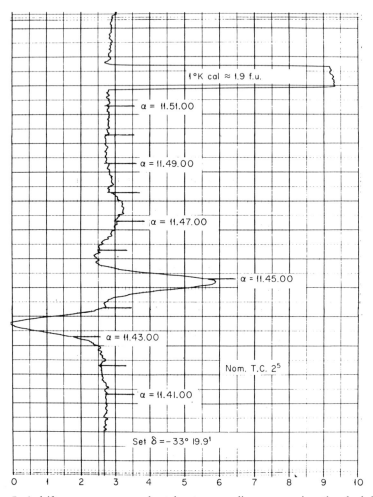

FIG. 7. A drift scan across a moderately strong radio source using the dual beam techniques. The telescope was fixed in declination (δ), and the right ascension (α) increased with time. Note that a weaker source is also present about $2\frac{1}{2}$ min east of the main one. This record was made at a wavelength of 11 cm with the 64-m (210-ft) telescope at Parkes by R. A. Batchelor, J. W. Brooks, and B. F. C. Cooper [*IEEE Trans. Antennas Propag.* **16**, 228 (1968)].

Even if we have a good signal-to-noise ratio, accurate calibrations, etc., there is still an ultimate limit to the accuracy of any observations caused by "confusion noise" in the integrated signal from very weak sources within the beam. Let the number of sources N with the flux densities greater than a given value S be given by the expression:

$$N = \text{const } S^x. \qquad (4.1.14)$$

The value of the index x depends upon the model of the universe adopted and is still uncertain. The constant value of -1.8 found empirically[11] seems reasonable except for sources of very weak flux density. Now, we wish to observe sources with flux densities greater than some limiting value S_{lim} for which there is an average of n sources greater than this flux density within one solid angle subtended by the beam. Then, the average number of sources within the beam with flux densities greater than some value S is

$$N = n\left(\frac{S}{S_{lim}}\right)^x, \qquad (4.1.15)$$

or the average number of sources within the beam with flux densities between S and $S + dS$ is

$$dN = nx\left(\frac{S^{x-1}}{S_{lim}^x}\right) dS. \qquad (4.1.16)$$

Assuming that the sources are distributed randomly, we should find that their number follows a Poisson distribution with a standard deviation of $(dN)^{1/2}$. The contribution of these sources to the background radiation will therefore have a standard deviation of $S(dN)^{1/2}$, or the total standard deviation of the background from all sources weaker than S_{lim} is

$$\Delta S = \left(\int_0^{N_{lim}} S^2\, dN\right)^{1/2} = \left(nxS_{lim}^{-x}\int_0^{S_{lim}} S^{x+1}\, dS\right)^{1/2}. \qquad (4.1.17)$$

This "noise" is always present, and the only means of improving the situation is to build larger antennas with their correspondingly smaller beam-widths and thus smaller values of n.

The evaluation of the integral in Eq. (4.1.17) is complicated because we do not know the exact value of x for the strong sources, and we have even less information on the distribution of the very weakest ones. Assuming that the detection of a given source is valid if its flux density is five times the standard deviation of the noise, von Hoerner[12] has calculated the number of sources which can be detected for some different models of the universe. He concludes that one source for every 75 beam areas should be safe for reliable detection.

The value for the minimum detectable signal of five times the rms noise level adopted by von Hoerner is a conservative limit. The noise fluctuations are, of course, statistical, but they should have a normal distribution. If so, there is only one chance in 500 that a point will deviate from the average by more than three times the standard deviation. This value seems sufficient for most applications.

[11] M. Ryle, *in* "Radio Astronomy Today" (H. P. Palmer, R. D. Davies, and M. I. Large, eds.), Chapter 18. Harvard Univ. Press, Cambridge, Massachusetts, 1963.
[12] S. von Hoerner, *Publ. Nat. Radio Astron. Observ.* **1**, 19 (1961).

The confusion noise can also be determined empirically by measuring the noise level of a series of records at a given position in the sky, The confusion noise remains constant no matter how many records are averaged together, whereas the contribution from the receiver noise will decrease as the square root of the observing time, or number of records averaged. If we measure the total noise TN for an average of m records at a given position, then we can solve the simultaneous equations

$$TN_1^2 = RN^2 + CN^2 \quad \text{and} \quad TN_m^2 = (1/m)RN^2 + CN^2 \quad (4.1.18)$$

for both the receiver noise RN and the confusion noise CN. Several groups have done this (e.g., Illinois[13] and Ohio State[14]), and it appears that a value of about 1 source per 25 beam areas is sufficient to give a chance of 1 in 20 that a source at the minimum detectable level of three times the rms noise level is fictitious. (In the literature on this subject it is often common to refer to the reciprocal of n or $1/n = \mu = 25$ beam areas per source.) It is also possible to use these empirical values of n in Eq. (4.1.17) to put some constraints upon acceptable models of the universe.[15, 16]

[13] J. R. Dickel, K. S. Yang, G. C. McVittie, and G. W. Swenson, Jr., *Astron. J.* **72**, 757 (1967).
[14] J. D. Kraus, R. S. Dixon, and R. A. Fisher, *Astrophys. J.* **144**, 559 (1966).
[15] R. Ringenberg, Ph.D. Dissertation, Univ. of Illinois (1969) (unpublished).
[16] A. Hewish, *Mon. Not. Roy. Astron. Soc.* **123**, 167 (1961).

4.2. Fundamentals of Spectral-Line Measurements*

4.2.1. Brief Historical Survey of Early Spectral-Line Measurements

The study of spectral lines in radio astronomy began in 1951 with the detection of neutral hydrogen from interstellar space at a wavelength of 21 cm. Even after the discovery of many other atomic and molecular lines in the years since 1951, the 21-cm line continues to be studied at many observatories, principally because of its ubiquitous nature in the galaxy and in external galaxies, and also because of the wealth of information contained in the velocity structure of the line.

All of the early work on the 21-cm line involved single-channel scanning-type receivers used in conjunction with the largest available single-dish antennas that would work at that wavelength. The typical superheterodyne receiver that was used employed a crystal mixer front end, a narrow-band i.f. (5–40 kHz), and a long output time constant (10–100 sec). The output was recorded directly on a strip chart. Because of the weakness of the line ($\sim 100°K$) compared to the receiver noise level of several thousand degrees, frequency switching was used from the start. The narrow-band channel was switched on and off the spectral region of interest, and the wanted difference signal was phase detected in a later portion of the receiver. In this process, the frequency of the local oscillator was required to be carefully stabilized, yet it had to be tunable through an appropriate region of spectrum (1–2 MHz) containing the spectral line. With the signal-to-noise ratios then available, it took one or more hours to scan the narrow-band filter through the line and thus to delineate the line profile on the strip chart. To provide a means of measuring accurate velocities for the peaks in the profiles, elaborate systems were devised to put frequency markers on the strip chart along with the profile. From the very beginning, spectral-line measurements have posed several important requirements:

  (i)   good pointing accuracy of the telescope;
 (ii)   excellent gain stability (as in continuum receivers);
(iii)   great frequency stability in the local oscillators.

Until recently, the use of single paraboloid antennas has continued to dominate the field of spectral-line measurements, although the use of 2-

---

* Chapter 4.2 is by D. R. W. Williams.

element interferometers is now common in both short baseline and very long-baseline (VLBI) work. For several reasons, however, it would appear that single dishes will continue to be useful both in line search and in survey work:

(i) A pencil-beam instrument is convenient for studying extended source galactic objects both in emission and absorption.

(ii) The single dish is simplest to use in the observations of lines that are widely distributed in angle in the galaxy, as well as for observing dust clouds, and it has been used in the study of nearby external galaxies.

(iii) With a single dish, only one front-end, a costly item, is required for each frequency range. Change from one frequency to another is relatively easy, and operation at several frequencies simultaneously is readily achieved.

(iv) The versatility of the single dish is especially apparent in the recent surge toward observations of centimeter- and millimeter-wavelength molecular lines which were all discovered with single-dish instruments of modest size but having very accurate surfaces.

The complexity of the radio-frequency spectrum currently being explored is best shown graphically. Figure 1 shows the number of lines which we might expect to detect in a typical HII region within the frequency range 4.5–5.0 GHz. As we improve our antenna-receiver systems to permit detection of small intensities we can expect a further increase in the number of known lines. The frequencies, intensities, and widths of spectral lines are discussed in detail in subsequent sections.

4.2.2. Special Equipment Requirements for Spectral-Line Work

Important areas for instrumental development were apparent from early observations of HI. For the antenna system an important requirement was precise pointing and tracking of the source during lengthy integrations. This led to the construction of the present sophisticated antenna systems which are completely computer controlled. Use of an on-line computer enables such factors as precession, antenna pointing corrections, atmospheric refraction, calibration, polarization mode, focusing, etc., to be programmed into every observation. Precise pointing can thus be achieved from a variety of input observational data such as epoch 1950 coordinates, galactic coordinates, or others. The observer can specify a variety of observational programs involving grids of points to be observed in special order, etc.

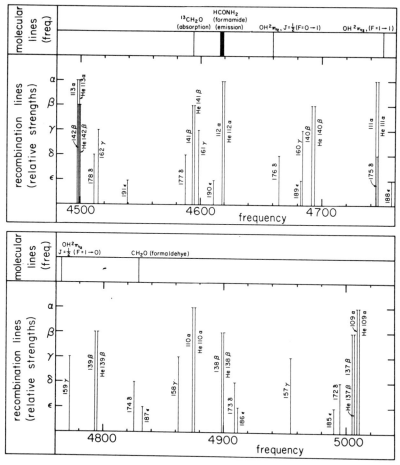

FIG. 1. Diagram of the radio spectrum of a typical HII region in the frequency range 4.5 to 5.0 GHz showing the frequencies of known molecular lines and hydrogen and helium recombination lines. The approximate relative intensities of the recombination lines are shown on a logarithmic scale.

Computer control of the receiver has been used to great advantage in:

(i)   the automatic setting of the frequency synthesizer so that a particular velocity component of a spectral feature can be preprogrammed to appear in a particular channel of a multichannel reciever;

(ii) the calibration of the observations so that every nth observation is a reference region and a calibration signal is applied every m seconds;

(iii) the recording of data automatically at prescribed integration periods.

The need for extremely low-noise receivers for spectral-line work has been apparent from the earliest observations. The limiting noise fluctuation of a receiving system is given by

$$\Delta T_{rms} = aT_{sys}/(\beta\tau)^{1/2}, \qquad (4.2.1)$$

where ΔT_{rms} is the root-mean-square noise fluctuation present in the observation, T_{sys} represents the system temperature, β denotes the receiver bandwidth, and τ denotes the time interval over which the observations are made. The constant a contains a factor of 2 for a symmetrical Dicke switch and also a weighting function of the order of unity, depending upon the shape of the bandpass of the receiver as discussed by Kraus[1] (also see Section 3.1.2).

In a continuum receiver the bandwidth β can be increased in order to improve the sensitivity per unit time. In a spectral-line receiver the bandwidth is that required to resolve the line. For a spectral-line receiver the sensitivity can be improved only by reducing T_{sys} or increasing τ. Since T_{sys} is the sum of the receiver noise T_R and the contribution from the antenna T_A, the solution is twofold:

(i) We may reduce T_R to its ultimate limit by employing a cooled parametric amplifier, a maser, or one of the exotic solid state devices now under development. Chapters 3.2 and 3.3 discuss in detail two such low noise amplifiers and their use. A recent survey of receivers for radio astronomy has been presented by Robinson.[2]

(ii) Special antenna feeds may be employed to reduce the spillover contribution to T_A. Such feeds provide a shaped illumination pattern across the paraboloid and cut off sharply at the edges. Minimum sky contribution to T_{sys} is achieved by use of a Cassigrain telescope. With such an instrument stray radiation spills over onto the cold sky.

In making the choice between the filter receiver (see Chapter 3.4) and the autocorrelation receiver (see Chapter 3.5) for a particular type of spectral-line measurement there are many factors to consider. The basic consideration of the baseline determination, the problems of which are produced principally by the front-end components of the receiver, are common to both methods of spectral analysis. The discussion of Section 4.2.6 is thus equally relevant to filter-type and autocorrelation receivers. Whereas early autocorrelation receivers were extremely costly to build and operate, requiring considerable computer time to perform the necessary Fourier transform, they have now evolved to the point where they are economically competitive with filter receivers. This situation has been brought about by the recent developments

[1] J. D. Kraus, "Radio Astronomy," p. 245. McGraw-Hill, New York, 1966.
[2] B. J. Robinson, *Annu. Rev. Astron. Astrophys.* **2**, 401 (1964).

in digital logic component and integrated circuits and by the availability of modestly priced on-line computers which make use of the fast Fourier transform (see Chapter 6.2). While present autocorrelators have instantaneous bandwidths up to 20 MHz, they do not as yet cover the wide bandwidths required in the millimeter wave range and beyond. The sensitivity of the autocorrelator is reduced by the factor of $\pi/2$ over that for the equivalent filter receiver under the assumption of one-bit quantizing, thus increasing the integration time by a factor of 2. The autocorrelator yields two information points per bandwidth for cosine weighting (this varies slightly with the form of weighting used) versus the single point for the equivalent filters overlapped at the half-power points, thus requiring twice the number of autocorrelation channels to cover the same spectral range. In addition, approximately 5 % of the transformed autocorrelation channels at either end are not usable because of the increased signal-to-noise ratio at the edges of the video filter. In principle, the frequency resolution of the autocorrelator can be increased indefinitely for the study of sharp spectral lines, limited only by the signal-to-noise ratio. Advantage has been taken of this feature in the measurements of the OH lines with an instrumental frequency resolution of only 190 Hz.

4.2.3. The Frequencies of the Radio Lines

For the majority of radio lines detected so far, accurate frequencies were available before detection occurred. If a reasonably accurate line frequency is available, one can limit the range of spectrum to be searched and make use of the lengthy integrations usually required for line detection. The frequencies of some radio lines can be calculated—the recombination lines, for example—whereas others are known from precise laboratory determinations. The rotational lines occurring generally in the centimeter–millimeter range can sometimes be calculated from a knowledge of the molecular constants of the species in question.

Vibrational–rotational frequencies of the molecules are frequently well known from laboratory determinations. Tables of spectral-line frequencies are given by Cord[3] in the National Bureau of Standards five volume series. The recombination line frequencies v of the elements are given by the Rydberg equation

$$v = cR_\infty Z^2 \left(\frac{M^+}{M}\right)\left\{\left(\frac{1}{N}\right)^2 - \left(\frac{1}{N+n}\right)^2\right\}, \qquad (4.2.2)$$

[3] M. S. Cord, N. S. Lojko, and J. D. Peterson, "Microwave Spectral Tables: National Bureau of Standards Monograph 70," Vols. 1–5. U.S. Govt. Printing Office, Washington, D.C., 1968.

where c is the velocity of light, 2.997925×10^{10} cm/sec, R_∞ the Rydberg constant for infinite mass, M the mass of the atom, M^+ the mass of the ion ($M^+ = M - m_e$, where m_e is the mass of the electron), Z the charge on the nucleus, N the principal quantum number of the Nth level, and n the order of the recombination line, for example, $n = 1$ for the α lines, $n = 2$ for the β lines, and $n = 3$ for the γ lines. The values for the constants are:

Hydrogen	^{1}H	$R_H = 109,677.6$ cm^{-1},	$cR_H = 3,288,052$ GHz
Helium	^{4}He	$R_{He} = 109,722.4$ cm^{-1},	$cR_{He} = 3,289,395$ GHz
Carbon	^{12}C	$R_C = 109,732.3$ cm^{-1},	$cR_C = 3,289,692$ GHz
∞ mass		$R_\infty = 109,737.3$ cm^{-1},	$cR_\infty = 3,289,844$ GHz.

where R_H, R_{He}, and R_C, the Rydberg constants for the reduced mass of these atoms, already include the M^+/M factor in Eq. (4.2.2).

Tables of computed recombination line frequencies have been published by Lilley and Palmer[4] for hydrogen with $n = 1$ to $n = 4$ and for helium with $n = 1$. The symbolism ^{12}C 110α designates the mass 12 isotope of carbon with $N = 110$ and $n = 1$.

A discussion by Townes[5] lists the frequencies of those lines that might be expected to be found in the galaxy and in planetary atmospheres. Many of these have now been detected. We list in Table I the frequencies of lines which have been observed in the galaxy and HII regions.

4.2.4. The Strengths of the Radio Lines

The calculation of the strengths of the radio lines is an important task. One may use such calculated strengths to estimate the possibility of detection of a line as yet unfound, or to calculate the number of atoms or molecules producing a detected line. A radio line may be visible in emission or absorption (or both). In either case, however, the absorption coefficient of the transition in question determines the opacity and hence the strength of the line. For an emission line the antenna temperature T_L of the line may be calculated from

$$T_L(v) = T_{ex}(1 - e^{-\tau(v)}) \, \Omega_S/\Omega_B, \qquad (4.2.3)$$

where $\tau(v)$ is the optical depth at frequency v, T_{ex} the excitation temperature, and Ω_S and Ω_B the solid angles of the emitting region and the antenna beam, respectively. In absorption, for the case in which the absorbing region covers

[4] A. E. Lilley and P. Palmer, *Astrophys. J. Suppl. Ser.* **16**, 143 (1968).
[5] C. H. Townes, in *IAU Symp. No. 4: Radio Astron.* (H. C. van de Hulst, ed.), p. 92. Cambridge Univ. Press, Cambridge, England, 1955.

the background source, the temperature of the absorption line ΔT_{abs} is given by

$$\Delta T_{abs}(v) = T_c(1 - e^{-\tau(v)}), \qquad (4.2.4)$$

where T_c is the antenna temperature of the continuum source derived from

$$T_c = (\eta/2k)SA, \qquad (4.2.5)$$

where η is the aperture efficiency of the antenna, k the Boltzmann constant, S the flux of the source, and A the aperture area of the antenna. For a given opacity in the line, if we assume local thermodynamic equilibrium with the excitation temperature T_{ex} small, the line should be more readily detectable in absorption against a bright continuum source than it is in emission. Hence, the OH molecule was searched for and discovered in the absorption spectrum of Cas A by Weinreb *et al.*[6] It was seen in emission only much later, and the character of the emission indicated strongly nonthermal excitation.

When a line is seen both in absorption in front of a source and in emission just off the source, Eqs. (4.2.3) and (4.2.4) may be used to determine the excitation temperature explicitly, provided the source is in local thermo-dynamic equilibrium. This procedure has been used principally for the 21-cm line for which a recent discussion has been given by Thompson *et al.*[7]

The quantity which is of principal concern in the calculation of the expected line strengths under conditions of thermal equilibrium is the opacity τ, as given by

$$\tau = \frac{8\pi^3 v^2 N |\mu_{12}|^2}{3ckT_{ex}\,\Delta v} \frac{g_1}{\sum g_i}, \qquad (4.2.6)$$

where N is the number of particles per square centimeter in line-of-sight, Δv the full linewidth at the half-intensity points (a small correction factor is required if the line is non-Guassian), g_1 the statistical weight of the upper level, $\sum g_i$ the sum of the statistical weights of the lower level, and μ_{12} the dipole-moment matrix element for the transition $2 \to 1$.

For the recombination lines observed in HII regions the ratio of the antenna temperature of the emission line, T_L, to the antenna temperature of the HII region, continuum temperature T_c, is given by Mezger[8] as

$$\Delta v_L(T_L/T_c) = 2.036 \times 10^4(v_L^{2.1}/T_{ex}^{1.15}), \qquad (4.2.7)$$

where Δv_L is the line width given in kHz and v_L the line frequency in GHz.

[6] S. Weinreb, A. H. Barrett, M. L. Meeks, and J. C. Henry, *Nature (London)* **200**, 829 (1963).

[7] A. R. Thompson, M. P. Hughes, and R. S. Colvin, *Astrophys. J. Suppl. Ser.* **23**, 323 (1971).

[8] P. G. Mezger and B. Hoglund, *Astrophys. J.* **147**, 490 (1967).

TABLE I. Frequencies of Detected Spectral Lines as of 1973[a]

Designation	Frequency (MHz)	Designation	Frequency (MHz)	Designation	Frequency (MHz)
CH₃OH (methyl-alcohol)	834.301	OH	13,441.371	X-ogen (unknown)	89,190
H (HI)	1420.4057	CH₂O	14,488.65	HNC (hydrogen isocyanide)	90,665
¹⁸OH (hydroxyl)	1639.460	HCNO (isocyanic acid)	21,981.7	OCS (carbonyl sulfide)	97,301.2
OH (hydroxyl)	1612.231	H₂O	22,235.22	OCS	109,462.8
	1665.401	NH₃ (ammonia)	23,694.48	C¹⁸O	109,782.2
	1667.358		23,722.71	HCNO	109,905.9
	1720.527		23,870.11	¹³CO	110,201.4
HCOOH (formic acid)	1638.805		24,933.47	CO (carbon monoxide)	115,271.201
			24,959.08	CH₃CN (methyl-cyanide)	110,383.5
CH₂S (thioformaldehyde)	3139.38	CH₃OH (methyl-alcohol)	25,018.14		110,381.4
			25,124.88		110,375.0
CH₂¹⁸O	4388.797		25,294.41		
		CH₂O	28,974		
		CH₃OH	36,169		

26

Molecule	Frequency	Molecule	Frequency	Molecule	Frequency
NH₂HCO (formamide)	4617.118	CH₃OH	48,372.47	CN (cyanide radical)	110,364.5
	4618.970		48,376.95		110,349.7
	4619.988	DCN (deuterated hydrocyanic acid)	72,414.6		110,330.7
OH	4660.242				113,492
	4750.656	CH₃OH	84,521.2	SiO (silicon monoxide)	130,268.4
CH₂O (formaldehyde)	4765.562	OCS (carbonyl sulfide)	85,139.0	CH₂O	140,839.53
	4829.659				145,602.97
CH₂NH (methanimine)	5289.00	CH₃C₂H (methylacetylene)	85,457		150,498.36
	5289.82			OCS	145,946.8
	5290.75	HCNO	87,925.4	CS (carbon monosulfide)	146,969.16
	5291.70	HCN (hydrogen cyanide)	88,630.416	H₂S (hydrogen sulfide)	168,762.7
OH	6035.085		88,631.847		
	6030.739		88,633.936		
OH	8135.868				
HC₃N (cyanoacetylene)	9097.036				
	9098.332				
	9100.279				

[a] All the above molecules are detected in emission only, except for the 4.8- and 14-GHz lines of formaldehyde which are in absorption. The H and OH lines are detected both in emission and absorption. The OH and H₂O are the known maser lines.

In general, atoms or molecules having transitions with long lifetimes compared to the collisional rate would be expected to come into equilibrium with the kinetic temperature of the surrounding gas. Under these conditions of local thermodynamic equilibrium (LTE), the Boltzmann equation gives the distribution of the number of atoms or molecules between the levels:

$$N_1/N_2 = (g_1/g_2) \exp(h\nu/kT_{ex}), \qquad (4.2.8)$$

where N_1 and N_2 are the numbers, and g_1 and g_2 the statistical weights of the upper and lower levels, respectively. The collision rate in the interstellar medium is determined principally by the hydrogen density since hydrogen is the most abundant element by a factor of over 1000, except for helium which is about one-tenth as abundant as hydrogen. The time t in seconds between collisions for hydrogen is given to a close approximation by

$$t = 10^{11}/N\sqrt{T}, \qquad (4.2.9)$$

where N is the number of hydrogen atoms per cubic centimeter, T the absolute kinetic temperature of the gas, and where the geometric cross section for the hydrogen atom has been used. Thus, for a region containing 10 hydrogen atoms per cm^3 at a temperature of $100°K$, the collision time is 10^9 sec, or about 30 yr. By comparison, the natural lifetime of the upper state of the 21-cm line is 1.1×10^7 yr; hence the populations of the levels of the HI ground state are completely controlled by the collisions. Many of the molecules detected in the millimeter wave region are in emission, presumably being collisionally excited in the denser dust clouds.

In contrast to the collisional case, atoms and molecules which have transitions with short lifetimes compared to the collisional time come into equilibrium with the radiation field. In the dark clouds, remote from ionizing radiation, the excitation temperature may be very low, probably below $10°K$ and falling in the limit to the $2.7°K$ blackbody temperature of the universal radiation. In such regions lines with small opacities ($\simeq 0.01$) become too weak to measure, antenna temperatures being in the milli-degree range. A detailed analysis of a molecule under both collisional and radiative excitation is given by Rogers and Barrett,[9] whose treatment yields the result that the excitation temperature of the molecule lies between the background radiation temperature and the kinetic temperature, depending upon the density.

The calculation of the expected line strengths of the spectral lines from interstellar space is complicated by many factors. The principal uncertainty is our lack of knowledge of the precise excitation mechanisms operative in the regions concerned. Frequently, the atoms and molecules are out of equil-

[9] A. E. E. Rogers and A. H. Barrett, *Astrophys. J.* **151**, 163 (1968).

ibrium with their surroundings. The prime example of this is the OH emission, where the overpopulation of the upper levels gives rise to coherent maser action which produces line ratios completely different from those predicted by theory. The attempt to explain this phenomenon alone has given rise to a wealth of proposals. So varied are the observational findings that it is likely that many of the proposed processes are at work, either singly or operating together.

The formaldehyde molecule provides another interesting case. Formaldehyde is found in absorption in dark clouds which contain no background continuum source. The 6-cm spectral line of the molecule must therefore be seen against the universal 2.7°K radiation. The populations of the energy states involved must be characterized by a temperature less than 2.7°K; the molecule is "refrigerated." An explanation of this phenomenon, providing for preferential population of the lower rotational level by collisions, has been proposed by Townes and Cheung.[10] The hydrogen recombination lines were also found (by Williams[11]) to have non-LTE line ratios in the higher order components. For a detailed discussion of the effects of non-LTE and high density clumping on the recombination lines in HII regions, see Hjellming et al.[12]

Calculation of expected line strengths under these anomalous conditions is clearly very uncertain, as is the inverse problem of interpreting line intensities in terms of atomic or molecular abundances. The processes of formation of the molecules in interstellar space are poorly understood. They may play important roles in determining whether a particular molecule is or is not observable and whether the molecule does or does not show anomalies.

4.2.5. The Calculation of Expected Linewidths

In the radio domain the natural width of a line is small, since the lifetimes of the states involved compared to the lifetimes of the states in the optical region are relatively long. Consequently, the width is principally determined by (a) the doppler width characteristic of the kinetic temperature of the gas and (b) the turbulent velocity. If the instrumental bandwidth is of the order of the resultant linewidth, then its contribution must also be considered. The relation between these quantities is given by

$$(\Delta v_{obs})^2 = (\Delta v_{bw})^2 + (\Delta v_{th})^2 + (\Delta v_{turb})^2, \qquad (4.2.10)$$

where

$$(\Delta v_{th})^2 = 457T/\lambda^2 M, \qquad (\Delta v_{turb})^2 = 10^4 V_T/\lambda^2, \qquad V_T = 2.36\sigma_T,$$

[10] C. H. Townes and A. C. Cheung, Astrophys. J. **157**, L103 (1969).
[11] D. R. W. Williams, Astrophys. Lett. **1**, 59 (1967).
[12] R. M. Hjellming, M. H. Andrews, and T. J. Sejnowski, Astrophys. J. **157**, 573 (1969).

and where Δv_{obs} is the observed width in kHz between the half-intensity points, Δv_{bw} the instrumental half-power bandwidth, Δv_{th} the thermal width, Δv_{turb} the contribution from the turbulent width V_T in km/sec, λ the wavelength in centimeters, T the absolute temperature, M the mass of the atom or molecule in units of atomic hydrogen mass, and σ_T the rms velocity in km/sec of the turbulence. As an example, for the neutral hydrogen line (21 cm) at $100°K$ with $V_T = 2$ km/sec internal turbulence, Eq. (4.2.10) gives an observed width Δv_{obs} of 14 kHz for zero instrumental broadening.

If lines from two emitters having differing masses but located in the same cloud are observed, then Eq. (4.2.10) permits separation of the thermal and turbulent components, hence a determination of T and V_T. Such separation has been performed for the case of Cas A (zero-velocity components), for the HI and OH lines, by Barrett et al.[13]

Equation (4.2.10) is not applicable to lines produced by coherent maser action. For the OH masers extremely high brightness temperatures $\gg 10^6$ °K have been observed from sources of small diameter. For the same sources linewidths of the order of 2 kHz are observed corresponding to a kinetic temperature of 48°K, an obvious inconsistency. Solution of the equation of transfer for a maser having a gain g leads to narrowing of the line by an amount given by

$$(\ln 2/\ln(g/2))^{1/2}. \qquad (4.2.11)$$

For a gain of 10^8 the line is narrowed by a factor of 5. Large values of maser gain are implied from this equation, and maser saturation almost certainly occurs in some emitting regions. In such regions in which saturation exists, Eq. (4.2.11) is no longer valid and the narrowing is much less.

The effect of instrumental smoothing becomes an important one for these narrow lines. The situation is exactly analogous to that of the smoothing of data by the antenna-beam pattern which has been discussed in detail by Bracewell and Roberts.[14] These authors show that restoration of the smoothed data to obtain the original information is only partially possible. They refer to the solution obtained as the principal solution. Bracewell[15,16] describes a method for simple restoration when using a Guassian-shaped analyzing bandpass described by the expression $(1/\sigma\sqrt{2\pi}) \exp(-x^2/2\sigma^2)$. The data are sampled at intervals of $\sqrt{2}\,\sigma$. Correction for the bandpass is then accomplished by subtracting the mean of the values on either side of the point to be corrected from the value at that point. This difference is then

[13] A. H. Barrett, M. L. Meeks, and S. Weinreb, *Nature (London)* **202**, 475 (1964).
[14] R. N. Bracewell and J. A. Roberts, *Aust. J. Phys.* **7**, 615 (1954).
[15] R. N. Bracewell, *Aust. J. Phys.* **8**, 54 (1955).
[16] R. N. Bracewell, *Aust. J. Phys.* **9**, 297 (1956).

applied to the original value to obtain the corrected value for that data point.

However, despite the availability of this and other computing techniques, published data are rarely corrected. If the observed width of a spectral line is indeed of the order of the bandwidth, it is generally agreed preferable to reobserve the line using a finer resolution than to apply a large correction term.

4.2.6. The Use of Switching Techniques in Spectroscopy

In spectral-line measurements some form of switching or differencing is invariably used to obtain adequate gain stability and an appropriately flat baseline. A flat (or at least a smooth) baseline is necessary for precise line measurements or whenever detection of weak features in the line structure is to be achieved.

Of the various forms of switching, the frequency switch is used extensively because of the ease with which good baselines can be obtained. Frequency switching has the principal advantage that the comparison window off the line has the same underlying continuum temperature as the spectral region of the line. The system is thus automatically balanced. In addition, there is a cancellation of some of the curvature that is inherent in the baseline.

The technique of switching against a load or horn pointed toward the sky also plays an important role in spectral-line work. The present rate of discovery of spectral lines makes it clear that the radio spectrum (particularly the centimeter–millimeter range) is rather densely filled with lines (see Fig. 1). As receiving systems become more sensitive, there exists the possibility that an unwanted line will appear in one of the windows and hence introduce uncertainty in the measurements. For this reason, at least one survey of recombination lines was made using load switching. Again, the load or sky switch is useful when lines have lengthy velocity tails which may not be covered with one frequency setting of the multichannel receiver or require too large a frequency switching range.

Load switching usually requires that an off-source reference region be observed in order to determine the true baseline level. Such reference region observations present difficulties because the off-source measurement has, in general, a different continuum temperature, which gives rise to a different baseline curvature.

Beam switches and polarization switches as applied to spectral-line work are discussed briefly in the subsequent sections.

4.2.6.1. Frequency Switching. The frequency switch is used extensively in both survey work and line searches, and we now discuss it in more detail. One frequency switching scheme, shown diagramatically in Fig. 2, is the

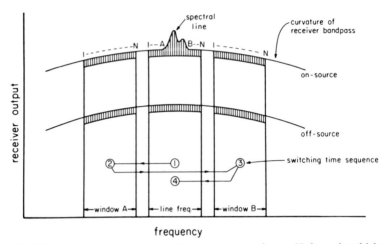

frequency

FIG. 2. Diagrammatic representation of measurements using an N-channel multichannel receiver and employing the double frequency switch. In the technique, comparison windows are time shared on either side of the line frequency.

double frequency switch first introduced in the OH measurements made with a receiver designed by Williams (see Weaver et al.[17]). In this technique, comparison is made to the spectrum viewed through two windows symmetrically displaced in frequency on either side of the line frequency. A switching pattern provides equal time on each position (line, window A, line, window B), with the sequence repeated at the switching rate. The frequency switching takes place close to the front end of the receiver, hence the bandpass of the following components acts as a constant multiplier. Thus, the only change in shape of the bandpass from window to line that must be considered is that ahead of the frequency switch, and is usually associated with the parametric amplifier, maser, or other amplifier combined with that of the antenna feed itself. If this bandpass is of simple mathematical form, then the effect of differencing in the way described above is to cancel the first derivative of the bandpass characteristic leaving only higher order terms. For example, if the curvature is of second order, only a constant term will remain in the data. A third-order curvature will leave only a small linear gradient, etc. A full mathematical discussion of this is given in Weaver and Williams,[17] where it is shown that, in general, an Nth order bandpass yields an $N - 2$ order term in the resulting baseline.

The presence of an underlying continuum source affects the magnitude of the bandpass curvature to a marked degree. If a comparison spectrum is

[17] H. Weaver, N. H. Dieter, and D. R. W. Williams, *Astrophys. J. Suppl. Ser.* **16**, 272 (1968); H. Weaver and D. R. W. Williams, *Astron. Astrophys. Suppl.* **8**, 1 (1973).

used in the data reduction, it is usual to take the off-source spectrum with sufficient artificial noise added to the antenna to equal the on-source temperature. This reference spectrum is then subtracted from the on-source spectrum to yield the "true" spectrum. Spectra obtained from the double frequency switch are normally of such quality that it is usually sufficient to take only the on-source spectrum and then subtract either a linear or quadratic least-squares baseline through channels 1 to A and B to N in the region outside the line. The detection of weak spectral features thus involves searching among the higher order terms in the baseline. It is often inadvisable to fit a curve higher than second order to the baseline under the area of interest.

The presence in the bandpass of ripples, which are of the order of the line-width being measured, constitutes a difficulty with this or any other method. Such ripples, if present, arise from the standing wave pattern in the feed produced by reflections from the central portions (central Fresnel zone) of the paraboloid, and have a period of $c/2F$, where F is the focal length and c is the velocity of light. These impedance variations in turn modulate the noise from the first stage to produce the observed undulations in the baseline. Various schemes have been proposed to eliminate them: use of circular polarization feeds, the use of a vertex cone, etc. Baseline irregularities can be minimized by these schemes, by careful matching of the feed at the line frequency, and by the use of high isolation components in the first stage of the front end of the receiver.

4.2.6.2. Load Switching. The load switch and the sky switch are also used extensively in spectral-line work for reasons discussed earlier; we now consider such schemes in more detail. The switching scheme is shown diagramatically in Fig. 3a. It is important to note that balancing white noise is added to the comparison load in sufficient amount to equal the observed source temperature. The load may be either a cold matched resistor in a cryogenic system or a sky horn looking at cold sky, etc. Another method of balancing the system is by the use of the gain modulator, as in continuum Dicke receivers discussed in Section 3.1.4. In either case, impedance variations with frequency across the band are likely to be quite different in the two halves of the switching cycle. Whereas the antenna feed may have rapid variations with frequency, the comparison load may exhibit quite a smooth function; consequently, the resulting baseline will contain these rapid undulations without the advantage of cancellation as in the frequency switch. Thus, the load switch may introduce baseline problems for small-signal work. (The large horn antennas may avoid this problem.) One possible practicable solution to the problem of avoiding an undesirable baseline is shown diagramatically in Fig. 3b. A reference region off the source is employed. Uncorrelated white noise is added to the antenna side of the switch to equal the on-source temperature, while noise is also added to the load side as in Fig.

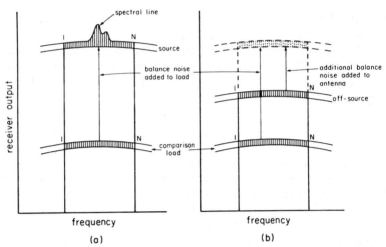

FIG. 3. Diagrammatic representation of measurements using an N-channel multichannel receiver employing a load or sky switch. In (a) (on-source spectrum), balancing noise is added to the load to equal the source temperature. In (b) (off-source reference spectrum), white noise is added to the load as in (a), and additional artificial balancing noise is added to the antenna to equal the on-source temperature. Spectra are taken on the source and at the reference position, the difference yielding the "true" source spectrum.

3a. The spectrum obtained with this configuration is then subtracted from the source spectrum to yield the "true" spectrum containing the wanted line.

4.2.6.3. Other Switching Schemes. An extremely useful switching scheme that has found application at the centimeter wavelength range is the beam switch. The comparison is made between the main beam and a beam displaced by a small angle from it. This procedure has the singular advantage that the comparison beam passes through an almost identical air mass and allows cancellation of sky fluctuation noise. Additionally, if the source has negligible continuum emission, the system is automatically balanced. However, it has the disadvantage of being limited to sources of small angular extent and excludes the study of extended sources.

Polarization switches provide a powerful technique with which to measure the Stokes parameters and Zeeman effect in spectral lines. An excellent discussion of wave polarization and the Stokes parameter is to be found in Kraus.[18] Numerous observational techniques have been employed; only references will be given here.

The general equations from which to derive the Stokes parameters are given by Ball and Meeks,[19] from whom we quote.

[18] J. D. Kraus, "Radio Astronomy," p. 108. McGraw-Hill, New York, 1966.
[19] J. A. Ball and M. L. Meeks, *Astrophys. J.* **153**, 577 (1968).

If $T(\phi)$ represents the antenna temperature observed with linear polarization at a position angle ϕ, and $T(R)$ and $T(L)$ represent the antenna temperatures with right- and left-circular polarization, then the Stokes parameters are defined as

$$S_0 = T(0°) + T(90°), \quad S_1 = T(0°) - T(90°), \quad S_3 = T(R) - T(L).$$
$$S_0 = T(R) + T(L), \quad S_2 = T(45°) - T(135°),$$

$$(4.2.12)$$

We consider right-circular polarization as rotation of the electric vector in a clockwise sense when viewed along the direction of propagation. This is the standard radio definition, which is opposite to the definition used in optics. The position angle θ of the major axis of the polarization ellipse is given by

$$\tan(2\theta) = S_2/S_1. \qquad (4.2.13)$$

The degree of polarization P is given by

$$P = (S_1{}^2 + S_2{}^2 + S_3{}^2)^{1/2}/S_0. \qquad (4.2.14)$$

The degrees of circular polarization P_c and of linear polarization P_L are defined as

$$P_c = |S_3|/S_0, \qquad (4.2.15)$$

$$P_L = (S_1{}^2 + S_2{}^2)^{1/2}/S_0. \qquad (4.2.16)$$

Finally, P_c and P_L combine to give P though the relationship

$$P_c{}^2 + P_L{}^2 = P^2. \qquad (4.2.17)$$

The OH lines were the first radio spectral lines which were found to be highly polarized and, hence, to require a full Stokes analysis. The polarization measurements of OH made by various authors include those of Coles et al.,[20] Palmer and Zuckerman,[21] and Manchester et al.[22] In some of the polarization work the individual polarizations are Dicke switched against a load and recorded separately for future reduction.

In the detection of the Zeeman splitting of the 21-cm line, Verschuur[23] describes a system in which he switched a digital correlator at a 1 Hz rate between left- and right-hand polarizations. These were recorded separately and later combined in the computer so that both the difference between the two polarizations, as well as one of the polarizations alone, was obtained.

[20] W. A. Coles, V. H. Rumsey, and W. J. Welch, Astrophys. J. **154**, L61 (1968).
[21] P. Palmer and B. Zuckerman, Astrophys. J. **148**, 727 (1967).
[22] R. N. Manchester, B. J. Robinson, and W. M. Goss, Aust. J. Phys. **23**, 751 (1970).
[23] G. L. Verschuur, Phys. Rev. Lett. **21**, 775 (1968).

4.2.7. Intensity Standardization in 21-cm Spectral-Line Work

Determination of the absolute intensity scale in 21-cm line measurements poses special problems because the HI clouds are, in general, larger than the antenna beam solid angle. The method outlined here is general: it is equally applicable to emitting areas of large angular extent or small angular extent. Briefly stated, the problem to be solved involves the conversion of the observed antenna temperature, T_A, for a region into a brightness temperature, T_B, in the sky. To make the transformation one must have knowledge of the source brightness distribution and the antenna properties. Further discussion of this problem is contained in Chapter 4.4.

There are various uses of standard regions whose brightness temperatures have been thus determined:

(i)   observations of the standard regions permit the brightness temperature scales of the various HI surveys to be intercompared;

(ii) observations of the standard regions provide a convenient way of checking the internal calibration of the equipment on a day-to-day basis;

(iii) such observations provide a way of calibrating an antenna–receiver system.

Three standard fields have been recommended by the I.A.U. (van Woerden[24]). The primary standard region $S8$ ($l = 207°$, $b = -15°$) has been used in several surveys; $S7$ ($l = 132°$, $b = -1°$) and S9 ($l = 356°$, $b = -4°$) are recommended as secondary standards. To estimate the brightness temperature associated with these regions, Williams[25] has made a study which involves mapping the regions to find the source distributions and also measuring in detail the antenna pattern of the 85-ft telescope used.

At the central point of any extended source, the antenna temperature $T(v)$ at a particular frequency v, from a source having a brightness distribution $T_B(\theta, \psi, v)$ when observed with an antenna having a normalized response $F(\theta, \psi)$ and gain G_0, is given by

$$T_A(v) = \frac{G_0}{4\pi} \int_{4\pi} T_B(\theta, \psi, v) F(\theta, \psi) \, d\Omega. \qquad (4.2.18)$$

Here, θ and ψ are the polar coordinates of the source region. Thus, by measuring the antenna gain pattern and G_0 by independent methods (or as discussed in Chapter 1.5 on antenna calibrations), one reduces the problem to that of solving the integral equation numerically for the various source distributions.

[24] H. van Woerden, *Trans. IAU Proc. General Assembly, 14th, Brighton, 1970*, Vol. XIV B, p. 217 (1971).

[25] D. R. W. Williams, *Astron. Astrophys. Suppl.* **8**, 505 (1973).

The results obtained for the three standard regions, as determined with the 85-ft telescope, are given in Table II. It was found in the mapping process that the integrated intensity was a more constant function of position than the peak brightness temperature, and it is concluded that the integrated intensity is the more desirable quantity to use for intercomparison purposes. Also, the integrated intensity is less sensitive to the choice of instrumental bandwidth.

TABLE II. Positions, Velocities, and Temperatures of the Standard Regions

	Galactic longitude l	Galactic latitude b	Velocity of peak (km/sec)	Velocity range used for integral (km/sec)	Integrated intensity (°K km/sec)	T_B, Peak (°K)
S7	132°	$-1°$	-50	$-56.3--45.8$	1168 ± 82	104 ± 7
S8	207°	$-15°$	$+7.0$	$-4.6-+21.75$	897 ± 66	72 ± 5
S9	356°	$-4°$	$+6.0$	$1.05-14.75$	953 ± 71	85 ± 6

The standard source S8 was used on a day-to-day basis during a recent galactic survey, in which it was found to be reliable to $\pm 1\%$ as measured against the internal noise calibration system. In order to achieve this accuracy, it has been found necessary to correct for the atmospheric extinction as a function of antenna attitude, and the intensities presented here are those computed for outside the earth's atmosphere, a necessary procedure since the observatories are located at various altitudes. The principal sources of uncertainty in the conversion of T_A to T_B, apart from systematic instrumental errors in internal calibrations, lie in our lack of knowledge of the antenna pattern outside the 4° area for which the beam was mapped, and also in our lack of knowledge of the assumed source distributions. As new data are collected from other sources with different instruments, it is expected that the errors will converge to give more accurate values for the brightness temperatures of these areas.

4.2.8. Measurements with Multichannel Filter Receivers

The basic method of spectral analysis with the multichannel filter receiver is through the use of a set of N contiguous bandpass filters of bandwidth β, spaced β apart. The filters are followed by N square-law detectors, the outputs from which are integrated, displayed, and recorded for further data processing. The versatility of the multichannel filter receiver is illustrated by the fact that it is used where high frequency resolution is desired (1 kHz), and also at millimeter wavelengths where wide bandwidths are required

(1–50 MHz per channel). A full discussion of multichannel filter systems, including integration devices, is given in Chapter 3.4. The ultimate sensitivity of the filter receiver is given by Eq. (4.2.1), where β is the individual channel bandwidth and the constant a contains the weighting functions depending upon the chosen shape of the filter response characteristic. This theoretical sensitivity is realized, provided that certain stability criteria discussed below are satisfied. The measurement of accurate line profiles also places a requirement on channel gain stability and linearity. These design requirements, of great concern in the design considerations of the filter receiver, are usually met without difficulty.

Multichannel filter receivers take a variety of forms. Some have crystal filter elements, others have inductance–capacitance filters. The number of channels may vary from fewer than 10 to over 100. Various frequency translation schemes are used to arrive at the final configuration of a set of contiguous channels. The filter widths of the individual channels may vary within one set, so as simultaneously to observe, with narrow filters, sharp features in the center of the line and, with wider filters, the low level wings of the line. Some receivers have channels spaced widely in frequency with respect to the bandwidth, the gaps being filled by moving the entire set in half-bandwidth steps. Despite these and other differences, the analysis in the following sections is equally applicable to all types of multichannel systems.

The individual elements of the multifilter system, for the purposes of the following discussion, are shown in Fig. 4; we assume that some form of

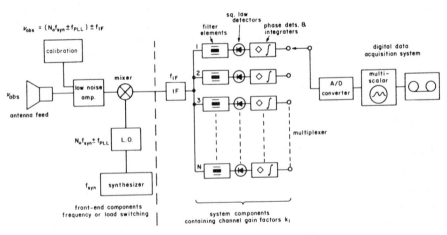

Fig. 4. Simplified block diagram of the multichannel filter receiver, showing the frequency scheme of local oscillator, the set of contiguous filters, and those components containing the normalized channel gain factors k_i.

frequency or load switching is employed. The observing frequency v_{obs} of the central channel is given by the expression

$$v_{obs} = (N_0 f_{syn} \pm f_{pll}) \pm f_{i.f.} , \qquad (4.2.19)$$

where N_0 represents the harmonic number of the synthesizer frequency f_{syn}. The frequency f_{pll} specifies the offset frequency in the phase-lock loop, and $f_{i.f.}$ denotes the frequency of the central filter channel referred to the first mixer. To simplify our discussion, we consider a signal from a narrow spectral line having excess temperature $T_{i, line}$ above the continuum T_c, and passing through the ith channel to produce N_i counts per unit time in the output data. Then, for a switched system,

$$N_i = G_i T_{i, line} + \Delta G_i T_c + n_i , \qquad (4.2.20)$$

where G_i is the transfer function of the entire multichannel system for the ith channel (we assume that the system is linear), ΔG_i is the differential transfer function for a switched system, and n_i is a channel offset parameter. The first term of Eq. (4.2.20) represents the modification of the spectral line by the transfer function, whereas the second term is the result of the differential transfer function operating on T_c. We consider G_i to be the product of two factors:

$$G_i = c_i k_i , \qquad (4.2.21)$$

where c_i is a smooth function of frequency related to the curvature of the bandpass of the front-end, and k_i is the individual channel gain factor, normalized so that the average channel gain is unity. Figure 5a shows the

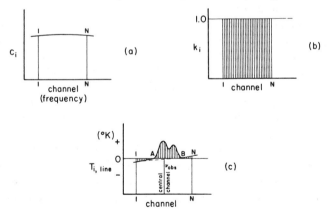

FIG. 5. (a) An illustration of the smooth component c_i of the transfer function G_i, (b) A representative set of channel gain factors k_i, (c) The resultant calibrated profile for a switched system, with the baseline shape exaggerated.

typical form taken by c_i, and Fig. 5b shows a representative set of k_i's. The latter would normally be set equal to within a few percent and then determined by calibration to better than 1 %. Whereas c_i may vary in shape as the receiver is tuned to a new line frequency, the k_i factors remain constant with time (or are very slowly varying functions of time), and may be determined once for a series of observations. Thus, the procedure is to apply a pulsed white noise calibration signal T_{cal} for a period, say 10 %, of the time on source. By such frequent calibration we remove the effect of gain changes in the receiver. The effect of changes in the bandpass upon the baseline, however, is not removed by calibration and is the subject of the next section.

We show in Fig. 5c the resulting calibrated spectrum, the shape of the baseline and its departure from zero in exaggerated form. This latter effect is a measure of the imbalance of the switched system and is due to the $\Delta G_i T_c$ term, the form of which depends upon the mode of switching employed. An ideal system would, of course, have the differential transfer function ΔG_i, and also its higher derivatives, equal to zero. However, in any real switched system these quantities depart from zero, resulting in the difficulties in determining the precise level of the instrumental zero. We refer the reader to our discussion in Section 4.2.6, which applies directly to the filter receiver in regard to the use of switching techniques, whether frequency, load, or other. We show that the effect of the front-end bandpass upon the baseline may be minimized by the application of various techniques. In particular, the symmetrical double frequency switch gives a cancellation of the first derivative of the passband, leaving only the smaller higher-order terms.

In the final analysis, it is usual to make a separate study of the baseline, and to infer its mathematical shape from a least-squares solution through channels 1 to A and B to N, which is then removed to obtain the "true" profile. As mentioned earlier, each channel also has its own zero point offset, n_i in Eq. (4.2.20), the stability of which must be such that it can be assumed constant throughout the observations. The fluctuation of n_i scaled in temperature units must be below the detection threshold by at least an order of magnitude. The observer may choose to determine these zero offsets in a separate measurement during which receiver noise is removed. In one of several alternative schemes, n_i can be effectively removed in the analysis by channel for channel subtraction of a reference region.

In a more elaborate reduction scheme described in detail by Weaver and Williams,[26] account is taken of not only the gain variations as a function of time, but also the changes in curvature of the front-end bandpass as a function of time. This is accomplished by the fitting of least-squares solutions

---

[26] H. Weaver and D. R. W. Williams, *Astron. Astrophys. Suppl.* **8**, 1 (1973).

to the frequent calibration data taken during a series of observations, and interpolating in time between them to find the value to be applied to each member observation. In the method, we assume that the bandpass shape is of simple mathematical form and changes only slowly with time. In the 21-cm measurements, application of the method gives internally consistent results from day to day of the order of 1 % in the determination of the temperature integral of a standard region of sky.

By careful application of the observing techniques described in Section 4.2.6, coupled with some of the more elaborate data reduction techniques, the multichannel filter receiver can be made to perform excellently under a variety of observing programs, whether it is used for low level search work or for general survey work. One example of multichannel filter performance, shown by Fig. 6, is a 12-hr integration on the 1715-MHz H 156α recombina-

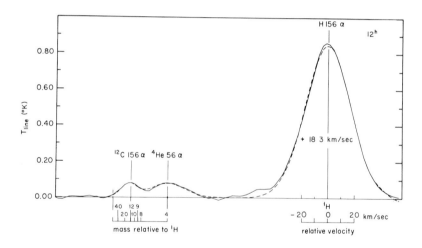

FIG. 6. The H 156α recombination line at 1715 MHz, from the HII region M17, as measured with a 100-channel filter receiver having a resolution of 30 kHz. Also resolved are the ⁴He 156α and the ¹²C 156α, the latter being anomalously strong. A residual second order baseline has been removed; consequently, the level of the baseline between the H and He lines is not precisely known. The profile represents a 12-hr integration with a system temperature of about 150°K, plus 50°K from the continuum source.

tion line from M17, which also contains the accompanying ⁴He 156α and ¹²C 156α lines, the latter being anomalously strong. The measurement was made using a 100-channel filter receiver with 30-kHz crystal filters and observations spaced every 15 kHz. A double frequency switch was used, and a reference region spectrum, off the source and with noise added, has been removed. A residual second-order baseline was further removed to give the final profile of Fig. 6. Consequently, the actual level of the baseline between the H and He lines is not precisely known. Also in this observation, as in others taken over an extended period of time, the local oscillator is frequently updated to take out the earth's diurnal and orbital motion so that a particular velocity feature always appears in a particular channel.

Another example, the CH_2O absorption in W51 observed with a similar filter receiver having a cooled parametric amplifier front end, is shown in Fig. 7. The observing bandwidths are 30 kHz and 10 kHz for the two spectra,

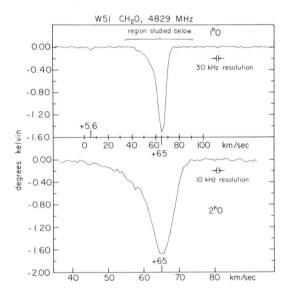

FIG. 7. The absorption spectrum of the source W51 observed in the formaldehyde CH_2O line at 4829 MHz, using a 100-channel receiver. The upper spectrum has a filter resolution of 30 KHz, the lower spectrum 10 KHz over the range indicated. The measurement was made with a 25.9-m (85-ft) telescope and a low-noise cryogenically cooled front end having a system temperature of 50°K.

and the integration times are 1 hr and 2 hr, respectively. A second-order baseline was removed by interpolation from each end of the profile under the region of the wide line. No reference region was used in this case.

For sharp spectral lines, as, for example, the OH maser-emission lines such as the W49 profiles shown in Fig. 8, it is essential that the observations be spaced no greater than one half-bandwidth apart. Multichannel filter receivers may have other spacings, in which case the procedure is to observe a series of spectra moving the local oscillator frequency in half-bandwidth steps to fill in the intervening data points.

In the case of lines having long wings, it is desirable to cover as much of the profile as possible at one time. Since this case implies a large number of channels, other techniques often have to be sought. For the OH lines in the galactic center, for example, the spectrum has extended wings, multiple velocity components, and the two OH lines involved (1665/1667 MHz) are a close doublet. Consequently, if the channels do not cover the entire spectral region, the observation must proceed piecewise in frequency with overlap.

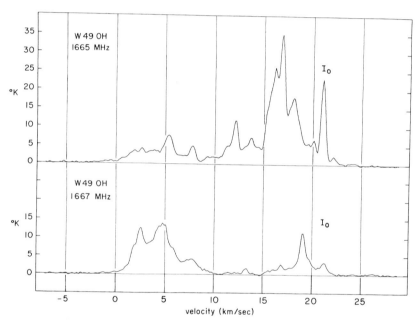

FIG. 8. Observations of the total intensity (I_0) 1665- and 1667-MHz OH maser lines from the source W49, observed with a 100-channel filter receiver having a frequency resolution of 2 kHz. The 200 data points in the spectra are obtained by shifting the LO frequency by one half-channel width (1 kHz) and interlacing the data. For sharp spectral lines such as these, of the order of the bandwidth, it is necessary to obtain data sampled at half-bandwidth spacings as required by the sampling theorem.

This has been done in the profiles in Fig. 9, in which six separate spectra have been fit together lengthwise. The observation was made using a sky switch and a reference region off the source to which balance noise was added. (See Section 4.2.6 for details of this method.) No other baseline has

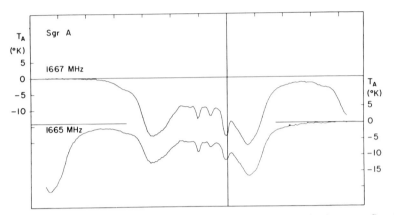

FIG. 9. The 1665- and 1667-MHz absorption spectrum of the galactic center Sgr A, observed with a 100-channel filter receiver having a resolution of 10 kHz. The observation was made using a sky switch, balance noise added, and a reference region off-source to which additional balance noise was added (see Section 4.2.6 for further discussion). Six separate spectra have been fit together lengthwise to obtain the profile of the close OH doublet. No other baseline has been removed from the data, and the residual uncertainty is less than a few tenths of a degree (°K) in the overlap region between the lines.

been subtracted from the data, and the uncertainty is less than a few tenths of one degree in the overlap region.

As an example of the usefulness of the multichannel filter receiver in survey work, we show in Fig. 10 a series of 81 21-cm profiles from a recent survey by Weaver and Williams.[26] The integration time was approximately 1 min per profile, employing a 150°K system temperature and a 10 kHz bandwidth. There are over 200 data points per profile, giving 5 kHz sampling spacing. The series of profiles were taken with the automated Hat Creek telescope, pointed in l and b coordinates corrected for precession and atmospheric refraction. The local oscillator synthesizer is set by the computer so that the chosen velocity with respect to the local standard of rest is in a particular (central) channel, and the integration proceeds automatically to yield the series of profiles.

The multichannel filter receiver has found application in the millimeter range of the spectrum where wide instantaneous frequency coverage is required; for example, at 100 GHz, 250 km/sec coverage, a minimal range to cover galactic velocities, corresponds to 80 MHz. Multichannel receivers having these properties have been used with great success in the centimeter and millimeter range for the discovery and measurement of the H_2O, NH_3, CO, CN, and other molecules.

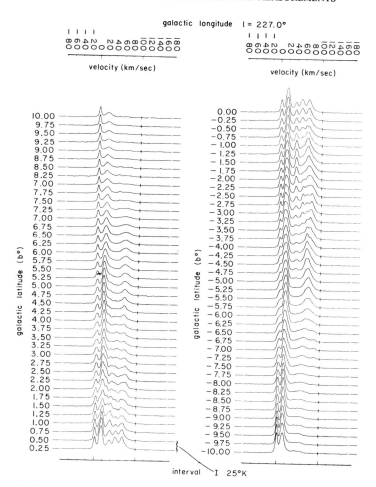

FIG. 10. A series of 81 21-cm profiles in galactic latitude, from $b = +10°$ to $-10°$ at longitude $l = 227°$, observed with a 100-channel filter receiver. The integration time is approximately 1 min per profile. The data were obtained from the 25.9-m (85-ft) automated telescope with over 200 data points, 10 kHz resolution, and sampled every 5 kHz. The frequency of each observation is preprogrammed so that the chosen velocity with respect to the local standard of rest occurs in a particular (central) channel. The vertical spacing between the profiles corresponds to 25°K in T_a.

4.3. Measurements with Radio-Frequency Spectrometers*

4.3.1. Introduction

Almost all spectrometers currently used in radio astronomy fall into two categories: multichannel filter banks and autocorrelators. Filter receivers are older, more common, and perhaps easier to understand. However, the use of the one-bit digital autocorrelator in radio-frequency spectrometers is now well established. Correlation systems have some advantages and some disadvantages in comparison with filter systems. For various practical reasons, correlators tend to be more versatile (i.e., they have a wider variety of spectral windows and resolution choices available) and more tractable if a large ratio of window width to resolution (i.e., a large number of channels) is needed. However, one-bit correlators have poorer noise performance for the same observing mode, and usually require the use of an electronic digital computer, at least to perform the Fourier transform.

Filter receivers are characterized by channels (samples) in frequency space, while correlators are characterized by channels in lag space but a continuous function in frequency space. It is incorrect, and sometimes misleading, to speak of correlator channels as if they were in frequency or velocity space. This error is prevalent because the Nyquist sampling theorem shows that the continuous spectrum from a correlation system is defined by a finite number of discrete points, and because digital computers can calculate the function at only a finite number of points.

It would be possible to design a filter reciver and a correlation receiver that would produce essentially identical output spectra. The filter receiver, in this case, would have approximately the same number of filters as the correlator has channels. To make such a system, the shape of the filter bandpass would be adjusted to be identical with the resolution shape (i.e., the response to a narrow spectral line) of the correlator. Then, for this case, the sampling theorem would apply to the filter receiver also, and $(\sin x)/x$ convolution, operating on the filter outputs, would produce the same function as the spectrum from the correlation system.

A correlation spectrum is a continuous function, and it is incorrect to plot it either as a series of points or as straight-line segments. In particular,

---

* Chapter 4.3 is by J. A. Ball.

46

straight-line segments between points spaced at the Nyquist interval for uniform weighting are very inaccurate for narrow spectral details. However, it is impossible to calculate the function at all points, and the recommended procedure is to use straight-line segments between points spaced closely enough to produce a good approximation to the continuous curve. One can also draw in a smooth curve by hand. The spacing for plotting purposes should be perhaps four to eight points per resolution width. $(\text{Sin } x)/x$ convolution will calculate additional points from points spaced at the Nyquist interval. Table I shows the worst-case plotting error resulting from straight-line seg-

TABLE I.   Worst-Case Plotting Error

No. points/Nyquist interval	Plotting error (%)
1	36
2	10
3	4.5
4	2.6
5	1.6

ments drawn between points spaced as indicated. This table is for a narrow spectral feature observed with uniform weighting.

It is more difficult to give recommendations for plots of spectra from filter receivers. For "square" filters, a histogram is appropriate; for other filter shapes, points may be the only honest display.

The noise performance of a one-bit correlator is poorer than that of a multibit correlator, or an equivalent filter receiver, because the process of representing the input signal only by its sign introduces quantization noise. A good discussion of this point has been given by Burns and Yao.[1] The noise is increased by a factor of about 1.5; however, this factor depends on the shape of the bandpass preceding the clipper and on the sampling frequency. Oversampling—that is, sampling at a rate more than twice the bandpass—reduces the excess noise somewhat. Multibit correlators with improved noise performance have been proposed. However, such correlators are more complex than the currently popular one-bit designs, and it is usually believed that increasing complexity should be such as to provide more channels and wider windows rather than a small improvement in noise performance.

The following sections discuss observing techniques generally applicable to either filter receivers or correlators, and some special techniques.

[1] W. R. Burns and S. Yao, *Radio Sci.* **4**, 431 (1969).

4.3.2. Switching Schemes and Baselines

The shape of the noise spectrum presented to the input of a spectrometer is largely determined by the shape of the i.f. bandpass. The signal to be observed usually appears as a small perturbation on top of this shape. The situation is thus analogous to the case in continuum observing where only a slight increase in system temperature occurs as the beam is scanned across the source. Both systems usually require some sort of switching scheme. The spectrometer case differs in that frequency switching is a possibility, and the one-bit correlator case differs in that the system is ideally insensitive to system gain. (A multichannel filter receiver can also be made to be insensitive to system gain.) Usually, the equivalent system noise temperature as a function of frequency across the spectral window and referred to the antenna (which is the relevant parameter for a clipped system) does not change appreciably over a period of many minutes or hours. By contrast, the system gain usually drifts more dramatically and more rapidly. These facts suggest that the switching cycle may be much longer with a clipped or gain-stabilized system, and with such long cycles, several new possiblities occur. In particular, the antenna may be moved between the "off" and the "on," the "off" and the "on" need not be the same length, and an "off" may be combined with more than one "on." This mode of operation has been termed the "unswitched" or "total-power" method of observing. Both these terms are incorrect; however, the latter will be used because it has become the accepted terminology.

One of the most persistent and difficult problems in spectral-line radio astronomy is the problem of obtaining good baselines. In a surprising number of observations, the limit on the detail that can be seen in a spectrum or the limit on the detectability of weak features is determined not by noise, but by instrumental baseline curvature. The various sources of instrumental baseline curvature can be divided into two categories: those that produce curvature by varying in time, and those that are independent of time. The problem of ameliorating baseline curvature of time-independent origin is essentially the problem of choosing a suitable comparison spectrum to switch against.

The perfect switching scheme would be one in which the spectral-line component in the signal is switched off and on without any other changes. Thus, the difference between the "on" and "off" would be precisely the quantity to be measured. Various practical switching schemes may be evaluated against this ideal.

One possibility is to switch to a room-temperature load. This scheme has two shortcomings. First, the noise temperature of the load will be near 300°K, and this is usually much higher than the noise temperature of the antenna. This produces a large change in level throughout the system, and often causes a change in bandpass, which leads to a sloping or curved baseline. Second, the impedance match of the load is often different from the match of the antenna, and the resulting change in bandpass will also produce a sloping or curved baseline.

The physical temperature of the load may be lowered by immersing it in liquid nitrogen or something similar; however, only very careful matching of both the antenna and the load will eliminate baseline curvature due to reflections.

Another popular scheme involves switching against a sky horn or an off-axis feed. The problem of reflections is similar; however, the noise temperature of such a comparison source is usually close to the noise temperature of the antenna, unless a strong continuum source is in the beam. It is important to insure that no confusing spectral signal is present in the power received by the comparison source. This proves difficult for certain experiments, such as those involving the 1420-MHz line of neutral hydrogen, which gives rise to emission over a large part of the sky.

Another switching possibility involves changing the frequency of the local oscillator (or one of the local oscillators) such as to move all spectral features outside the spectral window. Since the noise power from the front end is not independent of frequency, this scheme is recommended only for narrow spectral windows, i.e., where the spectral window is a small fraction of the bandpass of the front end. Symmetrical frequency switching, in which the comparison window is alternately above and below the signal window, is generally desirable. (See Section 4.2.6.) It can be shown that symmetrical frequency switching eliminates the effect on the baseline of the first three terms (through quadratic) in the Taylor-series expansion of the bandpass noise spectrum, whereas asymmetrical frequency switching eliminates only two.

When searching for new spectral features whose frequency is imprecisely known, one can sometimes make use of the comparison window as well as the signal window in frequency switching. A feature whose frequency places it in the comparison window will appear inverted but unattenuated with asymmetrical frequency switching, and inverted and half-size with symmetrical frequency switching. With some stipulations on the interpretation of the spectra, asymmetrical frequency switching can be used to search twice the frequency range with the same signal-to-noise ratio, and symmetrical frequency switching can be used to search three times the frequency range with half the usual signal-to-noise ratio in the two comparison windows. Other patterns are also possible.

With the total-power method of observing, an additional scheme is possible. The antenna may be pointed off the source for the " off." If the source has little continuum emission, and if the effects of the atmosphere can be neglected, then this scheme approaches the ideal very closely. If the same track across the sky is followed for the "off" as for the "on" (i.e., offset in right ascension by an amount equal to the time between the "off" and the "on"), then contributions due to atmospheric and ground pickup also cancel, provided only that they do not change in time. The total-power scheme and the method of taking the "off" with the antenna pointed off the source in angular position could also be (but apparently has not been) used with filter receivers.

With any of the switching schemes, instrumental baseline curvature may be associated with differences in power level between the "off" and the "on" caused by differences in continuum noise temperature. There are two popular ameliorants for this problem: noise injection and gain modulation. Noise may be injected during the part of the switch cycle (either "on" or "off") that has the lower noise temperature. This noise should be injected as closely as possible to the antenna, and should have a spectral distribution identical with the continuum noise it replaces. This is difficult in practice because the devices through which the injected noise passes usually do not have exactly the same bandpass as the equivalent bandpass for the continuum from the sky. Nevertheless, noise injected to balance the "on" and "off" usually improves the baseline. Note that if the comparison source is hotter than the antenna, so that noise must be injected during the "on," then the injected noise will raise the effective system temperature.

Gain modulation involves switching the gain somewhere in the system such that the noise level following that point will be the same for the "on" and "off." The gain change must be done in such a way that the bandpass remains unchanged. Gain modulation is usually less desirable than noise injection, but is recommended if suitable noise injection is not available.

Another source of instrumental baseline curvature involves drifts in time of some characteristic of the system. The switching cycle should be chosen to be short compared with the time scale for significant changes in the relevant system parameters, but long compared with the dead time or blanking time involved in the switch. For a clipped or gain-stabilized system, the shape of the bandpass noise at the input to the spectrometer is the relevant system parameter. Drifts in the absolute gain are unimportant, but drifts in the bandpass shape anywhere in the system are important. The time scale for such drifts ranges from milliseconds to hours, depending on the details of the design of the front end and the i.f., and depending on the fraction of the bandpass of the front end being used in the spectral window. There seems to be no good way to predict this time scale, and it needs to be determined by trial and error.

Another scheme that can be used to improve the baseline consists of subtracting a spectrum made by conventional symmetric switching off the source from a similar spectrum made on the source. However, this scheme can hardly be recommended except as a last resort, because it produces twice as much noise as symmetric switching alone.

Baseline curvature usually appears to improve as one goes to narrower frequency windows. However, this is largely due to the well-known effect of examining a smaller portion of a complex curve and so is, in some sense, illusory. A wider window will give a better idea of baseline curvature, and aid in evaluating the reality of marginal features and structure. With frequency switching, the actual shape of the baseline depends on the amount by which the frequency is switched, and this amount is usually changed along with the frequency window. So, with frequency switching, a real improvement in instrumental baseline curvature can sometimes result from going to a narrower window.

4.3.3. Baseline Fitting

A constant term in the baseline of a correlator spectrum is arbitrary, and the usual Fourier-transform formula chooses this constant such that the total spectrum has an integral of zero. In most filter receivers, this constant depends on the balance between signal and comparison. So, it is usually necessary to subtract at least a constant term from the spectrum, and higher-order baselines (usually polynomials) are sometimes subtracted in an attempt to compensate for instrumental baseline curvature.

The procedure involves fitting a polynomial in a least-square sense to those parts of the spectrum on which no spectral features are present, and then subtracting the polynomial from the whole spectrum. The idea is that instrumental baseline curvature is usually smooth and can be approximated by the first few terms of its Taylor-series expansion. Extreme caution is recommended when using this procedure with higher-order polynomials because such polynomials can either create or destroy spectral features!

In evaluating the reality of a possible new spectral feature, one approach is to first fit a polynomial to the entire spectrum. In effect, this assumes that no feature is present, only instrumental baseline curvature, and the polynomial is fitted to this curvature. If the feature persists nevertheless, then the fit can be redone, eliminating the region of the spectrum that contains the feature, to give an estimate of the true size and shape of the feature.

The order of the polynomial that is justifiable depends in part on the fraction of the spectral window occupied by spectral features. If the baseline regions are only a small fraction of the window, then only a lower-order polynomial should be used.

4.3.4. Noise Considerations

If the system permits the switching cycle to be long, then significant increases in observing efficiency can be achieved by taking the " off " before the source rises, for example, or by using the same " off " with several " ons " on the same or different sources.

Given that one " off " will be used with several " ons," the question arises as to how one should divide a given observing session among the " ons " and the " off " to produce minimum noise on the resulting spectra. This problem is analyzed below.

Let n be the number of " ons " to be used with each " off," t the time for each " on," α the ratio of the time for one " off " to t, and τ the total observing time for one cycle,

$$\tau = (\alpha + n)t. \tag{4.3.1}$$

We neglect the dead time or blanking time involved in the switch. We assume that τ and n are fixed by other considerations, and we wish to choose α (and t) for minimum noise.

The normalized noise for the " on " is $t^{-1/2}$, for the " off " $(\alpha t)^{-1/2}$ and for the " on " minus " off "

$$\left(\frac{1}{t} + \frac{1}{\alpha t}\right)^{1/2} = \left(\frac{\alpha + 1}{\alpha t}\right)^{1/2} = \tau^{-1/2}[\alpha + 1 + n + (n/\alpha)]^{1/2}. \tag{4.3.2}$$

We differentiate Eq. (4.3.2) with respect to α, set the result equal to zero, and solve for α, obtaining $\alpha = \sqrt{n}$. This means that we minimize the noise on each of the resulting spectra by spending $\sqrt{n}$ times as long on the " off " as on each " on." For example, we may divide a 5-hr observing session into a 1-hr " off " and sixteen 15-min " ons." For this example, the noise on the resulting spectra will be reduced by a factor of 0.625 compared with ordinary symmetric switching for the same total observing session. This scheme will be referred to as " optimum total power."

The noise obtained from the optimum-total power scheme may be compared with conventional schemes in Fig. 1, which shows the quantity a in the formula

$$\Delta T = a\gamma T_{\text{sys}}/(\beta\tau)^{1/2}. \tag{4.3.3}$$

In this formula, γ is the clipping correction for an autocorrelation system[1], T_{sys} is the system noise temperature, β is the resolution width, and ΔT is the rms noise on the spectrum. The formula for a may be derived from the unnormalized form of Eq. (4.3.2) with $\alpha = \sqrt{n}$. In Eq. (4.3.3), the time τ may be either $1/n$ of the total observing session (" ons " + " off "), in which case ΔT

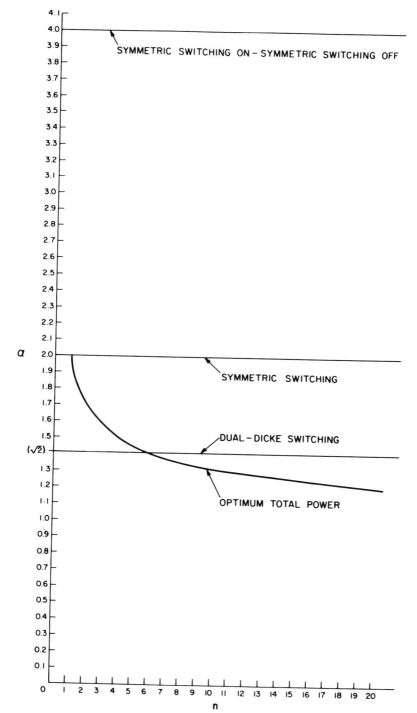

FIG. 1. Noise comparison of various commonly used methods of observing. The quantity a appears in the numerator of the noise equation, and n is the number of " ons " combined with each " off," or alternatively, the factor by which the " off " resolution has been widened by smoothing.

applies to each spectrum, or τ may equal the total observing session, in which case ΔT applies to the spectrum resulting from averaging the n runs.

So far, we have treated the resulting spectra ("on" minus "off") as if they were independent, but since the same "off" is to be used with several "ons," the noise in the spectra will be partially correlated, and if they are added together or averaged, the noise will be worse than expected. A recommended procedure for circumventing this problem involves stepping the frequency of the first local oscillator (LO) by at least a resolution width between "ons" during the observations, and then "unshifting" the spectra for averaging. Thus, the noise contribution in the "off" will be uncorrelated again. From the noise standpoint, this scheme of frequency stepping the LO is approximately equivalent to smoothing the "off" with a rectangular filter whose width is equal to the total shift, or to shifting the "off" before combining it with each "on." From the standpoint of instrumental effects, however, these schemes are distinct because the frequency stepping scheme is not sensitive to spectral structure in the i.f. bandpass as are these other schemes. For the frequency stepping scheme, one needs to have the bandpass of the system in front of the first LO smooth over the stepped frequency range, but the shape of the bandpass of the i.f. need not be smooth so long as it is constant in time.

It is worth noting that the spectral noise-power distribution in lag space for the resulting averaged spectrum is no longer flat because the "off" spectrum has, in effect, been smoothed so that the cut-off lag is smaller than for the "ons." Thus, the noise on the averaged spectrum will not fall as fast as $\beta^{-1/2}$ if it is convolved to a wider resolution. In the limit of β large compared with the total shift in the "ons," the noise actually becomes worse than would result from ordinary symmetric switching. The noise fluctuations contain a larger component at low lags (slow fluctuations in the spectrum) than is normally the case.

The above considerations may also be applied to a spectrum resulting from symmetric switching off the source subtracted from a similar spectrum made on the source. Also note that the scheme of fitting a polynomial to an "off" spectrum and subtracting the polynomial from the "on" is just a different (and usually drastic) method of smoothing the "off."

Schemes involving frequency stepping or smoothing usually result in a loss of spectral coverage at the edges of the spectral window approximately equal to the total shift. Another scheme that trades spectral coverage for improved noise performance is "dual-Dicke" switching. In this scheme, the LO frequency is switched by an amount equal to half the width of the spectral window such that the signal to be observed appears in the window during both the "on" and "off." The "on" minus "off" then contains the signal twice

—once inverted. The spectrum is then split, one half inverted, and averaged. The result is a spectrum with about half the frequency coverage but twice the effective integration time. Figure 1 shows how the noise performance of dual-Dicke switching compares with other schemes. Dual-Dicke switching is a case of asymmetrical frequency switching, and the comments given above regarding frequency switching apply.

4.3.5. The Spectral Resolution in a Correlator System

The user exerts a limited but useful control over the resolution shape of a correlator spectrum by choosing a weighting function for the Fourier transform and/or a convolving function for the spectrum (the two are mathematically equivalent). The situation is almost exactly analogous to the limited but useful control over the equivalent antenna pattern in mapping exerted by choosing the antenna feed illumination and/or a convolving function for the map. In general, as one reduces the resolution width or attempts to produce sharp-edged shapes, the sidelobes or spurious responses increase. Conversely, a very smooth response nearly free of sidelobes can be obtained, but with a wider resolution. It is always possible to convolve the spectrum by a function that produces a wider resolution, but not the converse.

Table II shows the resolution and the peak sidelobe level for some possible weighting functions. This table tells only part of the story, however, and the details of the shape of the spectral response may be important. Some of them

TABLE II.   Weighting and Resolution Functions

Name	Weighting function[a]	Resolution[b]	Peak sidelobe level, %
Uniform[c]	1	0.60	-22
Cosine	$\cos(\pi\tau/2\tau_m)$	0.82	-7
Triangle	$1 - \lvert\tau\rvert/\tau_m$	0.89	$+4.7$
Hanning (cosine²)	$\cos^2(\pi\tau/2\tau_m) = 0.5 + 0.5\cos(\pi\tau/\tau_m)$	1.00	-2.7
Hamming	$0.54 + 0.46\cos(\pi\tau/\tau_m)$	0.90	±0.67
Blackman	$0.42 + 0.5\cos(\pi\tau/\tau_m) + 0.08\cos(2\pi\tau/\tau_m)$	1.15	±0.12

[a] All weighting functions are zero for $\tau > \tau_m = M/F_s$, where F_s is the sampling frequency and M is the number of channels.
[b] The full width to half maximum in units of F_s/M.
[c] Uniform weighting corresponds to the "principal solution" of R. N. Bracewell, "The Fourier Transform and Its Applications." McGraw-Hill, New York, 1965.

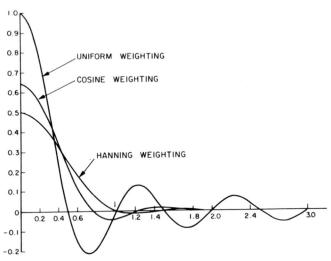

FIG. 2. Three popular resolution functions. The abscissa is in units of F_s/M. The Nyquist interval is 0.5 in these units.

have been plotted, for example, by Blackman and Tukey.[2] Three popular resolution shapes are plotted in Fig. 2.

All these weighting functions are unity for zero lag. This preserves the integral so that, except for baseline considerations, the spectral flux (the integrated flux density) is proportional to the sum of the temperature numbers spaced at the Nyquist interval; no more complicated integration formula should be used.

Note that the resolution function is the response to a narrow spectral line and also the function to be convolved with spectra from uniform weighting to produce the specified resolution shape.

There is a fairly dramatic difference between uniform weighting and any of the other popular weightings, both in terms of resolution width and side-lobe levels. Uniform weighting produces the highest signal-to-noise ratio for narrow (unresolved) features, and is thus recommended for searches for such features. For spectra in which there are no details comparable to the resolution width, the resolution shape is not important, only its width (in determining the noise). In searching for or studying narrow weak spectral features in the presence of strong features, or when studying narrow weak details in strong features, one should use a weighting function that produces low sidelobes.

[2] R. B. Blackman and J. W. Tukey, "The Measurement of Power Spectra." Dover, New York, 1959.

In searching for new spectral features, one may be tempted to make use of the matched-filter theorem, which states that the signal-to-noise ratio is maximized by matching the resolution width (and shape) to the sought-for feature. However, this theorem does not take into account other important considerations. First, it is always possible to convolve the spectrum by a function that produces a wider resolution, in case the sought-for feature is wider than the original resolution. But it is usually unnecessary to actually perform this convolution because the human eye is quite proficient at such averaging. And with filter receivers, it is undesirable to have a feature appear in only one filter because of the possibility that the filter may be faulty.

These considerations suggest that one should adjust the window to cover the desired frequency range, and check that the corresponding resolution is narrower than (or perhaps comparable to) the width of the sought-for feature.

ACKNOWLEDGMENTS

I thank C. Gottlieb, C. Heiles, A. E. E. Rogers, S. Weinreb, and B. Zuckerman for useful discussions. The total-power technique was originally suggested by M. Gordon and J. Carter at Haystack. I thank A. E. Lilley for providing the stimulating environment in which this work was carried out. Financial support came from the United States National Science Foundation through Grant GP-19717.

4.4. Measurements of Galactic 21-cm Hydrogen*

4.4.1. Introduction

Measurements of atomic hydrogen at a wavelength of 21-cm are performed either in emission or in absorption against a background source. The theory of line formation and summaries of known results are given by Simonson,[1] Kerr,[2] and the recent IAU Symposia Nos. 31, 38, 44, and 60.[3]

4.4.2. Emission Surveys

In the northern sky, future measurements will undoubtedly be oriented toward specific regions, obtaining data free of the deficiencies peculiar to the various catalogs already in existence. Survey data away from the galactic plane are sadly lacking in the southern sky. Extensive work is summarized in Tables I–IV, where we have included only those contributions in which the data are given in reasonably complete detail. It is hoped, but not guaranteed, that there are no serious omissions. Maps of total column density in the northern sky are given by Davies,[4] Tolbert,[5] and Heiles[6]; Fejes and Wesselius[7] have assembled both southern and northern data, and present a complete map.

In the earlier emission surveys it was customary to present the individual profiles measured at every position observed, and sometimes to combine them

[1] S. C. Simonson, *Astron. Astrophys.* **9**, 163 (1970).
[2] F. J. Kerr, *Annu. Rev. Astron. Astrophys.* **7**, 39 (1969).
[3] *IAU Symp. No. 31: Radio Astron. Galatic Syst.* (H. van Woerden, ed.). Academic Press, New York, 1967; *IAU Symp. No. 38: Symp. Spiral Structure Our Galaxy* (W. Becker and G. Contopoulos, eds.), Reidel Publ., Dordrecht, 1970; *IAU Symp. No. 44: External Galaxies Quasi-Stellar Objects* (D. S. Evans, ed.). Reidel Publ., Dordrecht, 1972; *IAU Symp. No. 60: Galactic Radio Astron.* (F. J. Kerr and S. C. Simonson, eds.). Reidel Publ., Dordrecht, 1974.
[4] R. D. Davies, *Mon. Not. Roy. Astron. Soc.* **120**, 483 (1960).
[5] C. R. Tolbert, *Astron. Astrophys. Suppl.* **3**, 349 (1971).
[6] C. Heiles, *Astron. Astrophys. Suppl.* **20**, 37 (1975).
[7] I. Fejes and P. R. Wesselius, *Astron. Astrophys,* **24**, 1 (1973).

---

* Chapter 4.4 is by Carl Heiles and G. T. Wrixon.

TABLE I. Extensive Surveys of the Galactic Plane

l^{II} Range (deg)	b^{II} Range (deg)	Grid size (deg)	Beam size (arc min)	Velocity resolution (km/sec)	Authors and Remarks
−6–120	$b = 0$	0.5	36	1.6	Burton[a]; contour map and profiles. See also references quoted therein.
190–299	−10–+6, typically	5 (in l)	14.5	7.5	Hindman and Kerr[b]; tracks perpendicular to the galactic plane. Contour maps.
200–30	−4–+4	5 (in l)	84	8.5	Kerr et al.[c]; tracks perpendicular to the galactic plane. Contour maps.
296–63.5	$b = 0$	0.1 (in l)	14.5	7.5	Kerr[d]; contour maps.
301–60	−2–+2	1 (in l)	14.5	7.5	Kerr[d]; tracks perpendicular to the galactic plane. Contours.
185–63	$b = 0$	1 (in l)	14.5	7.5	Kerr and Hindman[e]; contour maps.
0–240	$b = 0$	10 (in l)	35	4.4	Makarova.[f]
345–250	−5–+5	2.5	114×168	7.5	Muller and Westerhout[g]; profiles. Contour maps by Lindblad[h]; Gaussianized by Lindblad.[i]
10–270	−10–+10	0.25	35	2	Weaver[j]; profiles and contour maps. Also available on magnetic tape.
11–235	some −3–+3	0.08	10	1.5	Westerhout[k]; contour maps. Also available on magnetic tape.

[a] W. B. Burton, Astron. Astrophys. Suppl. 2, 261 (1970).
[b] J. V. Hindman and F. J. Kerr, Aust. J. Phys. Astrophys. Suppl. No. 18, 43 (1970).
[c] F. J. Kerr, J. V. Hindman, and C. S. Gum, Aust. J. Phys. 12, 270 (1959).
[d] F. J. Kerr, Aust. J. Phys. Astrophys. Suppl. No. 9 (1969).
[e] F. J. Kerr and J. V. Hindman, Aust. J. Phys. Astrophys. Suppl. No. 18 (1970).
[f] S. P. Makarova, Astron. Zh. 41, 608 (1964) [English transl.: Sov. Astron. Astrophys. J. 8, 485].
[g] C. A. Muller and G. W. Westerhout, Bull. Astron. Inst. Neth. 13, 151 (1957).
[h] Per Olaf Lindblad, Bull. Astron. Inst. Neth. Suppl. Ser. 1, 77 (1966).
[i] Per Olaf Lindblad, Bull. Astron. Inst. Neth. Suppl. Ser. 1, 177 (1966).
[j] H. Weaver and D. R. W. Williams, Astron. Astrophys. Suppl. 8, 1 (1973); 17, 1 (1974).
[k] G. W. Westerhout, "Maryland Green Bank Galactic 21-cm Line Surveys," 2nd ed. Univ. of Maryland, 1969.

TABLE II.   Extensive Surveys Not Confined to the Galactic Plane

Area covered (deg)	Grid size (deg)	Beam size (arc min)	Velocity resolution (km/sec)	Authors and Remarks
Sky north of $\delta = -32$	≤ 5 (in δ)	90×102	21, 5	Davies[a]; map of total column density.
$\|b\| \leq 20$, most of northern sky	10	120×180	2.4	Erickson and Helfer[b]; profiles.
$\|b\| \geq 10$, most of northern sky	0.3 (in l) 0.6 (in b)	35	2	Heiles and Habing[c]; contour maps. Also available on magnetic tape (see Heiles[d]).
$l = 210-30$, $b = -90-+90$	5	132	7.5	McGee et al.[e]; profiles. For other presentations see references quoted therein.
Sky north of $\delta = -29$	5	36	2	Venugopal and Shuter[f]; profiles.
$\|b\| = 10-25$, northern sky	10 (in l) 5 (in b)	36	2	van Woerden[g]; profiles. Gaussianized by Takakubo and van Woerden,[h] and analyzed by Takakubo.[i]
$l = \quad 0-270$ $b = -30-+30$	2.5 (in l) 0.25 (in b)	35	2	Weaver and Williams.[j]
$l = \quad 0-270$ $b = -90-+90$	10 (in l) 0.25 (in b)	35	2	Weaver and Williams.[j]

[a] R. D. Davies, Mon. Not. Roy. Astron. Soc. **120**, 483 (1960).
[b] W. C. Erickson and H. L. Helfer, Astron. J. **65**, 1 (1960).
[c] C. Heiles and H. J. Habing, Astron. Astrophys. Suppl. **14**, 1 (1974).
[d] C. Heiles, Astron. Astrophys. Suppl. **14**, 557 (1974).
[e] R. X. McGee, J. A. Milton, and W. Wolfe, Aust. J. Phys. Astrophys. Suppl. No. 1 (1966).
[f] V. R. Venugopal and W. L. H. Shuter, Mem. Roy. Astron. Soc. **74**, 1 (1970).
[g] H. van Woerden, K. Takakubo, and L. L. E. Braes, Bull. Astron. Inst. Neth. **16**, 321 (1962).
[h] K. Takakubo and H. van Woerden, Bull. Astron. Inst. Neth. **18**, 488 (1966).
[i] K. Takakubo, Bull. Astron. Inst. Neth. **19**, 125 (1967).
[j] H. Weaver and D. R. W. Williams, Astron. Astrophys. Suppl. **8**, 1 (1973); **17**, 1 (1974). (1975) (to be published).

TABLE III.   Surveys Directed toward "High Velocity" Hydrogen

Area covered (deg)	Grid size (deg)	Beam size (arc min)	Velocity resolution (km/sec)	Author and Remarks
Most of northern sky	2–30 (in l) 2 (in b)	35	2	Dieter[a]; contour maps. Discussed by Dieter.[b]
$l =$ 40–140 $b = -15-+40$	5 (in l) 1 (in b)	33	6.8	Habing[c]; contour maps.
$l = 228$–48 $b =$ 6–20	4 (in l) 2 (in b)	38	10.6	Kepner[d]; contour maps of Gaussian components.
$b = -40, +30-+90$ most of northern sky	5 (in l) 10 (in b)	36	2	Muller et al..[e] profiles.
$\delta = -5-+60$	1 (in δ)	10×40	20	Meng and Kraus[f]; maps and tables of cloud parameters.
$|b| \geq 15$, northern sky	≥ 5	36	3.4	Tolbert[g]; profiles, contour maps, and tabular material.
$l = 230$–350 $b =$ 10–40	2	120	16	Wannier et al.[h]
$b = 15$, northern sky	continuous (in α) 2.5 (in δ)	38	45	van Kuilenberg.[i,j]

[a] N. H. Dieter, Astron. Astrophys. Suppl. **5**, 21, 313 (1972).
[b] N. H. Deiter, Astron. Astrophys. **12**, 59 (1971).
[c] H. J. Habing, Bull. Astron. Inst. Neth. **18**, 323 (1966).
[d] M. Kepner, Astron. Astrophys. **5**, 544 (1970).
[e] C. A. Muller, E. Raimond, U. J. Schwarz, and C. R. Tolbert, Bull. Astron. Inst. Neth. Suppl. Ser. **1**, 213, (1966).
[f] S. Y. Meng and J. B. Kraus, Astron. J. **75**, 535 (1970).
[g] C. R. Tolbert, Astron. Astrophys. Suppl. **3**, 349 (1971).
[h] P. Wannier, G. T. Wrixon, R. W. Wilson, Astron. Astrophys. **18**, 224 (1972).
[i] J. van Kuilenberg, Astron. and Astrophys. **16**, 276 (1972).
[j] J. van Kuilenberg, Astron. and Astrophys. Suppl. **5**, 1 (1972).

TABLE IV. Detailed Surveys of Smaller Regions

Area covered (deg)	Grid size (deg)	Beam size (arc min)	Velocity resolution km/sec	Authors and Remarks
$l = 355-5$ $b = -5-+5$	0.25	36	4	Braes[a]
$l = 43-56$ $b = -4.5-+4.5$	0.5 some 1	36	2	Burton[b]
$l = 344-356$ $b = 0-10$	1	36	3.5	Cugnon[c]
$l = 230-290$ $b = -15-+15$	10 (in l) 5 (in l)	48	3	Davis[d]
$\lvert b \rvert \geq +80$	1	53	0.4, 6	Dieter[e]
$l = 120-170$ $b = 30-70$	3	180	22	Encrenaz et al.[f]
$l = 280-340$ $b = 32.5-57.5$	5 (in l) 2.5 (in b)	36	3.4	Fejes[g]
$l = 270-310$ $b = -3-+2$	1 (in b)	28	2	Garzoli[h]
$l = 195-208$ $b = -2-+5$	1	35	6	Girnstein[i]
$l = 200-212$ $b = -3.5-+3.5$	1	35	6	Girnstein and Rohlfs[j]
$l = 191-223$ $b = 15.4-22.4$	0.1	10	2	Gordon[k]
$l = 0-360$ $b = 30$	0.5	30 × 40	2.3	Grahl et al.[l]
$l = 100-140$ $b = 13-17$	0.17	10	1	Heiles[m]
$\alpha = 03^h14^m-04^h14^m$ $\delta = 21-37$	2 (in δ)	108 × 168	2.5	Helfer and Tatel[n]
$l = 132-247$ $b = -16.5-+13.5$	5 (in l) 2.5 (in b)	120 × 150	6.6	Höglund[o]
$l = 67-92$ $b = -10-+10$	5, some 2.5	102	3.3	Kaftan-Kassim[p]
$l = 355-5$ $b = -2-+2$	0.1	14.5	7.5	Kerr and Vallak[q]

(continued)

TABLE IV. (*continued*)

Area covered (deg)	Grid size (deg)	Beam size (arc min)	Velocity resolution (km/sec)	Authors and Remarks
$l = 165–195$ $b = -15–+5$	1	36	2	Lindblad[r]; less extensive average to $l = 240$, $\delta = -25$.
$l = 162–193$ $b = -11–+9$	0.5	36	2	Locke *et al.*[s]
$l = 192–216$ $b = -11––26$	3	102	3.3	Menon[t]
$l = 356–2$ $b = 0.5–5.0$	0.5	36	3.5	d'Odorico *et al.*[u]
$l = 161–179$ $b = -8–+9$	1, some 0.5	34	2	Raimond[v]
$l = 97–150$ $b = -1–+1$	0.16 (in l) 1 (in b)	12	1.6	Rickard[w]
$l \sim 120–170$ $b \sim 35–60$	2.5 (in δ)	10	1.8	Rickard[x]
$l = 8–22$ $b = -0.6–+0.9$	0.5, some 0.25	33	4	Rougoor[y]
$l = 150–170$ $b = -10––2.6$	0.5	36	2	Sancisi[z]; Sancisi and Wesselink[aa]
$l = 2–8$ $b = 33–39$	0.5, some 0.35	36	2	Sancisi[bb]
$l = 341–2$ $b = 12–26$	1	36	2	Sancisi and van Woerden[cc]
$l = 10–24$ $b = -8–+5$	0.5	36	2	Simonson and Sancisi[dd]; Simonson[ee]
$l = 36–49$ $b = -10––22$	1	36	3.4	Smith[ff]
$\alpha = 20^h04^m–20^h44^m$ $\delta = 38–46$	1	108×48	4.4	Sorochenko[gg]
$l = 120–240$ $b = -30–+30$	10 (in l) 0.5 (in b)	35	2.4	Velden[hh]
$\alpha = 02^h30^m–04^h00^m$ $\delta = 64–68$	0.25	10	1.5	Verschuur[ii]
$\alpha = 16^h00^m–17^h30^m$ $b = 69–73.5$	0.25	10	1.5	Verschuur[jj]

(*continued*)

TABLE IV. (continued)

Area covered (deg)	Grid size (deg)	Beam size (arc min)	Velocity revolution (km/sec)	Authors and Remarks
$l = 302–310$ $b = 2–12$	0.5	29	2	Vieira[kk]
$l = 96–110$ $b = 66–72$	1	36	2	Wesselink[ll]
$l = 70–86$ $b = 2–8$	1	84	2.6	Winnberg[mm]

[a] L. L. E. Braes, *Bull. Astron. Inst. Neth.* **17**, 132 (1963).
[b] W. B. Burton, *Astron. Astrophys. Suppl.* **2**, 291 (1970).
[c] P. Cugnon, *Bull. Astron. Inst. Neth.* **19**, 363 (1968).
[d] R. J. Davis, *Smithsonian Contrib. Astrophys.* **5**. No. 13 (1962).
[e] N. H. Dieter, *Astron. J.* **69**, 288 (1964); **70**, 552 (1965).
[f] P. J. Encrenaz, A. A. Penzias, R. Gott, R. W. Wilson, and G. T. Wrixon, *Astron. Astrophys.* **12**, 16 (1971).
[g] I. Fejes, *Astron. Astrophys.* **11**, 163 (1971).
[h] S. L. Garzoli, *Astron. Astrophys.* **8**, 7 (1970).
[i] H. G. Girnstein, *Veroff Univ. Sternw. Bonn.* No. 66, (1963).
[j] H. G. Girnstein and K. Rohlfs, *Z. Astrophys.* **59**, 83 (1964).
[k] C. P. Gordon, *Astron. J.* **75**, 914 (1970).
[l] B. H. Grahl, O. Hachenberg, and U. Mebold, *Beitr. Radio-astron. Bonn.* **1**, 1 (1968).
[m] C. Heiles, Catalog of Hydrogen-Line Data at Intermediate Latitudes. Nat. Radio Astron. Observatory, Green Bank, West Virginia, 1966; *Astrophys. J. Suppl. Ser.* **15**, 97 (1967).
[n] H. L. Helfer and H. E. Tatel, *Astrophys. J.* **129**, 565 (1959).
[o] B. Höglund, *Ark. Astron.* **3**, 215 (1963).
[p] M. A. Kaftan-Kassim, *Astrophys. J.* **133**, 821 (1961).
[q] F. J. Kerr and R. Vallak, *Aust. J. Phys. Astrophys. Suppl.* No. 3 (1967).
[r] Per Olaf Lindblad, *Bull. Astron. Inst. Neth.* **19**, 34 (1967).
[s] J. L. Locke, J. A. Galt, and C. H. Costain, *Astrophys. J.* **139**, 1066 (1964).
[t] T. K. Menon, *Astrophys. J.* **127**, 28 (1958).
[u] S. d'Odorico, R. Sancisi, and S. C. Simonson, *Astron. Astrophys.* **1**, 131 (1969).
[v] E. Raimond, *Bull. Astron. Inst. Neth. Suppl.* **1**, 33 (1966).
[w] J. J. Rickard, *Astrophys. J.* **152**, 1019 (1968).
[x] J. J. Rickard, *Astron. Astrophys.* **11**, 270 (1971).
[y] G. W. Rougoor, *Bull. Astron. Inst. Neth.* **17**, 381 (1964).
[z] R. Sancisi, *Astron. Astrophys.* **4**, 387 (1970).
[aa] R. Sancisi and P. R. Wesselius, *Astron. Astrophys.* **7**, 341 (1970).
[bb] R. Sancisi, *Astron. Astrophys.* **12**, 323 (1971).
[cc] R. Sancisi and H. van Woerden, *Astron. Astrophys.* **5**, 135 (1970).
[dd] S. C. Simonson and R. Sancisi, *Astron. Astrophys. Suppl.* **10**, 283 (1973).
[ee] S. C. Simonson, *Astron. Astrophys.* **12**, 136 (1971).
[ff] G. P. Smith, *Bull. Astron. Inst. Neth.* **17**, 203 (1963).
[gg] R. L. Sorochenko, *Astron. Zh.* **38**, 478 (1961).
[hh] L. Velden, *Beit. Radio Astron.* **1**, 172 (1970).

by hand and make contour maps. In the mid-1960's, Heiles[8] and, independently, Kerr[9] and Westerhout[10] introduced maps generated automatically by computer. This was made necessary by the large number of data points involved in their surveys. The making of such maps has now evolved to the making of computer-generated movies, shown at IAU Symposium No. 38 by Weaver and by Westerhout. These show, with increasing time, maps at successive galactic longitudes; the maps consist of contours of antenna temperature as a function of velocity and galactic latitude with constant longitude.

Modern observational astronomy is beset with the problem of large amounts of data,[11] but probably more so in this field than any other. The magnitude of the problem can perhaps be appreciated by considering the fact that the Berkeley survey of the northern sky, in its three parts,[12-14] covers more than 2.0×10^5 independent positions with at least 100 spectral points. The data are functions of three variables: two angles and velocity. Only when data are presented in the most appropriate (and succinct) form for the problem at hand can one hope to deduce anything other than the most elementary, trivial, and obvious conclusions. One should be able to project data onto any two-dimensional plane, with integration over appropriate limits in the third dimension if desired. However, the attainment of such flexibility requires much programming effort, and the production of maps requires much sorting and large amounts of computer time. Thus, hydrogen-line surveyors often become so disillusioned with their projects, after having spent years collecting data, that they prefer to turn, temporarily at least, to other research topics instead of interpreting their data. However, they are generally happy to provide computer tapes containing the data for anyone temerarious enough to be interested.

We now turn exclusively to the problems of measurement. Compared to

[8] C. Heiles, Catalog of Hydrogen-Line Data at Intermediate Latitudes. Nat. Radio Astron. Observatory Green Bank, West Virginia, 1966.

[9] F. J. Kerr, *Aust. J. Phys. Astrophys. Suppl.* No. 9 (1969).

[10] G. W. Westerhout, "Maryland Green Bank Galactic 21-cm Line Surveys," 2nd ed. Univ. of Maryland, College Park, Maryland, 1969.

[11] B. G. Clark, *Annu. Rev. Astron. Astrophys.* **8**, 115 (1970).

[12] N. H. Dieter, *Astron. Astrophys. Suppl.* **5**, 21, 313 (1972).

[13] C. Heiles and H. J. Habing, *Astron. Astrophys. Suppl.* **14**, 1 (1974).

[14] H. Weaver and D. R. W. Williams, *Astron. Astrophys. Suppl.* **8**, 1 (1973); **17**, 1 (1974).

[ii] G. L. Verschuur, *Astron. J.* **74**, 597 (1969).

[jj] G. L. Verschuur, *Astrophys. J.* **156**, 861 (1969); *Astron. Astrophys.* **1**, 473 (1969); **3**, 77 (1969).

[kk] E. R. Vieira, *Astrophys. J. Suppl. Ser.* **22**, 369 (1971).

[ll] P. R. Wesselius, *Astron. Astrophys.* **1**, 476 (1969).

[mm] A. Winnberg, *Ark. Astron.* **4**, 533 (1969).

molecular or recombination lines, the astronomer's choice of technique is severely restricted when it comes to observing the 21-cm line. This is because hydrogen exists all over the sky. Therefore, one cannot beam switch or use a reference horn; one is forced either to observe a reference load or to obtain a reference spectrum at a different frequency. Each technique has the problem, however, of establishing the zero level. Although the building of a well-matched cold load presents some technical problems, this approach is generally not considered because the load should be cooled to some tens of degrees Kelvin in order to match the antenna temperature. The alternative technique of frequency switching introduces additional problems because of receiver gain and system temperature variation with frequency. (See Sections 4.2.6 and 4.3.2.) These are usually accepted in order to avoid the difficulties associated with using a cold load. In fact, except for the work done at Bell Laboratories,[15,16] all of the work quoted in Tables I–IV was obtained by frequency switching. Switching is generally done symmetrically nowadays to reduce baseline problems. The switching interval should be large enough to avoid errors from galactic high-velocity hydrogen. Extragalactic hydrogen, usually at positive velocities, can produce additional errors.

4.4.3. Baseline Determination in Emission Measurements

The seriousness of the baseline uncertainty depends, of course, on the regions surveyed and the purposes to which the data will be put. Near the galactic plane, brightness temperatures are large so that baseline uncertainties are of little concern; away from the plane, temperatures are lower. These considerations have to some extent, influenced the care to which various observers have gone in the matter of baseline determination. For example, Westerhout,[10] in his galactic-plane survey, found sinusoidal baselines, with period $\simeq 1$ MHz and amplitude $1°K$ in antenna temperature, from special wide-band calibration observations. After correcting all the survey data, he found that the baselines were still accurate only to $\pm 1.5°K$ (by intercomparison of different data on adjacent regions). The Australians refer all zero levels to $\alpha = 22^h30^m$, $\delta = -30°$, "the coldest point discovered so far, and the emission can be taken as zero...."[9] The brightness temperature here is small over a region several degrees in diameter[17] (see also the Bell Laboratories measurements cited below). The Berkeley group has surveyed nearly the whole northern sky at closely spaced intervals in three surveys. In two of

[15] P. J. Encrenaz, A. A. Penzias, R. Gott, R. W. Wilson, and G. T. Wrixon, *Astron. Astrophys.* **12**, 16 (1971).

[16] P. Wannier, G. T. Wrixon, and R. W. Wilson, *Astron. Astrophys.* **18**, 224 (1972).

[17] J. V. Hindman and F. J. Kerr, *Aust. J. Phys. Astrophys. Suppl.* No. 18, 43 (1970).

these, the baseline is determined by linear interpolation in velocity and position to regions having less emission (presumably zero) than elsewhere (Dieter,[12] the high-velocity survey; Weaver and Williams,[14] the galactic-plane survey). The same procedure is also followed implicitly by those making drift scans of high-velocity hydrogen regions.[18]

The third Berkeley survey[13] covers galactic latitudes outside the strip bounded by galactic latitude |10°|. Baseline corrections were accomplished by the following rather elaborate procedure. The antenna temperature at each velocity was smoothed in angle to a beamwidth of about 10°. A smooth function was fit to these smoothed data. Deviations of individual observations from this fit are caused both by instrumental error and by small-scale structure. There are 100 channels, therefore 100 deviations for each observation; these frequency profiles were least-squares fit to determine baseline zero error and slope, as well as gain errors. After correction, the baselines and gains are all internally consistent within the survey; the baseline correction for the survey as a whole is then obtained from the zero level measurements of Wrixon and Heiles,[19] and the intensity scale is matched to the Dutch scale (see below). Baseline uncertainties were reduced to less than 0.1°K using this technique. The corrected data do not appear in the contour maps of Heiles and Habing,[13] but exist on the magnetic tape version. The details of this correction procedure are given by Heiles,[6] and the above description is somewhat oversimplified.

With their typical persistence and attention to detail, the Dutch have attacked this problem more rigorously, probably mainly because of their pioneering efforts in the field of high-velocity hydrogen. They chose a number of regions (see Burton[20]) which have weak, narrow emission around zero velocity, and which were thought to have none elsewhere (but see Wrixon and Heiles[19]); they interpolated their zero line across the emission features to obtain the instrumental baseline. Analysis of repeated observations showed that the baseline was a function of date, frequency, and telescope position; they least-squares fit the baselines to these three variables and subtracted the result from their survey profiles. This reduced baseline uncertainty from 1 to <0.3°K (see Muller et al.[21]).

The above are basically empirical determinations of baseline. A more fundamental (but correctable) contribution to baseline uncertainty is radiation entering the stray portion of the telescope's response pattern, first

[18] S. Y. Meng and J. B. Kraus, Astron. J. 75, 535 (1970).
[19] G. T. Wrixon and C. Heiles, Astron. Astrophys. 18, 444 (1972)
[20] W. B. Burton, Astron. Astrophys. Suppl. 2, 261 (1970).
[21] C. A. Muller, E. Raimond, U. J. Schwarz, and C. R. Tolbert, Bull. Astron. Inst. Neth. Suppl. Ser. 1, 213 (1966).

discussed by van Woerden.[22] Its influence became apparent by comparing profiles of calibration regions taken in different seasons. The seasonal variation of the relative Doppler corrections for line radiation entering the stray beam from different parts of the sky caused seasonal variations in baseline of several tenths of a degree. The correction procedure used by van Woerden[22] and Habing[23] is described by Raimond.[24] The stray beam was approximated by the sum of an isotropic pattern and a spillover pattern, each of which was assumed to contribute 12.5% of the total integrated directive gain. The 21-cm line data over that part of the sky above the horizon were convolved with the assumed stray pattern and the result (which depends on hour angle) was subtracted from the measured profile. This procedure eliminated the seasonal effects, but it is not used as standard practice by the Dutch astronomers. Tolbert[5] did not use it in his high-velocity survey (which is concerned with weak profiles), even though this effect is expected to provide as much as 0.4°K antenna temperature at galactic latitudes of about 60° where the spillover ring is tangent to the galactic plane.

Baseline uncertainties of a few tenths of a degree, accomplished by several of the above mentioned observers, are rather good for a conventional frequency-switched system. However, they are rather bad in absolute terms. For example, if one desires the projected column density of hydrogen—for example, to compare with soft x-ray data—he must compute the integrated area under the 21-cm line profile. A baseline uncertainty of 0.3°K over a 2-MHz bandwidth yields an uncertainty of about 3×10^{20} hydrogen atoms per cm², or more than 100% uncertainty in many high-latitude regions. However, one cannot expect to do better without extraordinary care in construction of equipment and in calibration.

An existing facility where such care has been taken is the horn-reflector antenna of the Bell Telephone Laboratories at Crawford Hill, Holmdel, New Jersey, a full description of which has been given by Crawford et al.[25] In this antenna, a horizontal mounted horn is used to illuminate a section of a parabolic reflector. This arrangement eliminates the multiple reflections of radiation which are often present in reflecting antennas and which can produce "ripples" on a 21-cm spectrum baseline.[26]

The absence of reflected radiation in the horn makes it an extremely broad band device with a low inherent return loss which is limited only by the transition to the horn. Due to its shielded construction, it has low sidelobes

[22] H. van Woerden, De Neutrale Waterstof in Orion. Dissertation, Groningen Univ. (1962).

[23] H. J. Habing, *Bull. Astron. Inst. Neth.* **18**, 323 (1966).

[24] E. Raimond, *Bull. Astron. Inst. Neth. Suppl. Ser.* **1**, 33 (1966).

[25] A. B. Crawford, D. C. Hogg, and L. E. Hunt, *Bell Syst. Tech. J.* **40**, 1095 (1961).

[26] A. A. Penzias and E. H. Scott, III, *Astrophys. J.* **153**, L7 (1968).

and backlobes and, consequently, a high beam efficiency, almost 92% at 21-cm.[27] The equivalent temperature of the antenna when pointed at the zenith is about 7°K, only about 2°K greater than would be obtained with a perfect antenna free of sidelobes and loss. The antenna has a spillover lobe whose amplitude is more than 35 dB below the main beam. The spillover lobe is confined to a small region of the sky about 70° away from the main beam, in the direction away from the feed. Since the antenna has an unlimited rotation altitude–azimuth mount, it is possible to observe any point in the sky at several hour angles, or 180° away in azimuth, to check whether any stray radiation is indeed entering the antenna.

The antenna is switched against a liquid-helium-cooled termination, thus eliminating many of the problems associated with frequency switching which were described above. Calibration is achieved using a low-loss, room-temperature variable attenuator located between the reference port of the radiometer switch and the cold load. This attenuator, also used to balance the system, has been calibrated. Thus, each measurement is calibrated against an absolute scale. Because all the problems associated with baseline determination have been taken care of in this system, flat baselines are obtainable. Also, as a consequence of using a maser as the first stage of rf amplification, a low system temperature of about 40°K is achieved, making highly sensitive studies of the hydrogen line possible.

The present authors have used the Crawford Hill antenna to examine a number of regions in the sky thought to be suitable for baseline calibration.[19] These regions were mostly selected from Heiles' survey[6] as having particularly low hydrogen-emission intensity; in addition, a number of so-called "zero-level" fields, currently in use for baseline calibration purposes by the Dutch and Australians, were examined. Figure 1 shows profiles for two of these regions which contain a small amount of hydrogen that is also restricted in velocity width. Many of the regions had smaller antenna temperatures but larger velocity spreads, often about 200 km/sec. The point at $\alpha = 22^h30^m$, $\delta = -30°$, is the region used by the Australians.

4.4.4. Calibration of Scale

The calibration of antenna temperature scale is, both in principle and in practice, no different for line work than for continuum work. The usual technique is to use a well-calibrated noise tube. Although there are technical problems associated with accurate calibration, we shall disregard such contributors to calibration inaccuracy in the following discussion. Atmospheric extinction should be allowed for in precise work; it amounts to less than 1% attenuation at the zenith.[28]

[27] A. A. Penzias, R. W. Wilson, and P. J. Encrenaz, *Astron. J.* **75**, 141 (1970).
[28] P. J. Encrenaz, A. A. Penzias, and R. W. Wilson, *Astron. Astrophys.* **9**, 51 (1970).

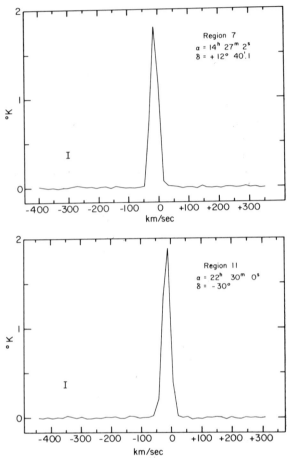

FIG. 1. Two of the regions measured by Wrixon and Heiles [*Astron. Astrophys.* **18**, 444 (1972)]. The profiles show antenna temperature as a function of velocity; brightness temperature is obtained by dividing by the beam efficiency, 0.92, if the hydrogen distribution is uniform over the area of the main beam. The velocity resolution is 15.8 km/sec. The bar on the left side of each profile shows its peak to peak noise.

The problem peculiar to the 21-cm line, of course, is the conversion of antenna temperature to brightness temperature. In the case of point sources, one simply refers to existing absolute measurements (for example, those of Baars and Hartsuijker[29]) of bright sources as primary standards. For hydrogen, however, the brightness temperature distribution must be integrated over not only the main beam, but particularly the near-in sidelobes and the remaining contributors to the stray factor as well. Naive use of a beam

[29] J. W. M. Baars and A. P. Hartsuijker, *Astron. Astrophys.* **17**, 172 (1972).

efficiency factor can lead to serious errors, especially for telescopes with poor surfaces or large diffraction lobes. An outstanding example of this is the pre-1970 version of the NRAO 91.4-m (300-ft) telescope, where surface irregularities caused nearly half of its response to 21-cm line radiation to come from the "error pattern," a region several degrees in diameter centered on the main beam.[10,30]

Historically, brightness temperature scales were determined independently at different observatories. The Dutch system, even now, is based on the assumption that the response of the 7.5-m Würzburg paraboloid to a pine-covered hill (with T_b assumed equal to $T = 290°K$) was 270°K.[31] The Dutch workers are still careful to point out that their profiles are really given in terms of "Dwingeloo units," approximately equal to °K derived ultimately from the pine woods calibration of Volders and Högbom.[32] The Australians, on the other hand, referred their first calibrations to measurement of the Sun with a standard horn. The agreement between the two was quite satisfactory, at least in published work.[33] For the 64-m (210-ft) Parkes telescope in Australia the beam efficiency is 0.8,[34] and the Australians assume all of the stray radiation arises from spillover.[9] This is undoubtedly a very good approximation since the telescope is under-illuminated for 21-cm line work, thus reducing response from diffraction sidelobes. The use of under-illumination also very considerably simplifies the conversion of antenna temperature to brightness temperature.

At Green Bank, Westerhout[10] has made a concerted attempt to correct his galactic-plane survey for the error pattern of the 300-ft NRAO telescope. The correction involves measuring the response pattern of the telescope down to −40 dB from the peak response, assuming the stray factor (but *not* including that portion arising from the error pattern) to be 19%. Brightness temperatures are then derived from antenna temperatures by a deconvolution procedure. The resulting scale is reported to be accurate to within 2% by Penzias *et al.*[27] who made an absolute determination of the integrated area under the hydrogen line profile with the horn reflector of Bell Laboratories, described above. However, one should be cautious in accepting this result because the brightness temperature varies significantly within the error pattern, thus rendering the validity of Westerhout's deconvolution procedure somewhat uncertain; furthermore, the integration under Westerhout's profiles may produce additional errors due to baseline uncertainty. Although we

[30] C. Heiles and W. Hoffman, *Astron. J.* **73**, 412 (1968).
[31] C. A. Muller and G. W. Westerhout, *Bull. Astron. Inst. Neth.* **13**, 151 (1957).
[32] L. Volders and J. A. Högbom, *Bull. Astron. Inst. Neth.* **15**, 307 (1961).
[33] F. J. Kerr, J. V. Hindman, and C. S. Gum, *Aust. J. Phys.* **12**, 270 (1959).
[34] J. V. Hindman, *Aust. J. Phys.* **20**, 147 (1967).

share the hope that Westerhout's scale[10] is indeed accurate to 2%, a more accurate determination should involve comparing profile shapes directly.

Comparison of the three scales shows that the Dutch temperatures are 7% higher than Green Bank's while the Australian temperatures are 18% lower.[10] A number of standard regions for intercomparison of intensity scales have been recommended by a subcommittee of Commission 40 of the IAU.[35] They were selected on the basis of their simply shaped profiles, and were thought to have little angular variation of brightness temperature. The subcommittee recommends S8 ($l = 207°$, $b = -15°$) as the primary standard; S7 ($l = 132°$, $b = -1°$) and S9 ($l = 356°$, $b = -4°$) as secondary standards. D. R. W. Williams (private communication) has mapped these regions, as well as some others which had been proposed earlier, and finds rather uncomfortably large variations in brightness temperature with position, although in some cases the gradient is nearly uniform. Unfortunately, observers have not published profiles of these regions, and, at present, intercomparisons have been circulated only informally.

4.4.5. Absorption Measurements

For a frequency-switched system, the antenna temperature in the 21-cm line, $T_a(v)$, is given by a somewhat modified form of the equation of transfer:

$$T_a(v) = \eta T_{ex} - T_a^c (1 - e^{-\tau(v)}), \tag{4.1.1}$$

where T_{ex} is the 21-cm excitation temperature, T_a^c is the continuum antenna temperature of a source lying behind the cloud, and $\tau(v)$ is the optical depth in the line at the frequency v. η is a factor less than unity which depends in part, on the relative solid angles of the hydrogen cloud, the radio source, and the telescope beam. This equation is, of course, an oversimplification; in general, there will be some continuum emission arising in front of, or even inside of, the cloud, and there will be several clouds in the beam. An improved equation is given by Kerr,[36] and a complete equation for OH (but also applicable to the 21-cm line) is given and discussed by Goss.[37]

There are two unknowns in the equation of transfer, T_{ex} and $\tau(v)$. If $\tau(v)$ is small, the antenna temperature is simply proportional to the number of hydrogen atoms; but, in general, one would also like to determine T_{ex} and $\tau(v)$ as well. This can be done if two measurements of the same cloud are obtained which have different values of T_a^c. This is the goal of most absorption studies. The studies with the procedure is, of course, the determination of the emission profiles that would be seen from only that hydrogen in front of, but in the

[35] H. van Woerden, *Trans. IAU* **16b**, 217 (1970).
[36] F. J. Kerr, *in* " Nebulae and Interstellar Matter " (B. M. Middlehurst and L. H. Aller eds.). Univ. Chicago Press, Chicago, Illinois, 1968.
[37] W. M. Goss, *Astrophys. J. Suppl. Ser.* **137**, 131 (1968).

absence of, the source (the "expected profile"). The fact that the hydrogen emission changes with position means that a profile taken off the radio source will differ intrinsically from the expected profile. Some idea of the extent to which the emission varies with position can be obtained by mapping the hydrogen in the neighborhood of the source, but the nagging uncertainty of the true, expected profile can never disappear.

A less ambitious (but more realistic) objective is to obtain a profile resulting from only the hydrogen lying in front of the source. This is not easy, however, because the telescope responds to the hydrogen surrounding the source as well as to the gas in front of the source. This surrounding emission must be subtracted out. One way to do this is to obtain emission profiles around the source in various directions, and to use the average as the probable contribution from hydrogen surrounding the source itself. This requires uniform gradients of hydrogen line intensity over regions equal to at least three telescope beamwidths in diameter. This requirement is probably never satisfied in practice, as shown, for example, by the interferometric measurements by Clark,[38] and the aperture synthesis measurements of Greison[39] and Elliot.[40]

This problem can be alleviated to some extent by realizing that one need not go completely off source to obtain the second measurement; one only requires some change in T_a^c. The amount of change required depends simply on the desired accuracy in the final result and the signal-to-noise ratio in the measurements. Radhakrishnan et al.[41] have used the 64-m (210-ft) Parkes telescope in Australia for absorption work by placing the source near the half-power point on the beam, where a small change in position produces a large change in T_a^c. Heiles (unpublished) has used the 42.7-m (140-ft) telescope at NRAO by taking two measurements, with the telescope directed at the source, one of which is defocussed (thus decreasing T_a^c).

A better method, suitable for small sources, is to use an interferometer at a baseline which is just short of beginning to resolve the source. The fringes cancel out the 21-cm emission very effectively, especially if observations are conducted over a range of hour angle. For small sources with continuum brightness temperatures $\gg T_{ex}$, the optical depth profile is thereby determined directly. Recent surveys have been performed in the northern sky by Hughes et al.,[42] and in the southern sky by Radhakrishnan et al.[41] The classic study of the strong sources was performed by Clark.[38]

[38] B. G. Clark, Astrophys. J. 142, 398 (1965).

[39] E. Greison, paper presented at URSI, Washington, D.C. (1972).

[40] D. Elliot, thesis, UCLA (1973) (unpublished).

[41] V. Radhakrishnan, W. M. Goss, J. D. Murray, and J. W. Brooks, Astro Phys. J. Suppl. Ser. 24, 49 (1972); V. Radhakrishnan, J. D. Murray, P. Lockhart, and R. P. J. Whittle, Astrophys. J. Suppl. Ser. 24, 15 (1972).

[42] M. P. Hughes, A. R. Thompson, and R. S. Colvin, Astrophys. J. Suppl. Ser. 23 323 (1971). See also R. D. Davies and E. R. Cummings, Mem. Roy, Astron. Soc. 170, 95 (1975).

A third method is to use polarized or time variable sources, either a time variable continuum source or a pulsar.[43] Successful absorption measurements have been performed on pulsars (e.g., by Gordon and Gordon [44,45]).(See Section 4.5.6.)

Although these methods do provide reliable profiles of optical depth, the problem of obtaining the expected profile required to solve the equation of transfer still remains. Despite the uncertainties, however, existing work— most particularly, the interferometric work referenced above— has indicated, with quite considerable force, that the temperature of interstellar hydrogen cannot be taken uniformly as 125°K, as was inferred from earlier work because of the apparent tendency of hydrogen emission profiles to saturate at this value. Instead, the derived temperatures cluster around lower values, and also at values higher than a lower limit of nearly 1000°K.[41,42] These results are in accord with recent theoretical developments concerning the two-phase character of the interstellar medium.[46,47]

A fourth method can sometimes circumvent the difficulty of obtaining an expected profile. Gardner and Whiteoak[48] measured the absorption due to linearly polarized component of Cent A. Since the polarized emission in this source varies slowly with angle, it can be assumed to be uniform within the telescope beam. However, most polarized sources are not so large in angular extent, which will cause problems similar to those discussed in the above paragraph. An additional difficulty in interpretation will occur if the source is not extragalactic or small in physical extent. For example, if the galactic synchrotron continuum radiation is used, the polarized emission may be generated at several locations relative to the HI along the line of sight. However, the most serious impediment to the use of this technique is the small degree of polarization of nearly all 21-cm continuum radiation. The instrumental pitfalls of measuring weakly linearly polarized line radiation are the same as for circularly polarized radiation, which are discussed in Section 4.4.7.

4.4.6. Baseline Determination in Absorption Measurements

For absorption work, the determination of accurate baseline is even more difficult than for emission work, but has somewhat different aspects. The problem of the ubiquity of hydrogen still remains. In the past, this has been somewhat alleviated by the fact that the optical depth τ in Equation (4.4.1) is

[43] C. Heiles and G. L. Verschuur, *Astrophys. Lett.* **3**, 21 (1969).
[44] K. J. Gordon and C. P. Gordon, *Astrophys. Lett.* **5**, 153 (1970).
[45] K. J. Gordon and C. P. Gordon, *Astron. Astrophys.* **27**, 119 (1973).
[46] G. B. Field, D. W. Goldsmith, and H. J. Habing, *Astrophys. J. Lett.* **155**, L149 (1969).
[47] R. M. Hjellming, C. P. Gordon, and K. J. Gordon, *Astron. Astrophys.* **2**, 202 (1969).
[48] F. F. Gardner and J. B. Whiteoak, *Astrophys. Lett.* **11**, 123 (1973).

proportional to the reciprocal of the excitation temperature of the line, and hence to the kinetic temperature.[49] Cold gas tends to be clumped, and so absorption spectra usually consist of a number of narrow components separated by velocity intervals containing very little absorption, at least in sources away from the galactic plane. These intervals have usually been taken as having zero absorption. The hot hydrogen is also of interest, especially recently, and it is just this component whose absorption can be apparent only in these intervals. As is often the case, the quantity of most interest is the most difficult to measure.

Thus, the true baseline determination, in addition to being subject to the usual grave difficulties outlined above in the emission case, will also depend on the continuum antenna temperature T_a of the source being observed. The amplitude of the baseline is dependent on T_a because the baseline shape is caused primarily by rf gain variations with frequency. These yield a baseline shape factor which simply multiplies the total system temperature. However, the receiver noise temperature itself is not independent of frequency. Thus, a better approximation to the baseline consists of a shape factor which depends on total system temperature. An additional contributor to such a factor is the back radiation occurring in reflecting antennas, which produces sinusoidal ripples in the baseline whose amplitude is proportional to T_a.[26] The baseline ripples can be alleviated to a considerable extent by combining observations taken with feed-focus positions differing by $\lambda/8$.

Finally, we have assumed that the i.f. imperfections have been perfectly eliminated through use of a reference spectrum. This is, in fact, not justified. The very act of moving the telescope off source to obtain the reference spectrum changes the total system temperature, and hence the i.f. signal level. M. Gordon (private communication) has shown that the NRAO Model II autocorrelator had an i.f. level-dependent bandpass characteristic. Such problems will, in general, occur with any radio spectrometer. For example, changing the input level to a multichannel receiver changes the operating points of the various detectors, and hence their relative gains. For these reasons, one might attempt to keep the input level to his spectrometer the same for both the source and reference spectra. This is often accomplished by inserting attenuation at the i.f. frequency. But this act in itself changes the i.f. frequency response by a small amount.

The best possible way to solve these problems is to obtain a reference spectrum on a source having the same intensity as the source being observed. Unfortunately, however, nature is not usually so kind as to provide one, especially one nearby in the sky. One can try " noise injection," which is simply to inject the proper amount of excess noise in the same way a noise

[49] G. B. Field, *Proc. IRE* **46**, 240 (1958).

tube is used for calibration. But it is difficult to ensure that the input noise level is frequency-independent, and, in any case, the sinusoidal baseline from back radiation (see above) cannot be duplicated in this manner.

In short, one must experiment with each particular system to discover its deficiencies, and calibrate accordingly. Successful examples of such experimentation are reported by Sandqvist[50] and Dent.[51] In Sandqvist's experiment, the constraints were particularly severe because he studied fine-scale structure in the galactic center using Lunar occulation; the system temperature varied in time scales of seconds due to motion of the moon.

The problem of baseline in absorption work is alleviated to large measure by use of an interferometer, since all instrumental properties of the two telescopes are uncorrelated. Here, the baseline uncertainty is strictly proportional to the fringe amplitude, i.e., the unresolved portion of the source intensity, which is usually much smaller than the total system temperature. However, the sinusoidal ripple due to back reflections, mentioned above, will still be present in an amount proportional to the full source intensity. In addition, reflections in the i.f. system (which includes variable delay lines) may produce additional problems.

4.4.7. Zeeman Splitting and Polarization

Verschuur[52] has presented the historical record of the attempts to detect Zeeman splitting of the 21-cm line, made mainly at Jodrell Bank. True success required long integration times with modern spectrometers. Verschuur[53] has summarized the present situation: the examination of a large number of positions has led to the unambiguous detection of splitting in only five of them. Verschuur is pessimistic about the possibilities for many more successful detections.

A magnetic field directed along the line of sight induces small changes in line frequency which depend on the sense of circular polarization of the radiation. The detection of this Zeeman splitting, therefore, requires the measurement of the profile in the two senses of circular polarization, and taking their differences. The magnetic fields so far detected are of the order of 10^{-5}, which induces a splitting of only about 20 Hz. Since this is much less than the width of any interstellar line, the difference profile will look like an " S " lying on its side, whose width is proportional to the width of the original profile, and whose amplitude is proportional to the line of sight component of the magnetic field divided by the width. The fact that the amplitude of the

[50] A. Sandqvist, thesis, Univ. of Maryland (1971) (unpublished).

[51] W. A. Dent, *Astrophys. J.* **165**, 451 (1970).

[52] G. L. Verschuur, *in IAU Symp. No. 31: Radio Astron. Galactic Syst.* (H. van Woerden, ed.), p. 385. Academic Press, New York, 1967.

[53] G. L. Verschuur, *Astrophys. J.* **165**, 651 (1971).

"S" is small (typically $<10^{-3}$ of the amplitude of the line) means, aside from the fact that long integration times are required, that care must be taken to eliminate systematic effects.

We shall not attempt to reproduce the comprehensive discussion of such effects given by Verschuur.[54] Similar considerations apply to measuring linear polarization in the 21-cm line itself, which arises from absorption of polarized background radiation, and has been detected by Gardner and Whiteoak[48] (see Section 4.4.5). Elimination of the simplest of these effects is accomplished by realizing that it is preferable to observe both polarizations simultaneously, using the same local oscillator, in order to ensure that the profiles in the two polarizations are precisely aligned. The most difficult effect to account for is the difference in beam patterns of the two circularly polarized feeds, which is largest near the edges of the main beam patterns. If the brightness temperature in the line varies with position, the difference profile will be affected. This is not particularly serious for the Zeeman studies of the strongest continuum sources in absorption, where the source antenna temperature is very large; but when attempting to measure splitting in emission regions where antenna temperatures are small, the relative contribution is much greater. Consequently, one should attempt to examine regions which are not only interesting from an astronomical viewpoint, but which also have little variation of emission with position.[53] Unfortunately, no Zeeman splitting has yet been detected for the 21-cm line in emission.

[54] G. L. Verschuur, *Astrophys. J.* **156**, 861 (1969); *Astron. Astrophys.* **1**, 473 (1969).

4.5. Pulsar Observing Techniques

4.5.1. Pulsar Radio Emission

Pulsars, as a class of galactic radio sources, have characteristics which are totally unlike other radio sources. Because of this, special observing techniques have been evolved which deserve separate discussion. First, a brief review of the major characteristics of pulsar radio emissions will be presented.

4.5.1.1. Pulsed Nature. The primary class characteristic of pulsars—that from which the name is derived—is the pulsed nature of their radio emission. An observation of a train of successive pulses from a typical pulsar reveals a great deal of detailed structure—some seemingly random, some definitely systematic—which varies considerably from pulse to pulse. The synchronous average of several hundred successive pulses results in a stable and repetitive "average pulse shape." This average pulse shape reflects the probability of occurrence of individual narrow pulses more than it represents any typical pulse, and therefore it should more properly be referred to as the "pulse window." Although emission occurs at varying phases with respect to the primary pulsation period P_1, it is confined to the narrow and well-defined pulse window. The fractional width of the pulse window is quite small, typically 5%, but ranges from 1 to 20% for the ~70 known pulsars.

Repeated determinations of the pulse window demonstrate that it is a very stable property of individual pulsars. No intrinsic temporal variations have been reported. The shapes of pulse windows range from simple Gaussian-like profiles to quite complex profiles consisting of up to five separate components. In only four instances do separate components of the pulse window exist at widely different phases (midphase pulses). The separation, width, and relative amplitudes of the components of a pulse window are generally slowly varying functions of the observing frequency.

The primary pulsation period P_1 ranges from 0.033 to 3.7 sec for the known pulsars, but is typically near 0.7 sec. The period P_1 is well-behaved, can be specified at a given epoch to a part in 10^{10} for most pulsars, and is gradually increasing at a rate $(P_{1,t})$ of, typically, a few times 10^{-15} sec/sec. A discontinuous behavior in the period and period derivatives has been observed for

---

* Chapter 4.5 is by G. Richard Huguenin.

the two shortest period pulsars, PSR 0531 + 21 and PSR 0833-45, which are in the Crab and Vela supernova remnants, respectively.

4.5.1.2. Polarization and Spectra. The individual pulses of radio emission from a pulsar are generally elliptically polarized, typically to a high degree. The polarization properties vary considerably from pulse to pulse, often in a systematic fashion. The long term (> 100 pulses) average polarization is a stable property of each pulsar. The average pulsar radiation is linearly polarized, with the position angle rotating systematically across the pulse window. A small amount of circular polarization has also been observed in some pulsars. Examples of the pulse window and the average polarization are shown in Fig. 1. The percentage polarization changes in a

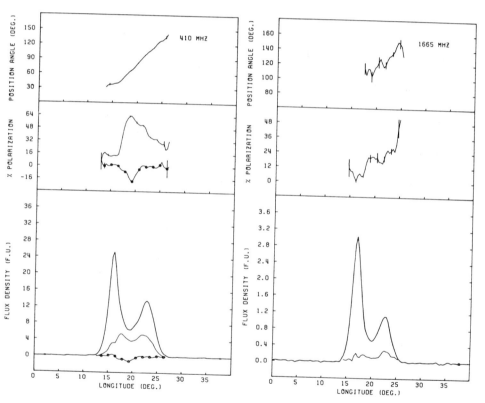

FIG. 1. Pulse window and average polarization properties of PSR 1133 + 16 [from R. N. Manchester, *Astrophys. J. Suppl. Ser.* **23**, 283 (1971)]. Bottom: total intensity, linear, and circular components. The circular component, defined in the sense LH–RH, is plotted with small circles every fourth point. Center: linear and circular percentage polarization with ±2σ error bars. Top: position angle of the linear component, also with ±2σ error bars.

systematic way with frequency; however, the position angle variation across the pulse window remains the same at all frequencies.

The large intensity variations exhibited by pulsars complicate the determination of their radio frequency spectrum. In particular, slow variations (those with time scales ranging from days to years) make it necessary to exercise extreme care in interpreting spectra. The spectra of pulsars are, typically, inverse power laws, with spectral indices considerably larger than unity and thus much larger than spectral indices of typical radio sources. Pulsars are therefore, primarily meter-wavelength objects.

4.5.1.3. Intensity Variations. The intensity of the radio emission from pulsars is observed to fluctuate over all time scales from tens of microseconds to tens of months—the extremes of time resolution thus far employed! While most of these intensity variations are believed to be intrinsic to the source, time variations resulting from interstellar and interplanetary scintillations are present under many observing conditions.

Although known to exist, intensity variations over time scales less than 1 msec have not been studied extensively because of great observational difficulties. Intensity variations of several orders of magnitude with time scales ranging from milliseconds to hours have been extensively studied. Random as well as periodic fluctuations in amplitude and phase are known to occur. Satisfactory theoretical explanations of these variations are still lacking. A daily sampling of intensity variations over a period of several years shows amplitude ranges of typically, 10 to 1, and characteristic time scales of variation of several tens of days.

4.5.1.4. Propagation Effects. The individual pulse of radiation is emitted simultaneously over a wide range of frequencies, and then propagates to the observer through the interstellar medium which contains many free electrons. Such a medium is dispersive, and the velocity of propagation v for waves of frequency f is given by

$$v = c(1 - f_p^2/2f^2), \qquad f \gg f_p, f_h, \qquad (4.5.1)$$

where c is the vacuum velocity of light, f_p the electron plasma frequency, and f_h the electron gyrofrequency.

Let us consider observations of the same pulsar made at two frequencies, f_1 and $f_2 < f_1$. The difference of times of arrival of a pulse at the two frequencies is given by

$$t_2 - t_1 = K(1/f_2^2 - 1/f_1^2)DM \quad [\text{sec}], \qquad (4.5.2)$$

where the dispersion measures DM (in cm^{-3} parsec) is defined as

$$DM \equiv \int_{\text{obs}}^{\text{puls}} n_e \, dl \quad [\text{cm}^{-3} \text{ pc}],$$

and the constant K is given by

$$K = (2\pi^2 e^2/\varepsilon_0 mc)(3.0856 \times 10^{18}) \quad [\text{cm pc}^{-1}].$$

The dispersion constant DC is defined as

$$DC \equiv (t_2 - t_1)/(1/f_2^2 - 1/f_1^2) \quad (= 4.149 \times 10^{15} DM),$$

and is the observed quantity. Generally, the dispersion constant DC can be observed with a precision which is better than our knowledge of the physical constants which make up the constant K. For this reason, the dispersion constant DC is often reported when high precision results are obtained.

Observations indicate that the dispersion of pulsar signals follows to very high precision the simple f^{-2} relationship given by Eq. (4.5.2). The absence of higher order (f^{-4}) terms implies that the pulsar signal does not propagate through any significant region containing a high electron density or a large magnetic field.

The propagation of the pulsar radiation through the interstellar medium effects the emitted radiation in ways other than simply dispersing the pulse arrival times. Specifically, irregularities in the interstellar medium cause small angle scattering which results in multipath propagation. The effects of multipath propagation are twofold: first, to induce continually changing frequency structure in the spectrum of the pulsar; and second, to broaden the received pulse in time.

An example of the scintillation-produced structure in the spectrum of pulsar PSR 0329 + 54 is shown in Fig. 2. The individual spectral features can be described statistically by a characteristic frequency bandwidth and a characteristic time span over which the observed intensities decorrelate. These quantities, known as the decorrelation bandwidth Δf_s and the decorrelation time τ_s, depend on observing frequency f (MHz) and dispersion measure DM approximately as follows:

$$\Delta f_s \simeq 3 \times 10^{-9}(f^4/(DM)^2) \quad [\text{MHz}], \qquad \tau_s \simeq 10(f/(DM)^{1/2}) \quad [\text{sec}] \quad (4.5.3)$$

In addition to amplitude scintillations, multipath propagation results in pulse broadening. Propagation along direct and scattered ray paths causes the emitted pulse to be asymmetrically broadened in time. The observed pulse profile is thus a convolution of the emitted pulse with an (approximately) exponential function whose characteristic time width τ_B is just Δf_s^{-1}. Thus, from Eq. (4.5.3):

$$\tau_B = \Delta f_s^{-1} \simeq 3 \times 10^5((DM)^2/f^4) \quad [\text{msec}]. \qquad (4.5.4)$$

The constant factors in Eqs. (4.5.3) and (4.5.4) are suitable for approximate calculations; however, large deviations can exist for individual pulsars.

An additional effect of propagation through the interstellar medium is

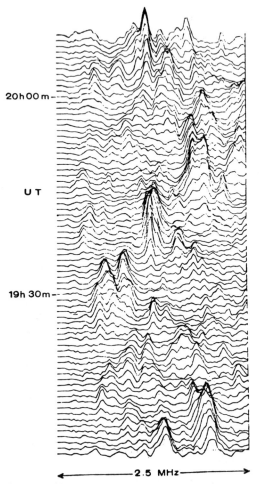

FIG. 2.   Scintillation produced spectral structure in PSR 0329 + 54 [from B. J. Rickett, *Nature* (*London*) **221**, 158 (1969)]. Successive mean pulse spectra from PSR 0329 + 54 (CP 0328) at 408 MHz at intervals of about 50 sec. The frequency resolution is 60 kHz, and the spectra include the effect of the receiver bandpass.

Faraday rotation. The position angle of the linearly polarized component of the pulsar emission is rotated an amount $\phi_2 - \phi_1$ between two observing wavelengths $\lambda_2 - \lambda_1$ according to the relation

$$\phi_2 - \phi_1 = RM(\lambda_2{}^2 - \lambda_1{}^2) \quad \text{[rad]}. \tag{4.5.5}$$

The rotation measure RM is defined as

$$RM = 0.812 \int_{\text{obs}}^{\text{puls}} n_e \mathbf{B} \cdot d\mathbf{l} \quad \text{[rad/m}^2\text{]},$$

where n_e is the electron density in cm^{-3}, $\mathbf{B}$ the magnetic field in microgauss, and l is measured in parsecs.

4.5.2. Sensitivity and Time Resolution

The necessity for submillisecond time resolution in the observations of the intrinsically weak and dispersed signals from pulsars always poses severe sensitivity limitations. For a specified time resolution Δt, the maximum bandwidth of Δf which can be employed is determined by the frequency drift rate $f_{D,t}$ (MHz/sec) resulting from dispersion. The dispersion-produced time smearing τ_D, due to a finite observing bandwidth Δf_o (MHz) at frequency f (MHz), is given by

$$\tau_D = \Delta f_o / |f_{D,t}| = 8.29 \times 10^3 DM(\Delta f_o/f^3) \quad [\text{sec}]. \quad (4.5.6)$$

For a specified time resolution, the maximum observing bandwidth Δf_o which can be employed is given by

$$\Delta f_o = |f_{D,t}|\tau_D = 1.205 \times 10^{-4}\tau_D(f^3/DM)$$
$$= 1.205 \times 10^{-4}(\tau_D/P_1)(P_1/DM)f^3 \quad [\text{MHz}]. \quad (4.5.7)$$

For a constant fractional time resolution τ_D/P_1, the observing bandwidth is then determined by the ratio of pulsar quantities P_1/DM, making short-period, high-dispersion pulsars the most difficult to observe. The time smearing due to dispersion can appear, occasionally, to be less than that indicated by Eq. (4.5.6) if the scintillation bandwidth Δf_s is smaller than the observing bandwidth Δf_o. In this circumstance, a single strong scintillation band can contain the dominant signal contribution in the observing bandwidth. Likewise, the simultaneous presence of two or more narrow scintillation bands within the receiver bandpass can produce complicated smearing effects.

Intensity fluctuations of time scale τ_s will be observed when the observing bandwidth $\Delta f_o \leq \Delta f_s$. In this situation, it is often quite worthwhile to be able to search in frequency for a strong scintillation band. Improvements in the observed signal-to-noise ratio of more than an order of magnitude can often be obtained by patient searching.

Ignoring scintillation effects, the rms fluctuations of the receiver output, in flux units, is given by

$$\Delta S_{rms} \simeq \frac{2kT_s/A_e + S_p}{(\Delta f_o \tau_o)^{1/2}}.$$

And, from Eq. (4.5.7),

$$\Delta S_{rms} \simeq \frac{64(DM/P_1)^{1/2}(2kT_s/A_e + S_p)}{\gamma^{1/2}f_o^{3/2}\tau_D}, \quad (4.5.8)$$

where T_s is the system temperature, [°K], A_e the antenna effective area [m^2], S_p the pulsar flux averaged over τ_0 in matched polarization, and $\gamma = P_1\tau_0/\tau_D$, the fractional time smearing due to dispersion. It is important to retain the pulsar flux S_p in Eq. (4.5.8), especially when τ_0 is in the submillisecond range since, over these very short time scales, the pulsar antenna temperature can occasionally far exceed the system temperature.

Another consequence of dispersion is the setting of a minimum observable time resolution independent of signal-to-noise considerations. Structures with time scales less than the reciprocal predetection bandwidth Δf_0 are not directly observable. Thus, $\tau_0 \geq \Delta f_0^{-1}$, and therefore the best time resolution attainable for a given pulsar occurs when $\Delta f_0 \tau_0 = 1$, or, from Eq. (4.5.6),

$$\tau_0 \geq 9.1 \times 10^{-2} DM^{1/2} f^{-3/2} \quad [\text{sec}]. \tag{4.5.9}$$

For the pulsar PSR 0950 + 08, which has the smallest measured dispersion, the best time resolution attainable is 20 μsec at 400 MHz, using a 50 kHz bandwidth. A larger bandwidth would produce more dispersion smearing, and a smaller bandwidth would increase the filter response time. Higher time resolution is theoretically attainable with narrower bandwidths by deconvolution of the filter response.[1]

4.5.3. Dispersion Removal

The signal-to-noise ratio attainable on existing telescopes with a simple total power receiver is not sufficient for the observation of individual pulses from a reasonable sample of pulsars. A device to remove the frequency dispersion and permit a larger effective observing bandwidth is clearly desirable. A number of such "dedispersers" have been employed for pulsar observations.

The individual pulse $s_T(t)$, as emitted at the pulsar, can be considered as a sum of its Fourier components. Thus, following the notation of Hankins,[2]

$$s_T(t) = \int_{-\infty}^{\infty} S(f) \exp(i2\pi ft) \, df. \tag{4.5.10}$$

Propagation over a distance z through the interstellar medium introduces a frequency dependent phase factor, $\exp(-ik(f)z)$, and the received pulse $s_R(t, z)$ becomes

$$s_R(t, z) \sim \int_{-\infty}^{\infty} S(f) \exp(i2\pi ft) \exp(-ik(f)z) \, df, \tag{4.5.11}$$

[1] T. W. Cole, *Astrophys. Lett.* **12**, 181 (1972).
[2] T. H. Hankins, *Astrophys. J.* **169**, 487 (1971).

where $k(f)$ is given, from Eq. (4.5.1), as

$$k(f) = (2\pi f/c)(1 - (f_p^2/2f^2)).$$

In theory, dispersion removal consists of multiplying the Fourier transform of the received signal by the factor $\exp(ik(f)z)$, and performing the inverse transform. In practice, varying approximations to this procedure are employed. There are basically two categories of dedispersing: coherent dedispersing, in which the phase relations between frequencies are preserved; and incoherent dedispersing, in which phases are ignored. Both categories of dedispersing would yield identical results if there were no frequency coherence present in the pulsar radiation. There are, as yet, no published results of tests for frequency coherence in pulsar radiation at radio frequencies.

For the study of all but the most detailed submillisecond structure, the simpler techniques of incoherent dedispersing suffice. The most practical dedispersing system in routine use is that of Taylor and Huguenin,[3] which employs a multichannel filter receiver and a tapped, variable delay line. A block diagram of the system is shown in Fig. 3. The desired time resolution

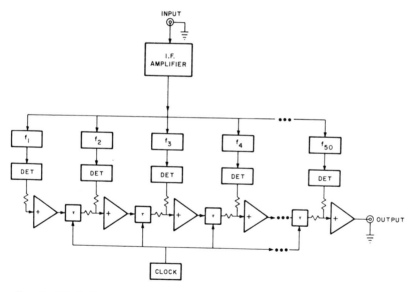

FIG. 3. Block diagram of the University of Massachusetts dedisperser. Boxes labeled $f_1, f_2, \ldots, f_{50}$ represent filters tuned to adjacent frequencies, in decreasing order; boxes labeled DET are square-law detectors; boxes labeled τ represent delay circuits, the timing of which is controlled by a programmable clock; large triangles represent operational amplifiers used as summing junctions.

[3] J. H. Taylor and G. R. Huguenin, *Astrophys. J.* **167**, 273 (1971).

τ_0/P_1 and observing frequency f define the bandwidth Δf_0 of each individual channel of the multichannel filter bank [Eq. (4.5.7)]. The detected output of each channel is entered into the appropriate tap of the variable delay line, with the output of the highest frequency filter connected to the input of the delay line, and the output of the lowest frequency filter connected to the output of the delay line. The delay per stage of the delay line is adjusted to equal the arrival time difference between adjacent channel centers, and is therefore different for each pulsar. The tapped, variable delay line used by Taylor and Huguenin[3] is an analog shift register which has been described by Orsten.[4] A digital shift register is to be used in a similar system under development at Arecibo. A bonus of the multichannel dedispersing device is the fact that broadband impulsive interference "spikes" are dispersed—thereby sharply diminishing their amplitude. The signal-to-noise ratio improvement for observations made with the multichannel dedisperser is $N^{1/2}$, or about 7 for a 50-channel system.

A swept-frequency postdetection dedispersing device has been used by Sutton et al.[5] for pulsar observations. Again, a multichannel filter receiver is employed. The local oscillator frequency of the receiver is carefully controlled to sweep in synchronism with the dispersed pulse. The oscillator sweep is synchronized in time so that the pulse window is centered in the middle of the multichannel filter bank. The detected outputs of adjacent individual channels of the properly synchronized system represent the pulse amplitude at adjacent phases within the pulse window. The frequency resolution of the filter receiver is thus translated into time resolution across the pulse window. The increase in signal-to-noise over a simple total power receiver is proportional to the square root of the ratio of the pulse tracking time to the time resolution between adjacent channels. The pulse tracking time is limited if each successive pulse is to be observed—a limitation not present in the delay line system. Operationally, the swept-frequency device is not as convenient to use as the delay line device because of its far greater complexity, and because of the requirement for precise synchronization. This sychronization requirement precludes the use of the swept-frequency device for the study of very weak pulsars.

The coherent, predetection dedispersing technique devised by Hankins[2] is accomplished off-line in a general purpose computer. Baseband signals are burst sampled, digitized, and recorded on magnetic tape for subsequent processing. The offline processing consists (conceptually) of: (a) Fourier transforming the burst of samples covering a single pulse: (b) applying the phase correction $\exp(ik(f)z)$ as indicated by Eq. (4.5.11); and (c) Fourier transforming the corrected frequency components to obtain the dedispersed

[4] G. S. F. Orsten, *Rev. Sci. Instrum.* **41**, 957 (1970).
[5] J. M. Sutton, D. H. Staelin, R. M. Price, and R. Weimer, *Astrophys. J.* **159**, L89 (1970).

pulse shape. Predetection bandwidths of 125 kHz at frequencies between 100 and 200 MHz have been employed by Hankins,[6] and have shown statistically significant pulses with unresolved durations of $(125 \text{ kHz})^{-1}$, or 8 μsec, and fluxes of $\sim 35,000$ f.u. The rather formidable amount of computer time necessary to realize coherent, predetection dedispersing has limited the widespread use of this very important tool in the study of the detailed pulsar emission process.

4.5.4. Period and Dispersion Determinations

Signal averaging techniques are used to great advantage in the study of average pulsar properties, such as the determination of average polarization properties and precise pulse arrival times. Average polarization properties over a wide range of frequencies are necessary for understanding the pulsar emission process, and for determining interstellar Faraday rotation. Time of arrival observations at different frequencies are required for the determination of the fundamental pulsation period and its time derivatives, for accurate positions, and for an accurate determination of dispersion measure.

The phase resolution "bin" of the signal averagers typically employed for pulsar timing observations is of the order of 10^{-3} of the fundamental period P_1. Interpolation to about 10% of the bin spacing is possible, and thus time resolution of $10^{-4}P_1$, or ~ 50–100 μsec, has become routine. In order not to smear the average pulse over integration times of the order of 10^3 to 10^4 sec, it is necessary to account for the changing doppler shift due to the observer's motions. This can be accomplished by integrating for shorter periods ($\sim 10^2$ periods) and making off-line corrections before compiling a grand average, or by updating the signal averager trigger rate every few hundred pulse periods.

The large variations of polarization within the pulse window make it necessary to observe with a single circular polarization (which is unaffected by ionospheric Faraday rotation), or more preferably, to sum orthogonal polarizations in order to obtain stable pulse shapes for accurate pulse arrival time applications. The cross correlation of carefully prepared pulse shape templates with the observed pulse profile usually permits arrival times to be determined to a small fraction (~ 10%) of a phase bin in the signal averager. Thus, daily average arrival times can typically be determined to within about 10–100 μsec—depending on the pulsar's period, dispersion measure, pulse shape, and the observed signal-to-noise ratio. Fluctuation in apparent arrival times can occur if the scintillation bandwith Δf_s [Eq. (4.5.3)] is smaller than the observing bandwidth Δf_0, and if the integration times are of the same order as the scintillation band lifetime τ_s. Clock systems capable of epoch specifica-

---

[6] T. H. Hankins, *Astrophys. J.* **177**, L11 (1972).

tion to better than 10 μsec are required for precision pulse arrival time observations.

The determination of pulsar dispersion measures is accomplished by measuring relative arrival times at two or more observing frequencies. Typically, a rough dispersion measure determination is made at two closely spaced frequencies to eliminate whole pulse period ambiguities. Observations are then made over a much wider frequency spacing to produce accurate dispersion measures. Observations at three or more frequencies are required for those pulsars that have pulse shapes which change with frequency—due to either intrinsic pulse shape variations or strong interstellar scattering. Dispersion-measure variations with time have been observed thus far only in the Crab pulsar, PSR 0531 + 21.[7]

4.5.5. Polarization Observations

The polarimeters employed in pulsar observations should be of the type which simultaneously produce the four signals necessary for a complete definition of the polarization state. Because of the inherently large variations in the polarization and intensity of individual pulses, and the consequent long times necessary to obtain stable averages, rotating feed type polarimeters are even less efficient than they are in the more usual continuous source application. Techniques similar to those normally used in the calibration of continuous source polarimeters are also employed in the calibration of pulsar polarimeters.

Accurate polarimetry requires carefully matched pre- and, especially, post-detection filters in each of four channels in order not to introduce differential time delays and therefore a spurious polarization. In practice, the post-detection filters must be matched to within 10% of the phase bin width of the signal averager. The predetection filters must not only have closely matched center frequencies and bandwidths, but they must also have similar skirt shapes because of the possible presence of strong scintillation bands near the edge of the passband.

The polarization of individual pulses has been studied, using conventional total-power systems for the strongest pulsars.[8,9] More recently, the University of Massachusetts group has employed a matched set of four dedispersers to study individual pulse polarization for a larger number of pulsars. Observing polarization with the University of Massachusetts type dedispersing

[7] J. M. Rankin and J. A. Roberts, *IAU Symp. No.* **46**: *The Crab Nebula* (R. D. Davies and F. G. Smith, eds.), p. 114. Reidel Publ., Dordrecht, 1971.

[8] D. A. Graham, A. G. Lyne, and F. G. Smith, *Nature (London)* **225**, 526 (1970).

[9] J. H. Taylor, G. R. Huguenin, R. M. Hirsch, and R. N. Manchester, *Astrophys. Lett.* **9**, 205 (1971).

hardware puts a limit on the total bandwidth which can be employed because of Faraday rotation effects. The differential rotation $|\Delta\phi|$ across an observing bandwidth Δf (kHz) is given, from Eq. (4.5.5), as

$$|\Delta\phi| = 1.03 \times 10^{-2} \, RM(\Delta f/(f/100)^3) \quad [\text{deg}]. \quad (4.5.12)$$

Dedispersing minimizes time smearing to that of a single channel Δf_i of the multichannel system, but Faraday rotation occurs across the total observing bandwidth $\Delta f_1 = N \, \Delta f_i$. The maximum number of channels N_p which can be used for polarization observations is independent of observing frequency, and depends on the fractional time smearing τ_D/P_1 of an individual channel and the differential rotation $|\Delta\phi|$ permitted across the total bandwidth. For typical values of $\tau_D/P_1 = 10^{-3}$ and $|\Delta\phi| = 5$ deg, the maximum number of channels is given as

$$N_p = 8 \times 10^{-4} \frac{|\Delta\phi|}{\tau_D/P_1} \frac{1}{P_1} \frac{DM}{RM} = \frac{4}{P_1} \frac{DM}{RM}. \quad (4.5.13)$$

Dedispersion and derotation could be accomplished in a "two-dimensional" version of the multichannel dedispersing polarimeter discussed above. Derotation could also be incorporated into the software dedispersing technique used by Hankins if the appropriate four polarization channels were digitized. Such complex systems have not yet been constructed.

Changing ionospheric electron content limits the integration times which can be employed in polarization observations—especially near sunrise and sunset. The averaging together, or even the comparison, of observations made on different days requires an ionospheric Faraday rotation correction. Total electron content observations of satellite-borne radio beacons have been employed for this purpose.[10]

4.5.6. Spectral Observations

Observations of the radio-frequency spectra of pulsars fall basically into two categories: those made over wide frequency intervals ($\Delta f/f > 10\%$) in order to establish the broad emission spectrum, and those made over narrow bands ($\Delta f/f < 10\%$) to study interstellar scintillation, interstellar absorption lines, and intrinsic fine-scale spectral structure. The specification of the gross features of the radio-frequency spectrum of a given pulsar is complicated in practice by slow intensity and spectral variations. Because of the presence of intensity variations over all time scales, from microseconds to years, it is important to properly sample the intensity waveform to insure freedom from aliasing errors and to minimize sampling noise. A program to properly elucidate the gross spectral feature of pulsars has not yet been completed,

[10] R. N. Manchester, *Astrophys. J. Suppl. Ser.* **23**, 283 (1971).

although a long-term program is underway at the Five College Radio Astronomy Observatory, where observations for several hours each day at two observing frequencies, 157 and 390 MHz, are obtained for analysis.

Pulsars have been reliably observed at frequencies as low as 40 MHz and as high as 8 GHz, although the vast majority of observations have been made between 80 MHz and 3 GHz.

Observations of the narrow band features in the spectra of pulsar emission have been obtained using both multichannel and digital autocorrelation spectrometers. The only practical system for observing the narrow band spectra of *individual* pulses is the multichannel spectrometer. The amount of information gathered by such a system is enormous and efficient ways of storing, analyzing, and displaying the data is of utmost importance.

The most versatile system for obtaining *average* spectra is the autocorrelation spectrometer. The autocorrelation spectrometer at Jodrell Bank was the first to be employed for pulsar spectral observations, and was used to detect the 21-cm line of HI in absorption in the spectrum of PSR 0329 + 54.[11] The NRAO and Jodrell Bank spectrometers have both been employed in the subsequent studies of λ21-cm HI, λ18-cm OH, recombination lines, and scintillation spectral observations.

The pulsed nature of the radio emission from pulsars simplifies the process of obtaining the comparison spectrum required for absorption studies. Thus, during the portion of the period ($\sim$5–10%) that the pulsar is emitting, the "signal" spectrum is obtained, and during the portion of the period (90–95%) that the pulsar is off, the reference spectrum is obtained. The basic switching is done by the pulsar itself, and it is only necessary to synchronize the spectrometer with the pulsar. The best signal-to-noise ratio is obtained when the "signal" spectrum window is matched to the dispersion-broadened width of the average pulse. Likewise, the reference spectrum should be taken over the entire interval between pulses. In practice, signal averaging techniques are used to synchronize the spectrometer with the pulsar.

4.5.7. Interferometric Techniques

Interferometric observations of pulsars are valuable for several reasons, particularly to determine positions independently of timing observations and to search for nonpulsed emission.[12] The optimum signal-to-noise ratio can be obtained when the interferometer correlators are gated on synchronously with the pulsar pulse. Likewise, by gating out the pulsar pulse and observing

---

[11] G. DeJager, A. G. Lyne, L. Pointon, and J. E. B. Ponsonby, *Nature (London)* **220**, 128 (1968).

[12] G. R. Huguenin, J. H. Taylor, R. N. Hjellming, and C. M. Wade, *Nature (London) Phys. Sci.* **234**, 50 (1971).

only between the pulses, a sensitive search for a nonpulsed component of pulsar emission can be conducted.

4.5.8. Search Techniques

Two rather unique characteristics of pulsar emissions are exploited in the search for new pulsars: their periodic pulsed emission, and the frequency dispersion of pulse arrival times. Searches have typically exploited one or the other of the characteristics, and have therefore been relatively more sensitive to different classes of pulsars. Some pulsars, as typified by the Crab Nebula pulsar, PSR 0531 + 21, emit occasional large pulses, and are best found by searching for individual dispersed pulses. Other pulsars, such as the Vela pulsar, PSR 0833-45, have a much smaller range of intensity fluctuations, and so searches exploiting their periodic pulsed nature can be more sensitive. An analysis of the sensitivity of various period analysis search techniques has been published by Burns and Clark.[13]

A majority of the pulsars cataloged to date have been found by searching analog chart records for pulsed emission. However, the most recent discoveries have used more sophisticated techniques. The recent Jodrell Bank survey[14] relies on the pulsed nature of pulsars by fast sampling the detected output of a single channel, and analyzing for periodic emissions. The planned University of Massachusetts survey[15] exploits, primarily, the dispersed nature of pulsed emission by sampling the detected outputs of a multichannel receiver, and looking for dispersed pulses. In order not to become inundated in huge piles of magnetic tape, both the Jodrell Bank and the University of Massachusetts searches employ on-line computers to do real-time processing of the received signals. The long-term variation of a pulsar's intensity by an order of magnitude or more, with time scales of tens of days, means that the same region of the sky must be repeatedly surveyed if all pulsars above a certain average flux are to be detected.

ACKNOWLEDGMENTS

The author wishes to acknowledge many useful discussions with his colleagues, especially J. H. Taylor, R. N. Manchester, A. Hartai, A. Rodman, and G. Orsten. This research was supported by NSF Grant GP-32414X.

[13] W. R. Burns, and B. G. Clark, *Astron. Astrophys.* **2**, 280 (1969).
[14] J. G. Davies, A. G. Lyne, and J. H. Sieradakis, *IAU Circ.* 2436 #(1972).
[15] J. H. Taylor, *Astron. Astrophys. Suppl.* (1973) (in press).

4.6. Lunar Occultation Measurements*

4.6.1. Lunar Occultations

Even the largest steerable pencil-beam antennas operated at their highest possible frequencies are limited to a primary resolution in excess of 1 arc min. However, by using such instruments not to study a source directly, but to investigate the behavior of the received power as the source is occulted by the Moon, it is possible, even at meter wavelengths, to achieve a resolution of the order of 0.1 arc sec.

In its orbit around the Earth the Moon moves in a plane which is inclined to the ecliptic at an angle of about 5.1°, and the normal to this plane precesses in space at such a rate that the intersection of the lunar orbit with the ecliptic ("the line of nodes") completes one revolution in 18.6 yr. Taking into account the size of the Moon and its parallax, it follows that occultations are confined to a strip of sky about 12° wide centered on the ecliptic, and that the sources in this strip undergo occultations in an 18.6-yr cycle. During this cycle, sources close to the ecliptic undergo two series of occultations, separated by about 9.3 yr, as the ascending and descending nodes pass through the region of sky in which they lie, while sources at and near 5° from the ecliptic will undergo only one (but, in general, more numerous) series of occultations. Only a few of these occultations will be visible from any one geographical location, but, by using instruments at different sites, it is possible to build up a large body of data on a particular source and to construct high resolution maps similar to those produced by aperture synthesis techniques.

In the following chapter, attention will be confined to the use of the occultation technique in the measurement of the positions and structures of small angular size radio sources with continuous spectra. However, it is also a powerful tool for the study of radio spectral-line emitting regions, and is one of the few optical methods of measuring stellar diameters as small as 0.001 arc sec. and detecting and measuring close binary systems. The considerations which follow are readily applicable to these situations.

4.6.2. Method of Observation

In principle, the beam of the antenna is simply directed to the source, and the received power is recorded using a highly stable receiver as the source first passes behind the Moon (immersion) and then reappears (emersion).

---

* Chapter 4.6 is by C. Hazard.

This procedure is satisfactory for intense sources and at low frequencies when the variation in received power due to the motion of the Moon through the beam is negligible compared to the source flux. For weak sources, particularly at high frequencies, the observations are better made by keeping the beam directed at that point on the Moon at which the occultation is predicted to occur, or, in the special case[1] where the beam is large compared to the size of the Moon, keeping the beam directed at the Moon's center. Since the latter procedure requires no knowledge of the source position and records all sources occulted during the period of observation, its use is not confined to predicted occultations but permits the occultation antenna to be used as a survey instrument. A single narrow beam directed at a point on the limb can also be used for survey work, but observations will then be restricted to a single phase of each occultation occurring along the section of limb within the beam.

To achieve the necessary gain stability and sensitivity, it has been usual to employ a switched Dicke-type receiver with a low-noise preamplifier, but, with modern techniques, it is now possible to construct adequate unswitched receivers with a consequent improvement in the signal-to-noise ratio. Observations should be carried out simultaneously on at least two widely separated frequencies. When more than one occultation of a source is observed, a different choice of frequencies then enables a detailed investigation of the variation of spectral index over the source. At each frequency, observations are made using a number of predetector bandwidths (defined by i.f. filters) as discussed in Section 4.6.14. For a preliminary analysis of these, the received power is recorded on a chart recorder using a time constant of 1 sec or greater, while for a more detailed analysis, a time constant of about 0.1 sec is employed, the received power being recorded at corresponding intervals in digital form on magnetic tape.

A serious problem for weak sources is the variation in output level due to tracking errors, and the consequent motion of the Moon in the antenna beam. These tracking errors are least important around 150 MHz where the Moon and sky temperatures are approximately equal, and also when the antenna beam is large compared to the Moon. At high frequencies, however, where the beam width may be only a few minutes of arc and the sky temperature close to zero, tracking errors limit the observations to only the more intense sources. They may be reduced by recording the difference in the power received in two beams kept fixed at diametrically opposed points on the limb, or by using an interferometric system designed to resolve the extended lunar component of the received power. However, an attractive aspect of the occultation technique is that it enables high resolution observations to be made with existing telescopes without extensive modifications or special facilities.

[1] C. Hazard, *Mon. Not. Roy. Astron. Soc.* **134**, 27 (1962).

4.6.3. The Moon as a Straight Diffracting Edge

The occultation technique is concerned only with sources that contain angular structure much smaller than the size of the Moon. The variation in received power as such a source passes behind the Moon depends on the size of the source relative to the size of a Fresnel zone at the Moon's distance, which, over the range of frequencies of interest ranges from a few seconds of arc to a maximum of about 20 arc sec. Provided the source size is much greater than the size of the Fresnel zone, diffraction effects can be neglected, and the shape of the occultation curve represents the strip integral of the brightness distribution in a direction perpendicular to the limb at the point of occultation. The true strip brightness distribution can then be recovered simply by differentiating the observed occultation curve. However, for sources containing angular structure measured in seconds of arc, the diffraction effects at the limb must be taken into account.

Relative to the size of a Fresnel zone the curvature of the Moon's limb is small, and, for an ideal Moon, the occultation curves may be considered as diffraction curves at a straight edge. An exact calculation of the diffraction effects would also take into account the true limb profile which is irregular, with deviations from the mean limb of up to 3 arc sec. Fortunately, except for grazing occultations, the part of the limb which contributes to the development of the diffraction pattern is mainly confined to a section a few Fresnel zones in extent on either side of the point of occultation. Over the range of frequencies of interest this is of the order of a few tens of kilometers, and over this length of limb the scale of the irregularities are such that the limb can usually be approximated by a straight edge parallel to the true limb. The diffraction fringes run parallel to this assumed right edge, which for the purpose of calculation is assumed to be infinite, and the structural information obtained from a single occultation curve is therefore confined to an effective strip distribution, with the elementary strips parallel to the limb at the point of occultation.

An assessment of the reliability of the source structure derived under the simplifying assumption of an ideal straight edge requires a knowledge of the displacement of the true limb from the adopted mean limb. Charts of the irregularities in the lunar profile have been given by Watts[2], from which it is possible to estimate the displacement of the true and mean limbs to an accuracy of about 0.2 arc sec, and also to estimate the effective slope of the true limb to an accuracy of around 1°. Where these charts show limb irregularities of the order of 1 or 2 arc sec occurring over a section of the limb less than or comparable in size to a Fresnel zone, great care must be taken in the interpretation of the observations.

[2] C. B. Watts, *Astron. Papers Amer. Ephemeris* XVII (1963).

4.6.4. Shape of the Occultation Curve of a Point Source

Consider the flux received by an antenna at O, from a source S, as it is uncovered by a screen M at a distance D which moves along the y-axis from ∞ to $-\infty$ (Fig. 1), where it is assumed that the antenna has a constant gain

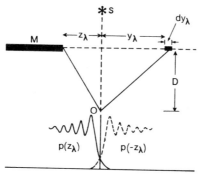

FIG. 1. Schematic representation of the generation of the occultation curve $p(z_\lambda)$, as the screen M at a distance D from the observer uncovers a source S. The occultation curve is the mirror image about the axis OS of the diffraction curve $p(-z_\lambda)$.

over the whole of the wavefront. If the field at O in the absence of the screen is A_0, corresponding to a flux $I_0 = A_0{}^2$, then the amplitude dA due to an element of wavefront at distance y from the line OS is

$$dA = kA_0 \exp(-i\pi y_\lambda{}^2 \lambda/D), \qquad (4.6.1)$$

where k is a normalizing factor and $y_\lambda = y/\lambda$. The power received when the Moon is at a distance $-z_\lambda = -z/\lambda$ from OS is therefore

$$p(z_\lambda) = k^2 A_0{}^2 \int_{-z_\lambda}^{\infty} \exp(-\pi y_\lambda{}^2 \lambda/D)\, dy_\lambda \int_{-z_\lambda}^{\infty} \exp(i\pi y_\lambda{}^2 \lambda/D)\, dy_\lambda.$$

Substituting $u^2 = 2y_\lambda{}^2 \lambda/D$, we obtain

$$p(v) = \frac{k^2 I_0 D}{2\lambda} \int_{-v}^{\infty} \exp(-i\pi u^2/2)\, du \int_{-v}^{\infty} \exp(i\pi u^2/2)\, du$$

$$= k^2 I_0 (D/\lambda) p_0(v), \qquad (4.6.2)$$

where

$$p_0(v) = \tfrac{1}{2}(C(v) + \tfrac{1}{2})^2 + \tfrac{1}{2}(S(v) + \tfrac{1}{2})^2, \qquad (4.6.3)$$

and $C(v)$ and $S(v)$ are the Fresnel integrals

$$C(v) = \int_0^v \cos(\pi t^2/2)\, dt \quad \text{and} \quad S(v) = \int_0^v \sin(\pi t^2/2)\, dt. \quad (4.6.4)$$

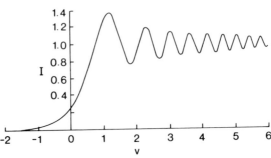

Fɪɢ. 2. The shape of the occultation curve of a point source. The horizontal scale is in units of v, the corresponding angular scale being given by $\theta = v(\lambda/2D)^{1/2}$, where λ is the wavelength and D the Moon's distance. For $\lambda = 1$ m, one unit of v corresponds to about 8 arc sec. $I = 1$ corresponds to the flux density of the unobstructed curve, while at the edge of the geometrical shadow $I = 0.25$.

The variation of $p_0(v)$ with v is shown in Fig. 2, the angular scale of the pattern, ϕ, being given by

$$\phi = z/D = v(\lambda/2D)^{1/2} \text{ (rad)} \approx 2 \times 10^5 v(\lambda/2D)^{1/2} \text{ (arc sec)}. \quad (4.6.5)$$

It can be seen from Fig. 1 that the occultation curve of a point source $p(z_\lambda)$ is the mirror image of the diffraction curve at a straight edge about the axis OS. The diffraction pattern $p(-z_\lambda)$ may therefore be considered as an antenna beam fixed to the edge of the Moon, and the occultation curve to represent the received power as this beam is swept across the source. Further-more, we may consider this beam to arise from an aperture extending from the edge of the Moon to ∞, which is illuminated in such a way as to simulate the phase delays introduced by the differing path lengths from the aperture to the observer and to produce a change in the mean level of the received power equal to I_0 as the source passes behind the Moon. From Eq. (4.6.2), this latter condition requires that $k = (\lambda/D)^{1/2}$, and, from Eq. (4.6.1), the appropriate aperture distribution $E(x_\lambda)$ is given by

$$E(x_\lambda) = (\lambda/D)^{1/2} \exp(-i\pi x_\lambda^2 \lambda/D) H(x_\lambda), \quad (4.6.6)$$

where $H(x_\lambda)$ is the Heaviside step function, x_λ $(= -z_\lambda)$ is measured positive from the edge of the Moon, and the height of the step in the occultation curve (I_0) is normalized to unity.

4.6.5. Time Scale of the Occultation Curve of a Point Source

The power received by an antenna directed at a point source throughout the occultation is shown schematically in Fig. 3, where it is assumed that the variation in the received power due to the motion of the Moon through the beam can be neglected. The time t taken to move 1 arc sec in a direction

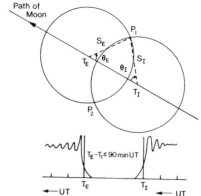

FIG. 3. Schematic representation of the geometry of an occultation. The circles represent the outlines of the Moon at the times of immersion (T_I) and emersion (T_E), when the occultation angles are θ_I and θ_E, respectively. These two times define two possible positions (P_1 and P_2) of the source. If the source is at P_1, the occultation curves at immersion and emersion correspond to the strip brightness distributions along $T_I P_1$ and $T_E P_1$, respectively. One division on the time scale of the occultation curves corresponds to about 1 min UT.

perpendicular to the limb, which gives the rate of motion of the source through the diffraction lobes, and the duration of the occultation ($T = T_{\text{immersion}} - T_{\text{emersion}}$) are given by

$$t = (b \cos \theta)^{-1} \quad \text{and} \quad T = (2s/b) \cos \theta, \qquad (4.6.7)$$

where b arc sec/sec is the apparent rate of motion of the Moon across the stellar background, θ, which will be called the occultation angle, is the angle subtended at the center of the Moon by the Moon's path and the source at the time of occultation, and $s \approx 900$ arc sec is the Moon's semidiameter.

As the Moon's apparent motion is determined both by its true motion and by its changing parallax throughout the occultation, s, b, and θ are, in general, different at immersion and emersion, but on the average, $b \approx \frac{1}{3}$ arc sec/sec. The maximum duration occurs for a central occultation and is of the order of 90 min. For such an occultation, the time taken to move through one unit of v at a frequency of 300 MHz ($\lambda = 1$ m, $v \approx 7.6$ arc sec) is about 20 sec. Both b and θ may be obtained by calculating the position of the Moon for two closely spaced times about the occultation time and from a knowledge of the source position. It will be noted that, for a noncentral occultation, θ varies throughout the occultation, thus producing a nonlinearity in the ratio of the time and angular scales across the diffraction pattern (see Section 4.6.15).

4.6.6. The Occultation Curve of a Source of Finite Size

4.6.6.1. Shape of the Occultation Curves.

As a source passes through the diffraction pattern at the Moon's limb, the received power will vary as in Fig. 2, unless its angular size is comparable to the lobe separation, in which case the pattern will be smoothed out, exactly as the fringes are smoothed out in an interferometer pattern. Figure 4 shows the calculated curves for

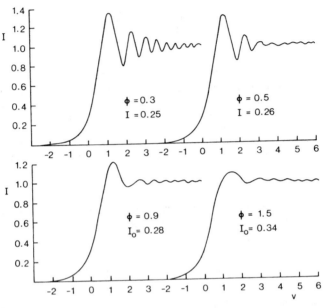

FIG. 4. Occultation curves of a uniform strip source calculated for different values of the angular size. ϕ expressed in units of v, where $v = (\lambda/2D)^{1/2}$, λ being the wavelength and D the Moon's distance (for $\lambda = 1$ m, one unit of v is about 8 arc sec). I is the relative flux density, while I_0 is the relative flux density at the edge of the geometrical shadow ($v = 0$).

different widths of a uniform strip source, showing how the lobe pattern varies with the source size. Provided the lobe pattern is visible, a simple inspection of the analog record of the occultation enables an immediate estimate of the equivalent source size without any knowledge of the circumstances of the occultation.

4.6.6.2. Estimate of Occultation Times.

The occultation times are the times at which the source lies on the Moon's limb and for a source where the diffraction lobes are visible, correspond closely to the points at which the received power falls to 25 % of the unobstructed value. Values of the relative flux density at the edge of the geometrical shadow for sources of larger angular size are indicated in Fig. 4; for sources of even larger size it is sufficiently accurate to measure to the 50% points of the occultation curves.

When several diffraction lobes are visible, more accurate occultation times may be obtained from the positions of the lobes and their known angular displacement from the diffracting edge, the relationship between the time and angular scales being obtained directly from the measured time scale of the occultation pattern and its known angular scale. For this purpose, it is adequate to estimate the Moon's distance from its tabulated apparent semi-diameter, and to ignore the effects of parallax.

4.6.7. Position Measurement

The determination of the source position from the observed occultation times requires a knowledge of the position and size of the Moon as seen by the observer, and, consequently, a calculation of its topocentric coordinates and semidiameter from the geocentric values tabulated in the Astronomical Ephemeris. The required relationships between the topocentric and geocentric coordinates are given in the Supplement to the Astronomical Ephemeris. It should be noted that the geocentric coordinates are tabulated not in Universal Time (UT), but in Ephemeris Time (ET), where ET = UT + ΔT. The value of ΔT varies with time, and is determined from observations of the Moon over an extended period. At the time of an occultation, only an approximate value is available for ΔT and therefore, for the ultimate in accuracy, all position calculations must be repeated at a later period.

Given a complete occultation curve, the times of immersion and emersion define two circles with radii equal to the corresponding radii of the Moon (corrected for limb errors), and hence two possible source positions (see Fig. 3).

If the beamwidth of the antenna is less than the size of the Moon, one of these positions can often be discarded immediately; otherwise the ambiguity can usually be removed using an approximate position obtained by another method.

When more than one occultation has been observed, the position is defined uniquely by the intersection of the limb positions defined by each occultation time but which must first be precessed to a standard epoch. It has been found possible from as few as four occultation times to obtain positions accurate to about ±0.2 arc sec.[3] A detailed analysis, however, is usually only applied after accurate occultation times have been obtained from the detailed structural analysis described below.

In some cases only a single occultation curve at immersion or emersion

[3] C. Hazard, J. Sutton, A. N. Argue, C. M. Kenworthy, L. V. Morrison, and C. A. Murray, *Nature (London) Phys. Sci.* **233**, 89 (1971).

will be available. The occultation time in combination with an alternative position then defines only a section of arc along which the source must lie. An accurate position can, however, in suitable cases, be determined from the occultation data alone. The principle of the method is as follows. The angular scale of the diffraction pattern is calculated in terms of v, and the time scale in terms of v derived from the observed occultation curve. If one unit of v corresponds to t_1, then, from Eqs. (4.6.5) and (4.6.7),

$$t = 5 \times 10^{-6} t_1 (2D/\lambda)^{1/2} = (b \cos \theta)^{-1},$$

and since b is known, this gives a value for θ which locates the source along the Moon's limb. The method is useful when θ is large, but is not suitable for central occultations or for extended sources.

4.6.8. Lobe Analysis and Model Fitting

Lobe analysis techniques are applicable when the source size is smaller than the first Fresnel zone, and the structural information therefore contained entirely in the diffraction lobes. It is also useful when interference on parts of the record precludes the use of the restoration techniques described below. The pattern outside the geometrical shadow approximates to a damped sinusoidal oscillation whose frequency increases with increasing v, and which can therefore be considered as a variable spacing interferometer whose resolution increases with increasing v. Since each lobe, or series of lobes, corresponds to a record which would have been obtained using an interferometer of the correct spacing, the well known theory of interferometers applies, from which it follows that the lobe amplitude, normalized relative to a point source, gives the amplitude of the corresponding Fourier component, and the lobe displacement, relative to that for a point source, gives its phase. The phase information, however, is usually irrelevant since the method is normally applied when the source appears unresolved, and all that is required is the estimated size of an assumed symmetrical model. It should be noted that while in an ideal variable spacing interferometer the signal-to-noise ratio remains constant with increasing resolution or lobe separation, in an occultation curve the signal-to-noise ratio decreases for higher frequency Fourier components. It is this decrease of the signal-to-noise ratio which sets a limit to the resolution which can be obtained using the occultation technique.

It may be shown that the lobe separation at a distance δv from the limb is given by $\delta v = 2/v$ units of v, which gives the angular lobe separation $\delta \phi = \lambda/x$. The spacing of the equivalent interferometer is thus equal to the distance of the limb of the Moon from the line joining the source to the observer.

The effective resolution at each point in the lobe pattern may, to a sufficient degree of accuracy, be obtained from the Moon's mean distance. The change in level as the source passes behind or emerges from the Moon is used to normalize the lobe amplitudes, and the height of the first lobe is used to infer the presence of any broader structure. The method requires neither a detailed knowledge of the theory of occultations nor of the circumstances of the occultation, and provides a rapid method of estimating the source size, which is particularly useful if no computer is available or the record is available only in analog form and must be laboriously digitized before processing in a computer. In particular, it may be noted that a minimum estimate of the source size is obtained by simply noting the extent of the observed lobe pattern.

If the source is complex and the components well separated, each component is analyzed separately. A double source with a projected separation less than one unit of v is revealed by beating in the outer lobes, the minimum of the beating pattern occurring when the patterns of the components are separated by half the lobe separation. When the component separation is such that the interference of the two diffraction patterns occurs in the region from 0 to $3v$, as in Fig. 5b,c, the record may be analyzed by a graphical curve-fitting procedure assuming various component fluxes, separations, and brightness distributions.

4.6.9. The Restoration Technique

Lobe analysis and model-fitting techniques are adequate, provided there is already some evidence of the source structure, the source is not too complex, and the lobe pattern is not too extensive. For complex sources, and particularly for sources with several components of different angular size, the most powerful method of analysis is the restoration technique first described by Scheuer.[4] A discussion of this technique requires a more complete understanding of the nature of the occultation curve than has been developed in the previous sections. The analysis is most conveniently carried out by considering the diffraction pattern to be the beam of a suitably illuminated antenna (Section 4.6.4).

4.6.9.1. The Fourier Transform of the Occultation Curve of a Point Source. The Fourier transform (FT) of the power pattern $p(\phi)$ of a linear antenna is given by the complex autocorrelation function of the field distribution across the aperture, i.e.,

$$\text{FT } p(\phi) = P(a) \propto \int_{-\infty}^{\infty} E(x_\lambda - a)E^*(x_\lambda)\, dx_\lambda, \qquad (4.6.8)$$

where $E(x_\lambda)$ represents the aperture distribution, $x_\lambda = x/\lambda$ is the distance in aperture plane in wavelengths, and $a = x_0/\lambda$ the spatial frequency of a Fourier

[4] P. A. G. Scheuer, *Aust. J. Phys.* **15**, 333 (1962).

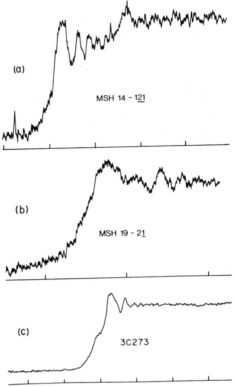

FIG. 5. Occultation curves of three double radio sources observed at Parkes at a frequency of 408 MHz. (a) The two components are well separated and the two diffraction patterns clearly visible; it is obvious that the stronger component has a significantly smaller angular size than the weaker component. (b) The components are separated by approximately 4 arc sec (0.7v), as can be seen from the beating in the lobe pattern at a lobe separation of about 1.5v. (c) The component separation is about 19 arc sec or about 3v.

component of $p(\phi)$. The Fourier transform of the normalized diffraction pattern is obtained by substituting for $E(x_\lambda)$ from Eq. (4.6.6), which gives

$$P(a) = (\lambda/D) \exp(-i\pi a^2 \lambda/D) \int_{-\infty}^{\infty} \exp(2\pi i a x_\lambda \lambda/D) H(x_\lambda - a) H(x_\lambda)\, dx_\lambda.$$

But $H(x_\lambda - a)H(x_\lambda) = H(x_\lambda - a)$ for $a + ve$ and $H(x_\lambda)$ for $a - ve$, from which it follows that

$$P(a) = \lambda/D \exp(i\pi a^2 \lambda/D \operatorname{sgn} a) \int_{-\infty}^{\infty} H(x_\lambda) \exp(2\pi i x_\lambda a\lambda/D)\, dx_\lambda,$$

where

$$\operatorname{sgn} a = \begin{cases} +1 & \text{for} \quad a + ve \\ -1 & \text{for} \quad a - ve \end{cases}$$

and, on substituting $\phi = x/D = x_\lambda \lambda/D$,

$$P(a) = \exp(i\pi a^2 \lambda/D \text{ sgn } a) \int_{-\infty}^{\infty} H(\phi) \exp(2\pi i a\phi) \, d\phi = Q(a) \text{ FT } H(\phi).$$

(4.6.9)

Hence

$$p(\phi) = q(\phi) * H(\phi),$$

(4.6.10)

and differentiating,

$$p'(\phi) = q(\phi) * H'(\phi) = q(\phi) * \delta(\phi) = q(\phi),$$

(4.6.11)

where

$$\text{FT } q(\phi) = Q(a) = \exp(i\pi a^2 \lambda/D \text{ sgn } a).$$

(4.6.12)

We therefore have the interesting result that the normalized diffraction pattern $p(\phi)$ is equal to the Heaviside step function $H(\phi)$ convolved with $q(\phi)$ ($= p'(\phi)$), the differential of the diffraction pattern, where $q(\phi)$ is pure imaginary, and represents the phase distortion introduced by the different path lengths from elements on the wavefront to the observer. Since $q(\phi)$ represents only a phase distortion, it follows that the diffraction pattern of a point source can be converted to a step function or the "geometrical diffraction" pattern by convolution with $q(-\phi)$.

4.6.9.2. The Occultation Curve of an Extended Source. As the diffraction fringes run parallel to the Moon's limb, it is necessary to consider only the case of an equivalent line source oriented perpendicular to the limb. For a source extended in two dimensions, the brightness distribution along this equivalent line source is the integral of the brightness distribution along strips parallel to the limb.

If $p(\phi)$ represents the angular diffraction pattern of a point source, then the observed response $f(\phi)$ to a point source is given by $p(-\phi)$, and the response to a line source with a brightness distribution $t(\phi)$ is the convolution of $p(-\phi)$ with $t(\phi)$. Hence,

$$f(\phi) = t(\phi) * p(-\phi).$$

(4.6.13)

Following the previous discussion, we may now convert to a geometrical occultation curve by convolution with $p'(\phi) = q(\phi)$, giving

$$f(\phi) * p'(\phi) = t(\phi) * p(-\phi) * q(\phi) = t(\phi) * q(\phi) * q(-\phi) * H(-\phi)$$
$$= t(\phi) * H(-\phi),$$

(4.6.14)

where the change in level or height of the geometrical occultation curve is equal to the height of the observed curve, namely, $\int_{-\infty}^{\infty} t(\phi) \, d\phi$. On differentiating Eq. (4.6.14), we have

$$f(\phi) * p''(\phi) = t(\phi) * \delta(\phi) = t(\phi).$$

(4.6.15)

Thus, given an occultation curve of a source of finite size, we may formally recover the true strip brightness distribution across the source by convolution with the double differential of the straight edge diffraction pattern, or cross-correlation with the double differential of the occultation curve of a point source $p''(-\phi)$.

It can be seen from Fig. 2, and is easily shown from elementary considerations, that both the amplitude and separation of the diffraction lobes for a point source decrease inversely as v. It follows that the double differential of $p(\phi)$ on the side remote from the geometrical shadow is an oscillating pattern of increasing amplitude extending to infinity.[4] It must, therefore, be made to converge before it can be convolved with $f(\phi)$, and to produce this convergence Scheuer[4] proposed convolution with a Gaussian beam $g(\phi)$, although in certain circumstances distributions other than Gaussian have certain advantages.[5,6]

From Eq. (4.6.15), and convolving with $g(\phi)$, we have

$$f(\phi) * p''(\phi) * g(\phi) = f(\phi) * c(\phi) = t(\phi) * g(\phi). \qquad (4.6.16)$$

This is the basic expression of the restoration procedure, and states that convolution of an occultation curve with the restoring function

$$c(\phi) = p''(\phi) * g(\phi) \qquad (4.6.17)$$

gives the true strip brightness distribution across the source as seen by a Gaussian beam of arbitrary width $g(\phi)$.

4.6.10. Effect of Finite Antenna Beam

In deriving Eq. (4.6.16) from Eq. (4.6.8), we have considered the observer to receive information with equal weight from all elements in the wavefront in the plane of the Moon. However, only those elements contained within the antenna beam will contribute to the received flux and they will be weighted according to the antenna gain in their direction, thus imposing a limit on the resolution which can be achieved.[7] Since the diffraction fringes run parallel to the diffracting edge, and any antenna can be reduced to an equivalent linear antenna, it is necessary to consider only the case of a long narrow antenna oriented with its long axis perpendicular to the edge. If the field pattern of this antenna in the plane of the Moon is $F(x_\lambda)$, then the field

[5] K. R. Lang, *Astrophys. J.* **158**, 1189 (1969).

[6] C. Hazard, *in* "Highlights of Astronomy" (C. De Jager, ed.). Reidel Publ. Dordrecht, 1971.

[7] C. Hazard, *in* "Quasi-Stellar Sources and Gravitational Collapse" (I. Robinson, ed.), p. 135. Univ. of Chicago Press, Chicago, Illinois, 1965.

distribution $E(x_\lambda)$ across the hypothetical antenna in this plane is modified by $F(x_\lambda)$ to become $E(x_\lambda)F(x_\lambda)$. To calculate the diffraction pattern as modified by the beam is then simply a matter of substituting this modified distribution for $E(x_\lambda)$ in Eq. (4.6.8) and following through the analysis in Section 4.6.9.1.

Unfortunately, the expression obtained for an arbitrary antenna beam permits no simple interpretation, apparently as a result of the coherent manner in which the antenna smooths the radiation. However, an approximate estimate of the effect of the beam on the resolution is easily obtained by considering a beam which has constant gain over an angle 2Ω but zero for larger angles. If such a beam is directed at the Moon's limb, then no Fourier components of higher spatial frequency than $\Omega D/\lambda$ can be present in the occultation curve. Since $\Omega \approx \lambda/d$, where d is the antenna size, the maximum resolution Ω_m is

$$\Omega_m \approx d/D \quad \text{(rad)}, \tag{4.6.18}$$

which is the angle subtended by the antenna at the Moon's distance. This is the fundamental resolution limit of the occultation technique using an antenna of width d, and, it will be noted, it is independent of the wavelength of observation. For a 300-m antenna it corresponds to an angular resolution of 0.16 arc sec, and for a 100-m instrument to 0.05 arc sec. For all existing instruments the effect of the antenna is therefore small, and may be considered as equivalent to convolution of the restored distribution and hence of the source brightness distribution [Eq. (4.6.16)] with the function $k(\phi)$, where for all practical purposes, $k(\phi)$ may be taken as Gaussian with a half-power width of d/D (rad).

4.6.11. Effect of Finite Receiver Bandwidth

If the antenna is connected to a receiver with a power gain $R(\lambda) = R(\lambda_0 + \Delta\lambda)$, where λ_0 is the adopted central frequency, the observed point source occultation curve $p_R(\phi)$ will be the sum of the patterns contributed by all elements in the bandwidth, each of which will be scaled as $\lambda^{1/2}$. For a rectangular bandwidth of width $\Delta\lambda_0$ this will produce a cancellation of the diffraction lobes, and therefore limit the resolution at a lobe spacing $\Delta\theta \simeq \phi(\Delta\lambda_0/\lambda)^{1/2}$, where ϕ is the width of the first Fresnel zone.[4] To calculate the resolution limit in the general case we note that if $P(a)$ is the Fourier transform of the diffraction pattern at λ_0, then, substituting $\Delta\lambda = \lambda - \lambda_0$ in Eq. (4.6.9), we find that, for the Fourier transform of the pattern $P_\lambda(a)$ at a wavelength λ,

$$P_\lambda(a) = \text{FT}\, p_\lambda(\phi) = \exp(i\pi a^2\, \Delta\lambda/D \, \text{sgn}\, a)P(a).$$

Hence,

$$\text{FT } p_R(\phi) = P_R(a) = P(a) \int_{-\infty}^{\infty} R(\lambda_0 + \Delta\lambda) \exp(i\pi a^2 \, \Delta\lambda/D \text{ sgn } a) \, d \, \Delta\lambda$$

$$= P(a)R(a)$$

and

$$p_R(\phi) = p(\phi) * \int_{-\infty}^{\infty} R(a) \exp(2\pi i a\phi) \, da = p(\phi) * r(\phi). \qquad (4.6.19)$$

For a symmetrical bandwidth,

$$R(a) = \int_{-\infty}^{\infty} R(\lambda_0 + \Delta\lambda) \exp(2\pi i \, \Delta\lambda \, a^2/2D) \, d \, \Delta\lambda$$

and

$$r(\phi) = r(-\phi) = \int_{-\infty}^{\infty} R(a) \cos(2\pi a\phi) \, da. \qquad (4.6.20)$$

Assuming a Gaussian pass band $R(\lambda_0 + \Delta\lambda) = \exp(-\Delta\lambda^2/2\sigma^2)$, with a width between half-power points $\Delta\lambda_0$ of $\sigma(8 \ln 2)^{1/2}$, then since $R(a)$ is the Fourier transform of $\exp(-\Delta\lambda^2/2\sigma^2)$ at a Fourier frequency $a^2/2D$, it follows immediately that $R(a) = \exp(-\gamma a^4)$, where $\gamma = \pi^2 \, \Delta\lambda_0^2/16D^2 \ln 2$ and

$$r(\phi) = \int_{-\infty}^{\infty} \exp(-\gamma a^4) \cos(2\pi a\phi) \, da. \qquad (4.6.21)$$

This function has been plotted by Scheuer[8], and is approximately equivalent to a Gaussian with a width between half-power points of $\beta_{\Delta f}$, where

$$\beta_{\Delta f} = 0.62(\lambda \, \Delta f/fD)^{1/2} \text{ (rad)} = 6.5\lambda^{1/2}(\Delta f/f)^{1/2} \text{ (arc sec)}, \qquad (4.6.22)$$

and where f is the frequency observation, Δf the width of the frequency response between the half-power points, and λ is measured in meters. From Eq. (4.6.16), the finite Gaussian bandwidth is equivalent to convolution of the source brightness distribution with a Gaussian beam of width $\beta_{\Delta f}$ between half-power points. At a frequency of 300 MHz and a fractional bandwidth $\Delta f/f = 0.01$, $\beta_{\Delta f} = 0.6$ arc sec. Bandwidths of shapes other than Gaussian have been considered by Scheuer and also by Lang.[5]

4.6.12. Effect of Receiver and Antenna Noise

An actual occultation curve consists not only of the pattern $f(\phi)$ due to the passage of the source behind the Moon, but also of a random component $n(\phi)$ contributed by the antenna and receiver noise. We now estimate the

[8] P. A. G. Scheuer, *Mon. Not. Roy. Astron. Soc.* **129**, 199 (1965).

limit which this noise contribution sets to the resolution which can be usefully attempted, and as a first step we will now establish the relationship between the signal-to-noise ratios on the observed and restored distributions.

4.6.12.1. Effect of the Restoration Procedure on the Signal/Noise Ratio. The restoration technique (see Section 4.6.9.2) involves convolution with $q(\phi)$ to convert the observed curve to the geometrical occultation curve of the same height, followed by differentiation and convolution with $g(\phi)$. Since convolution with $q(\phi)$ represents only a phase correction, and the phases of the Fourier noise components are in any case random, it follows that it leaves the signal-to-noise ratio unchanged, and it is necessary only to calculate the effect of differentiation and convolution with $g(\phi)$.

Consider the occultation of a source with flux I_0 and a Gaussian brightness distribution $t(\phi) = I_0 \exp(-\phi^2/2\sigma_s^2)/(2\pi)^{1/2}\sigma_s$ which produces on the occultation curve a step of height I_0 (see Fig. 6). Let the convolving Gaussian be

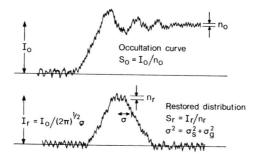

FIG. 6. Schematic representation of the restoration of the occultation of a source with a Gaussian brightness distribution of dispersion σ_S, using a convolving Gaussian with dispersion σ_g. The quantities n_0 and n_r are the rms noise levels before and after the restoration, and s_0 and S_r the corresponding signal-to-noise ratios.

represented by $g(\phi) = \exp-(\phi^2/2\sigma_g^2)/(2\pi)^{1/2}\sigma_g$, which has a width between half-power points of $\beta_r = \sigma_g(8\ln 2)^{1/2}$, and is normalized such that $\int_{-\infty}^{\infty} g(\phi)\,d\phi = 1$. Differentiation of the geometrical occultation curve and convolution with $g(\phi)$ will then give the distribution $t(\phi) * g(\phi)$, which is also Gaussian with a dispersion $\sigma = (\sigma_s^2 + \sigma_g^2)^{1/2}$, an area I_0, and a height $I_r = I_0/(2\pi)^{1/2}\sigma$.

It will be assumed that the output receiver time constant $\ll \sigma_g$ so that the noise $n(\phi)$ may be considered as white noise with a Fourier transform $N(v)$ which is constant over the range of interest. The variance of the noise on the observed and geometrical occultation curves is therefore $|N(v)^2|$, corresponding to an rms fluctuation n_0 when averaged over an angular interval α (arc sec) of $|N(v)^2|^{1/2}/\alpha^{1/2}$. After differentiation, the Fourier transform of the noise will be $2\pi i v N(v)$, and after convolution, $2\pi i v N(v)$ FT $g(\phi)$. The Fourier transform of the noise $A(v)$ on the restored distribution will therefore be

$$A(v) = 2\pi i v N(v) \exp(-2\pi^2 v^2 \sigma_g^2),$$

with a variance

$$C = \int_{-\infty}^{\infty} |A(v)|^2 \, dv = |N(v)^2| / 4\pi^{1/2} \sigma_g^3$$

and rms fluctuation

$$n_r = |N(v)^2|^{1/2} / 2\pi^{1/4} \sigma_g^{3/2} = 1.357 |N(v)^2|^{1/2} / \beta_r^{3/2}.$$

If we define the "signal" in the case of the observed distribution as the height of the step I_0 on the occultation curve, and the noise as the rms noise fluctuation n_0 when averaged over an angular interval $\alpha = 1$ arc sec, then the observed signal-to-noise ratio S_0 is

$$S_0 = I_0 / |N(v)^2|^{1/2}. \tag{4.6.23}$$

Taking the "signal" in the case of the restored distribution as the peak height I_r of the resultant Gaussian, the restored signal-to-noise ratio is given by

$$S_r = I_r / n_r = I_0 \, 2^{1/2} \sigma_g^{3/2} / \pi^{1/4} \sigma |N(v)^2|^{1/2},$$

which, for a point source where $\sigma = \sigma_g$, reduces to

$$S_r = I_0 (2\sigma_g)^{1/2} / \pi^{1/4} |N(v)^2|^{1/2} = 0.7 I_0 \beta_r^{1/2} / |N(v)^2|^{1/2},$$

and, on substituting for $|N(v)^2|$ from Eq. (4.6.23),

$$S_r = 0.7 \beta_r^{1/2} S_0 \tag{4.6.24}$$

(β_r measured in arc sec), an expression first obtained by von Hoerner[9] from numerical integration of a number of calculated restoring functions.

4.6.12.2. Resolution Limit in the Presence of Noise.   Consider the source $t(\phi)$ restored with successively narrower restoring functions. Since the width of the restored distribution is given by $\sigma = (\sigma_g^2 + \sigma_s^2)^{1/2}$, then, by reducing σ_g until σ is significantly greater than σ_g, we may obtain a measure of σ_s. This process is conveniently performed by normalizing such that for an unresolved source the restored distributions have a constant height, H_0, at all resolutions. The broadening due to the finite size of the source is then apparent as a decrease in the height of the restored curves with decreasing σ_g. We define the source as being resolved when the width of the convolving beam

[9] S. von Hoerner, *Astrophys. J.* **140**, 65 (1964).

$g(\phi)$ is equal to that of the source (i.e., $\sigma_s = \sigma_g$) that is, when the height of the restored distribution is $H_0/\sqrt{2}$, or $0.3H_0$ less than the height of an unresolved source. As the noise level of the restored curve increases with decreasing σ_g we may therefore define the noise resolution limit β_n as that value of σ_g at which the noise level would not permit the detection of a change in height of $0.3H_0$.

The height H_0 may be determined either by using large values of σ_g where the noise is negligible, or directly from the height of the step on the occultation curve. Assuming a well defined base line, the rms error in measuring directly the height of the restored curve is n_r, but is reduced to $n_r/\sqrt{2}$ if a curve fitting procedure is adopted. To ensure that the source is genuinely resolved we set $0.3H_0 \geqslant 2n_r/\sqrt{2}$, which corresponds to a signal/noise ratio $S_r = H_0/n_r \geqslant 4.7$, and reduces the chance that we are dealing with a spuriously resolved source to $<2.3\%$. Substituting for S_r from Eq. (4.6.24) then gives

$$\beta \geqslant 45/S_0^2 \quad \text{(arc sec)}, \tag{4.6.25}$$

which defines the minimum useful β_r for an occultation curve with signal-to-noise ratio S_0, and corresponds closely to the minimum of $52/S_0^2$ suggested by von Hoerner, who adopted a value of $S_r \geqslant 5$ from an inspection of the results of restoring several simulated distributions.

4.6.13. Minimum Useful Bandwidth

The signal-to-noise ratio S_0 of an occultation curve depends on the source flux, the antenna size, and the receiver bandwidth. The larger the antenna size and the bandwidth, then the larger the S_0 and, consequently, the smaller the β_r which can be employed. However, this is true only up to the point at which the resolution limits imposed by the antenna size and bandwidth themselves become operative. For a given antenna, the size d sets a fundamental resolution limit of d/D. The bandwidth limitation, however, is not of a fundamental nature since it is possible to subdivide the band into as many narrow channels as required and to restore the output from each channel separately, finally summing the restored distributions to give a signal-to-noise ratio equal to that which would have been obtained using the wider band.[9] Such a procedure is not possible with the coherent aperture distribution.

Subdivision of the bandwidth into a large number of arbitrarily narrow channels is usually not feasible, and for practical reasons it is useful to know the width of the narrowest single band which can be usefully adopted. One limit is clearly set by the antenna size and is given by

$$0.62^{1/2}(\Delta f/fD)^{1/2} \geqslant d/D \quad \text{or} \quad \Delta f/f = d^2/0.62\lambda D.$$

At $f = 300$ MHz, this gives $\Delta f = 0.001$, 0.01, and 0.1 MHz for a 30-, 100-, and 300-m telescope, respectively.

To estimate the minimum useful bandwidth set by noise fluctuations, we follow von Hoerner and rewrite Eq. (4.6.25) as

$$\beta_n = 45/S_1{}^2 \, \Delta f \quad \text{(arc sec)} \tag{4.6.26}$$

by substituting $S_0 = S_1(\Delta f)^{1/2}$, where S_1 is the signal-to-noise ratio of the occultation curve for a 1-MHz bandwidth. The minimum useful bandwidth Δf_0 and corresponding maximum resolution β_0 will then be given for $\beta_n = \beta_{\Delta f}$, where $\beta_{\Delta f}$ is the bandwidth limit. From Eqs. (4.6.22) and (4.6.26) we then have

$$\Delta f_0 = 3.6(f/\lambda)^{1/3}/S_1^{4/3}, \qquad \beta_0 = 12.5(\lambda/f)^{1/3}/S_1^{2/3}. \tag{4.6.27}$$

The signal-to-noise ratio S has been defined in terms of an integration over 1 arc sec, which, for a central occultation, corresponds to an integration time of $\simeq 3$ sec (see Section 4.6.5). If it is assumed that the observations are carried out at a frequency of f (MHz), using a telescope of diameter d (m), with an aperture efficiency of 60%, connected to a receiver with an effective input temperature of T_0 (°K), then, for a source with a flux density at 300 MHz of s_{300} (f.u.) and spectral index -0.5, we have the useful relations

$$\Delta f_0 \simeq 6T_0^{4/3}f^{4/3}s_{300}^{-4/3}\,d^{-8/3} \quad \text{and} \quad \beta_0 \simeq 30T_0^{2/3}s_{300}^{-2/3}f^{-1/3}\,d^{-4/3}. \tag{4.6.28}$$

Table I gives the values of Δf_0 and β_0 for a 100-m telescope operating at a frequency of 300 MHz, assuming $T_0 = 300$°K, a typical value for a receiving system operating at meter wavelengths.

TABLE I. The Minimum Useful Bandwidths and Corresponding Resolutions for Different Flux Levels Calculated for a 100-m Telescope Operating at a Frequency of 300 MHz and Connected to a Receiver with an Effective Input Temperature of 300°K.

Source flux (f.u.)	Signal-to-noise ratio for $\Delta f = 1$ MHz	Bandwidth (MHz)	Resolution β_0 (arc sec)
1	10	1.1	0.4
2	20	0.5	0.26
5	50	0.13	0.14
10	100	0.05	0.09
20	200	0.02	0.06

For sources > 20 f.u. the resolution limit is determined by the finite antenna size, which, for a 100-m instrument, sets a limit of $\simeq 0.05$ arc sec. For a 300-m telescope with a corresponding limit of 0.16 arc sec the antenna size becomes the limiting factor, even for sources as weak as 0.5 f.u., while for instruments smaller than 100 m the resolution is always limited by noise. The majority of radio sources are less than 10 f.u., and since β_0 depends only weakly on frequency, and T_0 can never be decreased significantly below 100°K because of the Moon's contribution to the input noise, it follows that the limiting resolution of the occultation technique using a single bandwidth is of the order 0.1 arc sec.

The resolution limits listed in Table I compare favorably with the resolution (~ 1 arc sec at 5000 MHz) now being achieved using aperture synthesis techniques, which, however, have the advantage of being able to select the sources to be studied rather than being confined to sources which happen to be occulted. On the other hand, not only can the occultation technique achieve this resolution at frequencies as low as 200 MHz, but it is a simple matter to carry out simultaneous observations at a number of widely different frequencies.

4.6.14. Choice of Operational Bandwidth

For a source of finite size there is an optimum bandwidth which depends on both the source size and flux, but which cannot usually be determined in advance since the majority of radio sources are complex, with components whose size and flux are unknown prior to the occultation. Given that subdivision of the available bandwidth into a very large number of channels is impractical, the following procedure is suggested. The output from each receiver is recorded at, say, four different bandwidths determined by a set of i.f. filters. The minimum bandwidth is chosen to permit maximum resolution, as discussed in the previous section, while the maximum is usually limited by local interference conditions. The intermediate bandwidths are chosen to give constant multiples of resolution, e.g., 125 kHz, 500 kHz, 2 MHz, and 8 MHz, thus ensuring at least a close approximation to the optimum conditions whatever the source structure may be.

4.6.15. Practical Restoration Procedure

The application of the restoration technique requires the use of a high speed computer, preferably provided with an automatic on-line graph plotter. A preliminary source position is first obtained, and then used to obtain an estimate of the parameters of the occultation, and hence the required relations between the time and angular scales of the occultation curves (Section 4.6.5). An inspection of the analog records permits an estimate of the resolution (β_r) which should be attempted, based either on an estimate of the source size or on the signal-to-noise ratio. The digital data may then, if necessary, be smoothed with a Gaussian time constant, with a width significantly less than the maximum resolution to be attempted; a suitable angular dispersion is $\sigma_t \simeq 0.15\beta_r$, with the smoothed data being sampled at corresponding intervals. The double differential of the diffraction pattern $p''(\phi)$ is then calculated in the computer at intervals of ϕ corresponding to the sample time σ_t, using a suitable series expansion of the Fresnel integrals, and then is convolved with a Gaussian of suitable chosen half-width to give the required restoring function $c(\phi)$. Convolution of $c(\phi)$ with the occultation curve then gives the restored brightness distribution, which is displayed directly on an angular scale using the automatic plotter. The width of this restored distribution between half-brightness points β is given by

$$\beta^2 = \beta_s^2 + \beta_g^2 + \beta_A^2 + \beta_{\Delta f}^2 + \beta_t^2, \qquad (4.6.29)$$

where β_s, β_g, and β_t are the corresponding widths of the source, the adopted Gaussian, and the smoothing time constant, respectively, and β_A and $\beta_{\Delta f}$ represent the effects of the finite antenna beam and finite receiver bandwidth. In practice, restorations are performed on a given source using successively smaller values of β_g until the source is resolved or the decreasing signal-to-noise ratio of the restored distribution makes further restoration useless. A sample restoration of the source 3C 273 is given in Fig. 7.

This is the basic method of restoration and is applicable to central or almost central occultations, or resolutions of the order of 1 arc sec. When the occultation angle is large, or for resolutions $\ll 1$ arc sec, account should be taken of the curvature and irregularities in the limb. The curvature produces a nonlinearity in the observed curve and a progressive phase shift between the outer diffraction lobes and the lobes of the restoring function. This non-linearity may be removed by rescaling the occultation curve before performing the convolution, using an accurately determined source position. Similarly, the limb irregularities may be taken into account by finding the true slope of the limb from Watts' limb profiles. An alternative and simpler procedure is based on the observation that the adoption of the wrong time scale for a source of small angular size produces an asymmetric restored distribution accompanied by a series of side lobes.[10,11] In arriving at a limit to the size of an unresolved source it is adequate to adjust the adopted time scale until these overshoots disappear. This process corresponds to adopting the best mean slope over the region of the limb near the occultation point, and provides an accurate method of determining the occultation angle α, and hence a source position, from a single occultation curve. For a resolved or complex source, however, a small error in the occultation angle merely produces a

[10] C. Hazard, S. Gulkis, and A. D. Bray, *Nature (London)* **210**, 888 (1966).
[11] C. Hazard, S. Gulkis, and A. D. Bray, *Astrophys. J.* **148**, 669 (1967).

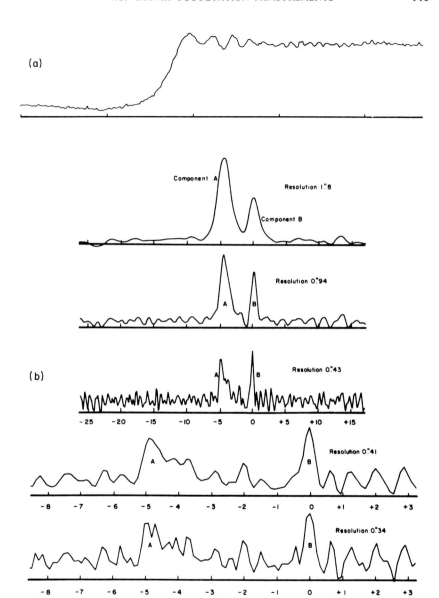

FIG. 7. An example of the restoration of an occultation curve of 3C 273: (a) the observed occultation curve, one division on the horizontal scale representing 1 min UT; (b) the derived distributions at different resolutions. The effective resolution is indicated on each curve. All curves are normalized so that an unresolved source would have constant height at all resolutions. The horizontal scales are indicated in seconds of arc. From *Nature* (*London*) **210**, 888, (1966).

variation in the apparent shape of the source. Great care must therefore be exercised in the interpretation of fine details in the source structure. When the source is complex, it often occurs that a different scaling factor is required for the various components. If these components are widely separated along the limb but pass almost simultaneously by the Moon, incompatibility in the scaling factors provides a serious limitation on the resolution which can be achieved. Similar problems caused by confusion of unrelated sources in the beam are not a problem at the flux density levels so far reached, even with the 1000-ft telescope at Arecibo.

Experience of analyzing many occultation curves suggests that irregularities in the Moon's limb are only important for resolutions <0.5 arc sec, except in unfavorable cases where the source passes close to a particularly large and sharp irregularity.

When the occultation times obtained from the high resolution analysis lead to a significant revision in the adopted source position and the experimental scaling technique was not applied, it may be necessary to recalculate the occultation parameters and repeat the restoration procedure. For sources of diameter <1 arc sec it is possible to estimate the occultation times to about 0.1 arc sec ($\simeq 0.3$ sec). The main source of error in determining the position of the Moon's limb is then due to uncertainties in the limb corrections and the Lunar Ephemeris, which are typically of the order 0.3 arc sec.

4.6.16. Occultation Surveys

For the majority of large instruments, the time available for occultation studies is limited and observations are usually confined to predicted occultations, either of catalogued sources or faint sources found by surveying in advance the calculated path of the Moon. Observations can, however, be extended to much fainter sources by carrying out an occultation survey.

A single-beam telescope can be used for survey work as described in Section 4.6.2, provided its beam is large compared with the size of the Moon, and the first such survey was carried out using the Jodrell Bank 76.2-m (250-ft) telescope at a frequency of 230 MHz. However, at higher frequencies, an instrument sufficiently large for survey work has a beam size less than the Moon's diameter, and needs to be designed to permit multiple beaming, the beams ideally being arranged to cover the whole of the Moon's limb. The only such instrument in operation, at Ootacamund, India, is a 12-beam system with an effective collecting area equal to that of a 60% efficient, 138-m diameter paraboloid, and operates at a frequency of 325 MHz with a bandwidth of 4 MHz.[12] For this instrument, the signal-to-noise ratio S_0 of a step in the received power as a source of flux s (f.u.) passes behind the Moon,

---

[12] G. Swarup et al., Nature (London) Phys. Sci. **230**, 185 (1971). Results for fifty sources observed with the Ooty telescope are given by M. N. Joshi et al., Astron. J. **78**, 1023 (1973), together with references to earlier Ooty observations.

using an integration time t (sec) and assuming an effective input temperature of $300°K$, is

$$S_0 \simeq 15st^{1/2}. \tag{4.6.30}$$

Taking the rate of motion of the Moon as $1/3$ arc sec/sec, this permits a resolution β of about $t/3$ arc sec, and the use of a fractional bandwidth $\simeq (t/20)^2/\lambda$. Given a sufficiently stable noise baseline, the maximum value of t, and hence the limiting flux, is determined by confusion between sources in each beam occulted within a time interval t (sec). Each limb intersects a total length of limb of about 10 arc min, and the effective area of each beam for confusion considerations is approximately $600t/3$ (sq arc sec). Taking the confusion limit as one source per 20 beam areas, the limiting source density is about one source per t (sq arc min). Recent source counts indicate that for sources between 0.01 and 1 f.u. the number of sources N per square degree, with a flux density greater than s (f.u.) at a frequency of 325 MHz, is given by $N \simeq 0.3s^{-1.1}$, and the flux at the confusion limit (s_c) by

$$0.3s_c^{-1.1}/3600 \simeq 1. \tag{4.6.31}$$

Adopting $S_0 = 5$ as the minimum useful signal-to-noise ratio, Eqs. (4.6.30) and (4.6.31) give the maximum useful value of t to be 186 sec and the minimum flux of the occultation survey as

$$s_{min} = 0.02 \text{ (f.u.)}.$$

This limiting flux, at which the resolution is about 1 arc min, corresponds favorably with the limit of other survey techniques, although the minimum of 18.6 yr required to cover the 12° strip along the ecliptic is much longer than would be taken to cover a comparable area of sky using, say, the 1.6-km (1-mile) Mills Cross at Molonglo, Australia. To reach this flux limit requires a baseline stability of $0.04°K$ over periods of several minutes, which is likely to be difficult to attain with a tracking instrument due to variations in input temperature produced by ground radiation, etc., particularly as repeated observations over a given region to check the reliability of weak sources cannot be made at will. While the occultation survey has, therefore, little advantage over existing survey instruments at the lower limit of its flux range, its great superiority is obvious for stronger sources, giving positions accurate to about 1 arc sec and structural information with a resolution of $\leqslant 3$ arc sec, even for sources as weak as 0.1 f.u.

4.6.17. Refraction Effects

4.6.17.1. Ionospheric Refraction. Since refraction in the earth's ionosphere displaces both the position of the source and the position of the Moon by the same amount, it enters into occultation work only as a second order correction. If χ is the angle of refraction which is assumed to occur at height $h(\sim 300$ km), its effect is to produce an error in the apparent zenith angle of the source of $\Delta z \approx \chi h/D$ (see Fig. 8) $\simeq 10^{-3}\chi$. The refraction angle χ is a func-

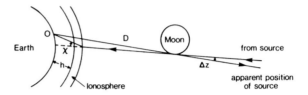

FIG. 8. Illustration of the effect of the Earth's ionosphere on the apparent source position. The refraction χ, which is assumed to occur at a height h, produces an apparent source displacement $\Delta z \approx \chi h/D$.

tion of both the frequency $(\propto f^{-2})$ and the zenith angle z, but at frequencies about 200 MHz it is less than 2 arc min for $z \geqslant 10°$, corresponding to $\Delta z <$ 0.1 arc sec. Provided position measurements are restricted to frequencies above 200 MHz, ionospheric refraction can therefore be neglected, but at frequencies below 100 MHz, and particularly at low zenith angles, it is necessary to take it into account.

Tropospheric and atmospheric refraction occurring much closer to the observer are negligible in all practical cases.

4.6.17.2. Refraction in the Solar Corona and a Possible Lunar Ionosphere. Refraction in the solar corona, assuming a gradually decreasing electron density, will lead to a systematic displacement of the source positions towards the Sun equal to the angle of refraction. Similarly, refraction in a lunar ionosphere would lead to an apparent displacement of the source towards the Moon's limb, with the result that the duration of the occultation is increased due to the immersion occurring earlier than expected, and the emersion later. Refraction in the solar corona has been discussed by Buckingham,[13] and it follows from his result that for an electron density at the Earth's distance of 10 cm^{-3}, a ray passing 20° from the Sun would be deviated by only of the order of 0.1 arc sec for a frequency as low as 100 MHz. It is, therefore, of no importance for position measurements at higher frequencies, particularly as observations of small angular size sources closer than 20° to the Sun will in any case be disturbed by scintillations due to coronal irregularities. The interest is rather that a comparison of occultation times of small angular size sources at high frequencies with the corresponding times at frequencies $\leqslant 50$ MHz may permit a measurement of the coronal refraction, and hence a measurment of the electron density in the outer

[13] M. J. Buckingham, *Nature (London)* **193**, 538 (1962).

corona. Similar measurements may also eventually permit an estimate of the excess electron density near the Moon above that of the surrounding interplanetary medium. The present lower limit, based on occultation of 3C 273, is $\leqslant 0.3$ arc sec at a frequency of 136 MHz,[14] which sets an upper limit to the excess density of 10^{-2} cm^{-3}.

[14] C. Hazard, M. B. Mackey, and A. J. Shimmins, *Nature (London)* **197**, 1037 (1963).

4.7. Scintillation Measurements*

4.7.1. Introduction

4.7.1.1. Scintillations. Radio sources having angular diameters less than about 1 arc sec show fluctuations of intensity, having a time scale of a few seconds, when observed within 90° of the Sun. It has been shown that the fluctuations are produced by scattering of the radiation from irregularities of plasma density in interplanetary space, and the phenomenon is known as interplanetary scintillation.[1] In addition, fluctuations of intensity of the radiation from pulsars, having a time scale of a number of minutes, have been shown to result from a similar mechanism operating in the interstellar medium, which is known as interstellar scintillation.[2]

4.7.1.2. Uses. Observations of the intensity fluctuations have proved useful in three main ways:

(a) to determine the electron density variations and scales of the plasma irregularities producing the scintillation.[3] Enhancements of interplanetary scintillation have been correlated with the presence of sector structure in the solar wind[3a];

(b) to study the angular structure of radio sources in the range 0.01–1 arc sec (from interplanetary scintillation)[3,4];

(c) to estimate the velocities of the irregularities by relating the time scale of the fluctuations to the spatial scale of the diffraction pattern as it drifts over the ground.[5,6]

The usefulness of the observations is increased if three spaced receivers are employed, when it becomes possible to determine the velocity and direction of the motion from cross-correlation analyses.

4.7.1.3. Outline. In this chapter we shall outline the diffraction effects which occur when radiation traverses an extended medium containing

[1] A. Hewish, P. F. Scott, and D. Wills, *Nature (London)* **203**, 1214 (1964).

[2] B. J. Rickett, *Mon. Not. Roy. Astron. Soc.* **150**, 67 (1970).

[3] M. H. Cohen, E. J. Gundermann, H. E. Hardebeck, and L. E. Sharp, *Astrophys. J.* **147**, 449 (1967).

[3a] Z. Houminer and A. Hewish, *Planet. Space Sci.* **20**, 1703 (1972).

[4] L. T. Little and A. Hewish, *Mon. Not. Roy. Astron. Soc.* **134**, 221 (1966).

[5] P. A. Dennison and A. Hewish, *Nature (London)* **213**, 343 (1967).

[6] R. D. Ekers and L. T. Little, *Astron. Astrophys.* **10**, 310 (1971).

* Chapter 4.7 is by L. T. Little.

randomly distributed phase-changing irregularities, and we shall show how the information listed above may be derived from a study of these effects. A detailed discussion of the determination of pattern sizes and wind velocities from simultaneous observations using spaced receivers will not be given but is contained in the literature.[7-12] The diffraction theory has been treated theoretically by many authors (e.g., Mercier,[13] Pisareva,[14] Budden,[15] Uscinski,[16] and the reviews by Ratcliffe[17] and Salpeter[18]), and the reader is referred to their papers for further details.

4.7.2. Diffraction Theory

4.7.2.1. Diffraction by Thin Layers. Initially we shall consider the behavior of a plane wave of unit amplitude, incident normally upon a thin layer of thickness L, containing the diffracting irregularities. We shall assume, as in the case of practical interest, that scattering angles are small. After passage through the layer the phase of the radiation is changed by $\phi(x, y)$, and the complex amplitude becomes

$$f(x, y) = \exp\{i\phi(x, y)\}.$$

If the fluctuations of electron density $\Delta N(x, y, z)$ within the layer are contiguous and have an autocorrelation function $\rho_{\Delta N}(\xi, \eta, \zeta)$, then it may be shown[19] that the two-dimensional autocorrelation function of the phase fluctuations $\rho_\phi(\xi, \eta)$ is given by

$$\phi_0{}^2 \rho(\xi, \eta) = r_e{}^2 \lambda^2 L \langle \Delta N^2 \rangle \int_0^\infty \rho_{\Delta N}(\xi, \eta, \zeta) \, d\zeta, \qquad (4.7.1)$$

where ϕ_0 is the rms phase deviation, r_e the Thomson radius, ΔN the rms deviation of electron density, and L the thickness of the layer. We shall assume that the autocorrelation function of the density fluctuations has the Gaussian form

$$\rho_{\Delta N}(\xi, \eta, \zeta) = \exp\{-(\xi^2 + \eta^2 + \zeta^2)/a^2\}.$$

[7] B. H. Briggs, G. J. Phillips, and D. H. Shinn, *Proc. Phys. Soc. London.* **B63**, 106 (1950).
[8] G. J. Phillips and M. Spencer, *Proc. Phys. Soc. London.* **B18**, 481 (1955).
[9] L. T. Little and R. D. Ekers, *Astron. Astrophys.* **10**, 306 (1971).
[10] B. H. Briggs, *J. Atmos. Terr. Phys.* **30**, 1789 (1969).
[11] J. W. Armstrong and W. A. Coles, *J. Geophys. Res.* **77**, 4602 (1972).
[12] J. Galt and A. G. Lyne, *Mon. Not. Roy. Astron. Soc.* **158**, 281 (1972).
[13] R. P. Mercier, *Proc. Cambridge Philos. Soc.* **58**, 382 (1962).
[14] V. V. Pisareva, *Astron. Zh.* **35**, 112 (1958) [English transl.: *Sov. Astron. Astrophys. J.*].
[15] K. G. Budden, *J. Atmos. Terr. Phys.* **27**, 883 (1965).
[16] B. J. Uscinski, *Phil. Trans. Roy. Soc. London.* **262**, 609 (1968).
[17] J. A. Ratcliffe, *Rep. Progr. Phys.* **19**, 188 (1956).
[18] E. E. Salpeter, *Astrophys. J.* **147**, 433 (1967).
[19] E. N. Bramley, *Proc. Inst. Elec. Eng. Part B.* **102**, 533 (1955).

The distance for this function to fall to $1/e$ characterizes a scale length a for the irregularities. Then the autocorrelation function of the phase fluctuations is also of Gaussian form, and their scale length is also a. The mean square phase deviation is given from Eq. (4.7.1) by

$$\phi_0{}^2 = \sqrt{\pi} r_e{}^2 \lambda^2 \langle \Delta N^2 \rangle a L. \qquad (4.7.2)$$

Alternative forms of correlation functions which have been considered are those whose spectra are of power law type $P(S) \propto S^{-n}$, as predicted from some theories of plasma turbulence. The modifications to the theory required for these forms are discussed in the literature.[20-25]

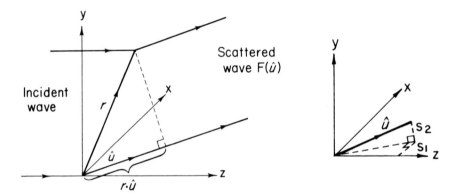

FIG. 1. The Fraunhofer diffraction pattern $F(\hat{u})$ of the phase changing screen $f(\mathbf{r})$.

After passage through the layer the incident wave is scattered into an angular spectrum. Using the notation of Fig. 1, the component of the spectrum traveling in the direction defined by a unit vector $\hat{u} = (l, m, n)$ may be determined from the theory of Fraunhofer diffraction as

$$F(\hat{u}) \propto \iint f(\mathbf{r}) \exp(ik\hat{u} \cdot \mathbf{r}) \, dx \, dy,$$

where the position vector $\mathbf{r}$ is confined to the xy-plane, $\mathbf{r} = (x, y, 0)$. The direction of $\hat{u}$ is defined by the angles S_1 and S_2 shown in Fig. 1. Since the

[20] R. V. E. Lovelace, E. E. Salpeter, L. E. Sharp, and D. E. Harris, *Astrophys. J.* **159**, 1047 (1970).

[21] J. V. Hollweg, *J. Geophys. Res.* **75**, 3715 (1970).

[22] R. Buckley, *Planet. Space Sci.* **19**, 421 (1971).

[23] J. R. Jokipii and J. V. Hollweg, *Astrophys. J.* **160**, 745 (1970).

[24] D. N. Matheson and L. T. Little, *Planet. Space Sci.* **19**, 1615 (1971).

[25] L. T. Little and D. N. Matheson, *Mon. Not. Roy. Astron. Soc.* **162**, 329 (1973).

scattering angles are very small, we make use of the approximations $l = S_1$, $m = S_2$, and

$$n = \{1 - (S_1{}^2 + S_2{}^2)\}^{1/2} \approx 1 - S^2/2,$$

where $S^2 = S_1{}^2 + S_2{}^2$. Hence,

$$F(S_1, S_2) \propto \iint f(x, y) \exp\{-ik(S_1 x + S_2 y)\} \, dx \, dy$$

so that $F(S)$ and $f(\mathbf{r})$ form a Fourier transform pair. At the exit from the layer ($z = 0$) the components of the angular spectrum add up in amplitude and phase to reproduce $f(x, y)$ so that

$$f(x, y) = \iint F(S) \exp(ik\mathbf{S} \cdot \mathbf{r}) \, dS_1 \, dS_2 \, .$$

Then the wave at $z = 0$ has fluctuations of phase but not of amplitude. To determine how the amplitude varies with distance z from the screen, we note that each component within the angular spectrum now travels away in a direction specified by $\mathbf{S}$. Since the waves with different $\mathbf{S}$ travel with different velocity components, their relative phases change, and intensity fluctuations are developed which increase with distance from the screen. At a distance z the amplitude $f(x, y, z)$ is the sum of the component waves with the appropriate phases inserted (see Fig. 1). So, we obtain

$$f(x, y, z) = \iint F(S) \exp\{ik(lx + my + nz)\} \, dS_1 \, dS_2$$

$$= \iint F(S) \exp ik\{\mathbf{S} \cdot \mathbf{r} + (1 - S^2/2)z\} \, dS_1 \, dS_2 \, .$$

The phase deviation $\phi(x, y)$ may also be expressed as an angular spectrum:

$$\phi(x, y) = \iint \Phi(S) \exp(ik\mathbf{S} \cdot \mathbf{r}) \, dS_1 \, dS_2 \, . \tag{4.7.3}$$

For an autocorrelation function $\rho_\phi(\delta) = \exp(-\delta^2/a^2)$, the power spectrum is

$$|\Phi(S)|^2 = \phi_0{}^2(\pi a^2/\lambda^2) \exp(-\pi^2 a^2 S^2/\lambda^2),$$

and the angular width to $1/e$ is given by $2\theta_0 = 2\lambda/(\pi a)$. The behavior of the fluctuations now depends on the value of ϕ_0. If $\phi \ll 1$, multiple scattering may be neglected and first order perturbation theory applied, whereas if $\phi_0 \gg 1$, a full multiple scattering theory must be worked out. In the interplanetary medium, at radio wavelengths, ϕ_0 is generally less than one when the observations are made far from the sun, but greater than one close to it, so that both possibilities need to be considered when interpreting observations. For the interstellar medium, ϕ_0 is greater than unity for typical radio frequencies.

4.7.2.2. Weak Scattering $\phi_0 < 1$. When $\phi \ll 1$, $f(\mathbf{r}) = \exp[i\phi(x, y)]$ may be expanded for small $\phi(x, y)$; i.e.,

$$f(x, y, 0) = 1 + i\phi(x, y), \tag{4.7.4}$$

and the autocorrelation function of complex amplitude $\rho_f(\delta)$ is the same as that of the phase fluctuations $\rho_\phi(\delta)$.

Substituting Eq. (4.7.3) into Eq. (4.7.4),

$$f(x, y, 0) = \iint \{\delta(\mathbf{S}) + i\Phi(\mathbf{S})\} \exp(ik\mathbf{S} \cdot \mathbf{r}) \, dS_1 \, dS_2$$

so that the angular spectrum

$$F(\mathbf{S}) = \delta(\mathbf{S}) + i\Phi(\mathbf{S}).$$

The angular spectrum contains a large unscattered component $\delta(\mathbf{S})$ together with small unscattered components $\Phi(\mathbf{S})$. The diffraction pattern is formed by the interference of the scattered components with the unscattered component. If we form the intensity $I = ff^*$ and its autocorrelation function $\rho_{\Delta I}(\delta)$, then by neglecting second order terms, it is possible to derive

$$\langle \Delta I^2 \rangle \rho_{\Delta I}(\boldsymbol{\delta}) = 4 \iint \Phi^2(\mathbf{S}) \sin^2(kS^2 z/2) \cos(k\mathbf{S} \cdot \boldsymbol{\delta}) \, dS_1 \, dS_2.$$

The scintillation index $m = \langle \Delta I^2 \rangle^{1/2}$ is given by

$$m^2 = 4 \iint |\Phi(\mathbf{S}|^2 \sin^2(kS^2 z/2) \, dS_1 \, dS_2,$$

and the power spectrum of the scintillation by

$$M^2(S_1, S_2) = 4 |\Phi(\mathbf{S})|^2 \sin^2(kS^2 z/2).$$

Thus, the power spectrum of the fluctuations of intensity is that of the phase fluctuations multiplied by the Fresnel zone function $\sin^2(kS^2 z/2)$.

Far from the screen, in the Fraunhofer region, where $k\theta_0^2 z/2 \gg 1$ (i.e., $z \gg a^2/\lambda$), the $\sin^2$ term oscillates rapidly as S varies and may be replaced by its average value $\tfrac{1}{2}$. Then,

$$m^2 = 2 \iint |\Phi(\mathbf{S})|^2 \, dS_1 \, dS_2 = 2\phi_0^2$$

so that

$$m = \sqrt{2}\phi_0 = \sqrt{2}\pi^{1/4} r_e \lambda \langle \Delta N^2 \rangle^{1/2} (aL)^{1/2} \propto \lambda \tag{4.7.5}$$

and

$$M^2(S) = 2 |\Phi(\mathbf{S})|^2.$$

Close to the screen, in the Fresnel region, where $k\theta_0^2 z/2 \ll 1$ (i.e., $z < a^2/\lambda$), the $\sin^2$ term may be expanded to give

$$m^2 = 4 \iint |\Phi(\mathbf{S})|^2 (kS^2 z/2)^2 \, dS_1 \, dS_2 \tag{4.7.6}$$

so that

$$m = (\sqrt{8}/\pi)\phi_0(z\lambda/a^2) \propto \lambda^2. \tag{4.7.7}$$

Since, for the interplanetary medium, m scales accurately as λ over the range of frequency 80–2700 MHz,[26] at least for observations made at distances greater than $100R_\odot$, it appears that the Earth is adequately in the Fraunhofer region. The scale of the diffraction pattern is then the same as that of the irregularities, and observations of the scintillation index m allow an estimate of $\langle \Delta N^2 \rangle^{1/2}$ to be made from Eq. (4.7.5) if a and L can be determined.

Direct measurements of electron density variations made from space probes[27, 28] have indicated the presence of considerable large scale structure $\sim 10^6$ km in the interplanetary medium. The Fresnel filter function $\sin^2(kS^2z/2)$ largely attenuates any contribution from such low wave numbers, but there has been debate[22-24, 29-31] as to whether the irregularities accounting for the observed scintillations are simply the tail of this large scale structure or exist as a separate distribution. The close proportionality between m and λ argues in favor of the latter conclusion. Otherwise, one would expect m to increase more sharply than λ, since more of the low frequency structure would contribute to the scintillations at longer wavelengths.

If the irregularities are not contiguous but widely spaced, then ΔN will be underestimated.

4.7.2.3. Strong Scattering $\phi_0 > 1$. When $\phi_0 \gg 1$ rad, the physical situation differs from weak scattering, since much of the energy is scattered out of the primary wave. Energy is rescattered out of the singly scattered components, and the angular spectrum becomes broader than in the single scattering case. The unscattered component is heavily attenuated, and the diffraction pattern is formed by the interference of the scattered waves with each other rather than just with the primary wave. Due to the broadening of the angular spectrum, the diffraction pattern contains structure finer than that of the phase blobs. Bramley[19] has shown that the autocorrelation function of complex amplitude is given by

$$\rho_f(\zeta, \eta) = \exp\{-\phi_0^2(1 - \rho_\phi(\zeta, \eta))\}$$

so that, for $\rho_\phi(\zeta, \eta) = \exp(-(\zeta^2 + \eta^2)/a^2)$, we may expand ρ_ϕ for small ζ, η to give

$$\rho_f(\zeta, \eta) = \exp\{-\phi_0^2(\zeta^2 + \eta^2)/a^2\}.$$

[26] A. Hewish, *Astrophys. J.* **163**, 645 (1971).
[27] D. S. Intriligator and J. H. Wolfe, *Astrophys. J. Lett.* **162**, L187 (1970).
[28] T. W. J. Unti, M. Neugebauer, and B. E. Goldstein, *Astrophys. J.* **180**, 591 (1973).
[29] W. A. Coles, B. J. Rickett, and V. H. Rumsey, preprint (1974).
[30] W. M. Cronyn, *Astrophys. J.* **161**, 755 (1970).
[31] B. J. Rickett, *J. Geophys. Res.* **78**, 1543 (1973).

Thus, the fluctuations of complex amplitude are smaller by a factor ϕ_0 than the scale of the phase blobs. The angular power spectrum $|F(S)|^2$ is the Fourier transform of $\rho_f(\zeta, \eta)$. Hence

$$|F(S)|^2 = \exp\{-\pi^2 S^2 a^2/\lambda^2 \phi_0^2\},$$

and the radiation is scattered through an angle $2\theta_0 = 2\lambda\phi_0/\pi a$.

Close to the screen ($z \ll a^2/\lambda$), the equations of geometrical optics may be applied to determine the scintillation index and scale of the pattern. The results derived are the same as that for the corresponding situation in weak scattering [Eqs. (4.7.6) and (4.7.7)] and the scale of the pattern of intensity is of the same order as that of the phase blobs.

Far from the screen ($z \gg a^2/\lambda$), the scattered components interfere with each other in a manner which may be regarded as random. Under these conditions, where the angular spectrum is said to be randomly phased, it may be shown[32] that

$$\rho_{\Delta I}(\zeta, \eta) = \rho_f(\zeta, \eta)^2. \qquad (4.7.8)$$

After leaving the screen, no further scattering takes place. Thus, $|F(S)|^2$ does not change with distance from the screen and neither does $\rho_f(\zeta, \eta)$. Hence,

$$\rho_{\Delta I}(\zeta, \eta) = \exp\{-(2\phi_0^2/a^2)(\zeta^2 + \eta^2)\}.$$

The scale of the intensity fluctuations is therefore reduced by a factor $\sqrt{2}\phi_0$ compared to that of the phase blobs. The random phasing of the interfering components implies that the amplitude has a Rayleigh distribution and the scintillation index is unity. Mercier[13] has given a formula for the scintillation index m which is valid for all values of ϕ_0 in the Fraunhofer region:

$$m^2 = 1 - \exp(-2\phi_0^2). \qquad (4.7.9)$$

A criterion for the development of large intensity fluctuations is that scattered waves from different phase blobs should be able to interfere; i.e., if $z\theta_0 > a$.

For interstellar scintillations the fluctuations are strong, and the scintillation index is generally unity. Hence, we are in the Fraunhofer region with $\phi_0 > 1$ for the interstellar irregularities: observing the scale of the pattern gives a/ϕ_0 rather than a, and measuring the scintillation index does not yield ΔN. The way to overcome this difficulty will be discussed in the last section of this chapter.

4.7.2.4. Extended Scattering Medium. In the discussion so far, it has been assumed that the irregularities are located in a thin layer so that the

[32] E. N. Bramley, *Proc. Inst. Elec. Eng. Part 1* **98**, 19 (1951).

medium may be approximated by a phase changing screen. Since both the interstellar and interplanetary medium have a finite extent, it is natural to ask how valid this approximation is. In the interplanetary medium, the refractive index fluctuations decrease with distance r from the Sun according to the law[33] $\Delta n \propto r^{-2.5 \pm 0.5}$. This rapid decrease implies that the effective scattering region for any observation is confined to the vicinity of the point of closest approach of a line of sight to the Sun, and means that the thin screen approximation is a good one. To estimate the effective thickness of the layer, using the geometry of Fig. 2, one would replace the $\langle \Delta N^2 \rangle La$ term in

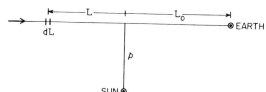

FIG. 2. Solar geometry.

Eq. (4.7.2) by a profile function $\langle \Delta N^2 \rangle (L) a(L) \, dL$ so that the expression for ϕ_0^2 is then

$$\phi_0^2(p) = \int_{-L_0}^{\infty} \pi^{1/2} r_e^2 \frac{\langle \Delta N^2 \rangle (r) \, a(r) \, dL}{v^2}.$$

In practice, for a radial law $\Delta N \propto r^{-2.5}$, the expression

$$\phi_0^2(p) = \frac{\pi^{1/2} r_e^2 \langle \Delta N^2 \rangle (p) \, a(p) p}{v^2}$$

is adequate to determine $\langle \Delta N^2 \rangle^{1/2}$ from measurements of ϕ_0. The effective thickness of the layer is of the order of the solar elongation. A detailed treatment of this effect has been given by Readhead.[33a]

For the interstellar medium, however, the location of the irregularities is not well known at all, but the most reasonable assumption is that they are distributed uniformly along the line of sight. The theoretical treatment of the development of intensity fluctuations in such a medium requires much mathematics but has been worked out by Uscinski.[16] (We note here that solutions put forward by Chernov[34] and Tatarski[35] are in error.) The physical principles are similar to those discussed for the thin screen, and the broad distinctions between weak and strong scattering and between the Fresnel and

[33] L. T. Little, *Astron. Astrophys.* **10**, 301 (1971).

[33a] A. C. S. Readhead, *Mon. Not. Roy. Astron. Soc.* **155**, 185 (1971).

[34] L. Chernov, "Wave Propagation in a Random Medium." McGraw-Hill, New York, 1960.

[35] V. I. Tatarski, "Wave Propagation in a Turbulent Medium." McGraw-Hill, New York, 1961.

Fraunhofer regions still remains; subject to replacing the thickness of the layer L and the distance from the screen by the same parameter z, the equations quoted in the above discussion remain valid apart from small numerical factors.

4.7.3. Interplanetary Scintillation and Radio Source Structure

4.7.3.1. Angular Diameter Measurement. A very important application of interplanetary scintillation is to the measurement of radio source structure in the range 0.01 to 1 arc sec. The method has three advantages:

(a) It may be applied to any source which is detectable with a sufficiently large signal to noise ratio, and is thus particularly useful for sources of low flux density which lie below the limits of detectability of long baseline interferometers.

(b) Unlike interferometric methods, high resolving power is obtained at low frequencies. However, the apparent angular diameters of the sources themselves may be increased by interstellar scattering at these low frequencies (see Section 4.7.4).

(c) A diameter survey of a large number of sources may be made relatively easily.

The disadvantage is that only a limited number of different source models may be fitted to the available data. The method allows the fraction of a source which is present as a small diameter component to be determined, and the angular diameter of the small component to be estimated.

4.7.3.2. Relation between Scintillation and Source Dimensions. If a point source at P produces a diffraction pattern $\Delta I(x, y)$ on the ground, then the same source displaced by angles θ and ϕ at P_2 will produce a similar pattern shifted by distances $z\theta$ and $z\phi$. An extended source, which will be represented by an intensity distribution $p(\theta, \phi)$, where $\iint p(\theta, \phi)\, d\theta\, d\phi = 1$, will produce a pattern which is the result of superimposing the patterns due to each element of the source. The point-source diffraction pattern tends to become smeared out, and the scintillation index is reduced for a source of finite angular diameter. The diffraction pattern $\Delta I_s(x, y)$ for the extended source is given by

$$\Delta I_s(x, y) = \iint p(\theta, \phi)\, \Delta I(x - z\theta, y - z\phi)\, d\theta\, d\phi.$$

Introducing new variables $\zeta = z\theta$ and $\eta = z\phi$ we have

$$\Delta I_s(x, y) = z^{-2} \iint p(\zeta/z, \eta/z) f(x - \zeta, y - \eta)\, d\zeta\, d\eta, \qquad (4.7.10)$$

and if

$$p'(\zeta, \eta) = z^{-2} p(\zeta/z, \eta/z),$$

it is apparent that the integration represents the convolution of $p'(\zeta, \eta)$ and $\Delta I(x, y)$.

If we apply the convolution theorem to Eq. (4.7.10), we have the result

$$M_s(Q_1, Q_2) = P(Q_1, Q_2)M^*(Q_1, Q_2),$$

where $P(Q_1, Q_2)$ and $M_s(Q_1, Q_2)$ are the Fourier transforms of $p'(\zeta, \eta)$ and $\Delta I_s(x, y)$, respectively, and (Q_1, Q_2) are spatial wave numbers related to the angles (S_1, S_2) of Section 4.7.2 by $Q_1 = S_1/\lambda$ and $Q_2 = S_2/\lambda$. The effect of the finite source diameter on the spatial power spectrum M_s^2 is given by

$$M_s^2(Q_1, Q_2) = M^2(Q_1, Q_2)|P(Q_1, Q_2)|^2. \qquad (4.7.11)$$

This expression is convenient for studying the effect of the varying angular structure of the source upon the power spectrum of the diffraction pattern. The autocorrelation function $\rho_{\Delta I_s}(\zeta, \eta)$ is given by the Fourier transform of M_s^2.

If we could determine M^2 by observing a point source, and then M_s^2 by observing the finite source, it would, in principle, be possible to determine the amplitude of the source visibility function $|P|$. In practice, this has not been done because M_s and M cannot be measured simultaneously.

For a given radio source, the two most readily measured quantities are the scintillation index m_s and the autocorrelation function of the intensity fluctuations in time $\rho_{\Delta I_s}(\tau)$, which is measured as the fluctuating pattern drifts past the receiver:

$$m_s^2 \rho_{\Delta I}(\tau) = \langle \Delta I_s(t) \, \Delta I_s(t + \tau) \rangle.$$

If the solar wind is blowing with a constant velocity v in, say, the x direction, then

$$\Delta I_s(t) = \Delta I_s(x - vt, y),$$

and the temporal and spatial spectra are related by

$$\rho_{\Delta I_s}(\tau) = \rho_{\Delta I_s}(\zeta/v, 0) \qquad \text{and} \qquad M_s^2(v) = M_s(vQ_1, 0).$$

Variations of the wind velocity, however, destroy a straightforward transformation of the spatial spectra into time spectra. One unambiguous quantity is the scintillation index, and to study the effect of the source size on this quantity we define a scintillation visibility to be

$$V = \langle \Delta I_s^2 \rangle^{1/2} / \langle \Delta I^2 \rangle^{1/2},$$

which shows how the index is reduced by the finite source structure.

From Parseval's theorem,

$$\langle \Delta I(x, y)^2 \rangle \propto \iint \Delta I^2(x, y) \, dx \, dy \propto \iint M^2(Q_1, Q_2) \, dQ_1 \, dQ_2$$

so that

$$V^2 = \frac{\iint M^2(Q_1, Q_2) |P(Q_1, Q_2)|^2 \, dQ_1 \, dQ_2}{\iint M^2(Q_1, Q_2) \, dQ_1 \, dQ_2} \qquad (4.7.12)$$

4.7.3.3. Scintillation for Different Source Models. We can now make use of relations (4.7.11) and (4.7.12) in order to ascertain qualitatively how V and M_s^2 depend on the angular structure of different model sources. We shall assume that $\rho_{\Delta I}(\zeta, \eta)$ is circularly symmetrical and of the form

$$\rho_{\Delta I}(\zeta, \eta) = \exp(-\{\zeta^2 + \eta^2\}/\eta_0^2)$$

characterized by a scale ζ_0. Thus, taking the Fourier transform of $\rho_{\Delta I}(\zeta, \eta)$,

$$M^2(Q_1, Q_2) = \exp\{-\pi^2 \zeta_0^2 (Q_1^2 + Q_2^2)\}.$$

If we consider a circular Gaussian source whose intensity distribution may be written

$$p(\theta, \phi) = (1/\pi \varepsilon_0^2) \exp\{-(\theta^2 + \phi^2/\varepsilon_0^2)\},$$

then

$$P(Q_1, Q_2) = \exp\{-\pi^2 z^2 \varepsilon_0^2 (Q_1^2 + Q_2^2)\},$$

and substituting in relations (4.7.11) and (4.7.12) gives

$$M_s^2(Q_1, Q_2) = \exp\{-\pi^2 (\zeta_0^2 + 2z^2 \varepsilon_0^2)(Q_1^2 + Q_2^2)\} \qquad (4.7.13)$$

and

$$V^2 = [1 + 2(z\varepsilon_0/\zeta_0)^2]^{-1}. \qquad (4.7.14)$$

From (4.7.13) and (4.7.14) we see that as the angular radius ε_0 is increased, the scintillation index is decreased by a factor $[1 + 2(z\varepsilon_0/\zeta_0)^2]^{1/2}$, and the scale of the pattern is increased by a similar factor. This increase represents the blurring of fine structure in the diffraction pattern which is associated with the decrease of scintillation visibility: The finite source diameter acts as a filter to suppress the high frequency spatial components in the diffraction pattern.

Similar calculations have been carried out for other idealized source models, and some curves of V versus $\zeta_0/z\varepsilon_0$ are plotted in Fig. 3 for such models. It can be seen that for simple models such as (a)–(c), there is little difference in the shapes of the visibility curves. A source whose extent in one dimension exceeds that in another gives a visibility curve much the same as a symmetrical source of intermediate diameter, but it is important to note that scintillation occurs only for a source which is sufficiently narrow in both dimensions; a negligible extent in one direction will not, by itself, give rise to scintillation.

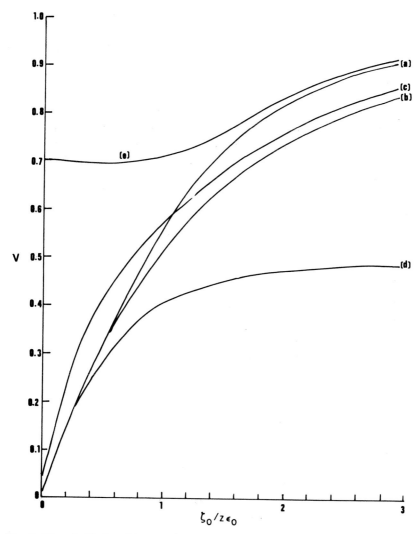

FIG. 3. The scintillation visibility $v(\xi_0/z\varepsilon_0)$ for idealized source models [L. T. Little and A. Hewish, *Mon. Not. Roy. Astron. Soc.* **134**, 221 (1966)]: (a) symmetric Gaussian: radius (to $1/e$) ε_0; (b) extended Gaussian: semiaxes (to $1/e$) $2\varepsilon_0$, $\frac{1}{2}\varepsilon_0$; (c) line source: semiaxes $2\varepsilon_0$, 0; (d) core-halo source: 50% core, radius $\frac{1}{2}\varepsilon_0$, 50% halo, radius $\gg \xi_0/z$; (e) double point source, separation $2\varepsilon_0$.

In the interplanetary medium, the scale a of the irregularities decreases closer to the Sun. In addition, the closer to the Sun the greater the scattering, and, depending on the frequency, the medium becomes phase thick ($\phi_0 > 1$)

so that the scale of the diffraction pattern is decreased still further ($\zeta_0 = a/\sqrt{2}\phi_0$).

Thus, as a radio source approaches the Sun, a series of observations of its scintillation can be reduced to yield V as a function of ζ_0/z. A source of diameter $2\varepsilon = 2\zeta_0/z$ will appreciably reduce the scintillation, and, since the observed values of ζ_0 range from about 5 to 250 km, substituting $z = 1$ A.U. shows that the inherent resolution of the method is in the range 0.01–1 arc sec The difficulty of discriminating between the different source models in Fig. 3 limits the amount of detail which may be derived, but the "fine component" model has proved particularly useful. This model consists of a small diameter component of intensity I_1 and angular radius ε_0, together with a very much larger component of intensity I_2 which does not scintillate. The visibility of this models is

$$V = [I_1/(I_1 + I_2)][1 + 2(z\varepsilon_0/\zeta_0)^2]^{-1/2}.$$

This model does not imply a "core-halo" structure, but allows for the presence of both a small component (< 1 arc sec) and larger scale components (> 1 arc sec), which many sources seem to possess.

It is unfortunate that the method has not allowed "double" structures to be detected, although this is possible in principle: a point double of separation $2\varepsilon_0$ along the x-axis has a visibility function $|P|^2 \propto \cos^2(2\pi Q_1 z\varepsilon_0)$, and the power spectrum of the scintillations from such a source would be given by

$$M_s^2(Q_1, Q_2) = M^2(Q_1, Q_2) \cos^2(2\pi Q_1 z\varepsilon_0).$$

The zeros in this function could, in principle, be detected in the temporal frequency spectrum. However, they are easily washed out by varying wind velocities and the finite thickness of the interplanetary medium.

4.7.3.4. Estimating the Diffraction Parameters ζ_0 and V.

4.7.3.4.1. Scintillation Index   The scintillation index m is easily calculated from records of scintillating sources made from day to day, but to compute V we must divide m by the scintillation index for an ideal point source observed in the same position, relative to the Sun, as the radio source under investigation. In practice, the strongest scintillators are used to calibrate the interplanetary medium. At frequencies below about 500 MHz, 3C119 and 3C138[3, 36] are the best calibrators, whereas at higher frequencies, P1148-00 appears to be the most suitable.[37] Typical curves of scintillation index versus elongation for these sources are shown in Fig. 4.

[36] L. T. Little and A. Hewish, *Mon. Not. Roy. Astron. Soc.* **138**, 393 (1968).
[37] G. Bourgois, *Astron. Astrophys.* **2**, 209 (1969).

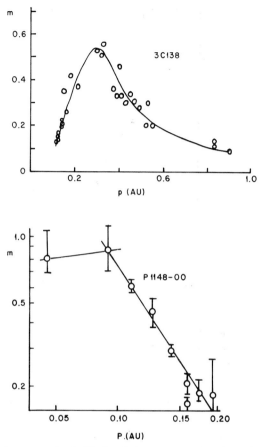

FIG. 4. Typical curves of scintillation index versus solar elongation for calibrating sources: 3C138 at 178 MHz [L. T. Little and A. Hewish, *Mon. Not. Roy. Astron. Soc.* **134**, 221 (1966)]; P1148–00 at 2695 MHz [G. Bourgois, *Astron. Astrophys.* **2**, 209 (1969)].

For $p < 0.3$ AU the cutoff in the scintillation of 3C138 (due to its finite angular diameter) means that it cannot be used as a calibrating source in this region. However, if the calibration scintillation index curve is known at one frequency, it may be scaled to another frequency using relation (4.7.9), $m^2 = 1 - \exp(-\phi_0^2)$, together with the fact that ϕ scales accurately as λ over a wide range of frequencies. For this purpose, it is convenient to plot a frequency independent $\phi_0 v$ curve as shown in Fig. 5, made up from observations of m at numerous frequencies.

4.7.3.4.2. Pattern Scale The parameter whose estimation is most difficult is ζ_0. Three different methods have been employed to derive ζ_0.

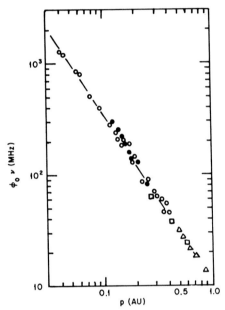

FIG. 5. Plot of $\phi_0\nu$ versus solar elongation [A. Hewish, *Astrophys. J.* **163**, 645 (1971)]:
○, 2695 MHz [G. Bourgois, *Astron. Astrophys.* **2**, 209 (1969)]; ●, 611 MHz [M. H. Cohen,
E. J. Gundermann, H. E. Hardebeck, and L. E. Sharp, *Astrophys. J.* **147**, 449 (1967)];
□, 178 MHz [L. T. Little and A. Hewish, *Mon. Not. Roy. Astron. Soc.* **134**, 221 (1966)];
△, 81 MHz.

(a)   Direct estimates of ζ_0 may be made by cross correlating the intensity
fluctuations obtained at two or more spaced receiving sites.[7, 8] Observations
have been made in this way at 81.5 MHz[5] and at 2700 MHz[6], covering the
range of solar elongation $0.03 < p < 1.0$ AU. For these observations, ϕ_0
was less than 1 rad so that the values of ζ_0 represented a, the scale of the
irregularities, and these have been plotted in Fig. 6. From this curve it is
possible to derive $\zeta_0 = a/\sqrt{2}\phi_0$ at a given elongation, for frequencies at
which $\phi > 1$, by dividing the values of a in Fig. 6 by the ϕ_0 given in Fig. 5.
So we see that both m for an ideal point source and ζ_0 may be obtained at a
given frequency from the curves in Figs. 5 and 6. Clearly, we would expect
these curves to vary with time throughout the solar cycle, and when doing a
source diameter survey it would be best to recompute them. However, sur-
prisingly little variation in ϕ_0 has been detected over the period 1964–1969.
On the other hand, the scattering angle θ has been measured during two solar
cycles,[38, 38a] and shows an increase at solar maximum which is well correlated
with sunspot number.

[38] S. E. Okoye and A. Hewish, *Mon. Not. Roy. Astron. Soc.* **137**, 287 (1967).
[38a] D. N. Matheson and L. T. Little, *Nature (London) Phys. Sci.* **234**, 29 (1971).

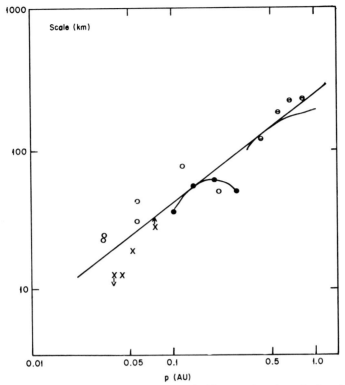

FIG. 6. Scale a of the small scale plasma irregularities as a function of solar elongation [L. T. Little, *Astron. Astrophys.* **10**, 301 (1971)]: ○, spaced receiver results at 2295 MHz ⊖, spaced receiver results at 81.5 MHz [P. A. Dennison and A. Hewish, *Nature* (*London*) **213**, 343 (1967)]; X, results derived from angular scattering at 81.5 MHz; ● results derived from angular scattering at 38 MHz [S. E. Okoye and A. Hewish, *Mon. Not. Roy. Astron. Soc.* **137**, 287 (1967)].

(b) Observations with interferometers at low frequencies out to 0.3 AU have allowed the angular spectrum of the scattered radiation to be measured in the region where $\phi_0 > 1$. If the radiation is scattered through an angle $2\theta_0$, then it follows from the results of Section 4.7.2 that $\zeta_0 = \lambda/\sqrt{2\pi\theta_0}$. Measurements of θ_0[38] have shown that the pattern is elongated along a radius to the Sun by a factor of 2 to 1. The results in Fig. 6 represent the mean radius of the characteristic ellipse.

(c) If the velocity of the solar wind is assumed, the fluctuations in time observed at a single receiver can be directly converted into fluctuations in distance as the diffraction pattern drifts over the ground. The temporal frequency spectra can be directly converted into spatial frequency using the relation

$$M^2(v/V) = M^2(Q_1, 0). \tag{4.7.15}$$

Workers at Arecibo[3] have assumed a simple Parker solar wind model, and hence obtained

$$M_s^2(v/v, 0) = M^2(v/v, 0)|P(Q_1, 0)|^2.$$

The Arecibo workers used the high frequency cutoff in the spectrum due to the source visibility $|P|^2$ in order to estimate the angular diameters of the scintillating components of the sources. They compare the observed spectra [Eq. (4.7.13)] with those of the calibrating sources observed in an equivalent position relative to the Sun. This method runs into difficulty at distances less than $30R_\odot$ from the Sun since the irregularities there have a high random component of velocity[6] which destroys the transformation of Eq. (4.7.15). In particular, the Gaussian frequency spectrum is changed into an exponential shape by the turbulent velocities, and it is necessary to use values of ζ_0 which have been obtained more directly.

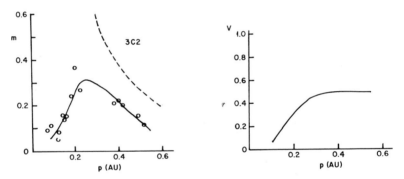

FIG. 7. Scintillation index m and visibility V curves for 3C2 at 178 MHz. The broken line shows m for a point source.

4.7.3.5. An Example of the Method. As an example of the method, consider the derivation of the structure of 3C2 at 178 MHz. In Fig. 7, its scintillation index is plotted as a function of distance from the Sun, and its visibility curve is also shown. We see that for $p > 0.4$ AU the visibility has a constant value 0.5, but by $p = 0.17$ AU it has decreased by a factor of 2. At this point, $\zeta_0 = 25$ km using Figs. 5 and 6, and $2\theta_0 = 2\zeta_0/z \sim 0.1$ arc sec. Interpreting this on the fine component model, the result indicates that 50% of 3C2 is contained in an angle 0.1 arc sec, but that the other 50% is > 1 arc sec.

Similar information has been derived in this way for hundreds of scintillating sources. The possibility of deriving such information from the probability distribution of intensity of the scintillations has been discussed by Bourgois.[39]

[39] G. M. Bourgois, *Astron. Astrophys.* **21**, 33 (1972).

4.7.4. Interstellar Scintillation and the Bandwidth Effect

If the rapid pulse-to-pulse variations in flux from pulsars are smoothed with a time constant of about 30 sec, longer period fluctuations having a time scale of tens of minutes may be distinguished. Studies of these fluctuations have shown that they have an exponential distribution of intensity, and Rickett,[2] by observing them with different bandwidths, has shown that they result from a scintillation mechanism operating in the interstellar medium.

Components in the angular spectrum scattered into an angle θ by the irregularities in the interstellar medium will arrive with a geometric delay $\Delta t \sim Z\theta^2/2c$ relative to the unscattered component (Fig. 8). There are two consequences of this delay:

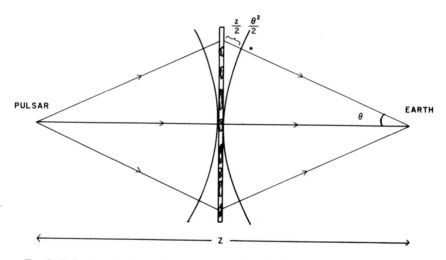

FIG. 8. Delay in radiation received at an angle θ_0 due to scattering by interstellar plasma.

(a)   The radiation is coherent over a time $\sim 1/B$, where B is the observing bandwidth. For scintillation to be observed, the interfering components within the angular spectrum must be coherent. This will be the case only if $1/B > Z\theta_0^2/2c$.[40] Rickett observed pulsars with increasing bandwidths and measured a decorrelation bandwidth $1/B_D = Z\theta_0^2/2c$ when the fluctuations were decreased by a factor $\frac{1}{2}$ compared to their narrow band value. From the results of Section 4.7.2, we see that

$$\frac{1}{B_D} = \frac{Z\theta_0^2}{2c} = \frac{Z}{2c}\frac{\lambda^2\phi_0^2}{\pi^2 a^2} = \frac{re^2}{2\pi^2 c}\lambda^4\frac{Z^2\langle\Delta N^2\rangle}{a} \qquad (4.7.16)$$

[40] L. T. Little, *Planet. Space Sci.* **16**, 749 (1968).

To establish the origin of the fluctuations, Rickett plotted B against the dispersion measure $\int N \, dZ$ on a logarithmic scale for a number of different pulsars, and obtained a slope of -2, which would be expected from Eq. (4.7.16) if $\Delta N \propto N$. To discover the allowed values of ΔN and a he used the bandwidth result together with the two conditions for observing strong scintillations:

(1)   that the medium must be phase thick ($\phi_0 > 1$), i.e.,

$$r_e^2 \lambda^2 \langle \Delta N^2 \rangle aZ > 1 \qquad \text{[from Eq. (4.7.2)]};$$

(2)   that the Earth is in the Fraunhofer region, i.e., if

$$Z\theta_0 > a \qquad \text{or} \qquad Z^3 \lambda^4 r_e^2 \langle \Delta N^2 \rangle / \pi^2 > a^3 \qquad \text{(from Section 4.7.2)}.$$

Plotting these limits on the ΔN, a-plane for CP 0328 (of known Z) he obtained the results shown in Fig. 9, from which he deduced that $\Delta N \sim 4.7 \times 10^{-5}$

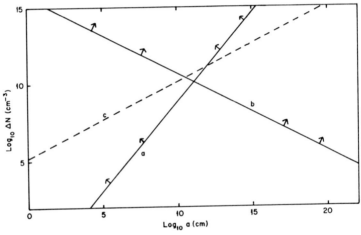

FIG. 9. Parameters of the interstellar plasma irregularities ΔN and a on logarithmic scales [B. J. Rickett, *Mon. Not. Roy. Astron. Soc.* **150**, 67 (1970)]. The lines represent the limits discussed in the text: (a) components of spectrum must interfere; (b) medium must be phase thick; (c) bandwidth relation.

cm^{-3} and $a \sim 10^{11}$ cm. Typically, the scattering angle $2\theta_0$, which is proportional to λ^2, has a value ~ 0.1 arc sec at 100 MHz, and radio sources of low galactic latitude will appear to have this diameter unless they are observed with sufficiently narrow bandwidths. In fact, the broadening of these sources at 81.5 MHz has been studied by interplanetary scintillation in order to show how the interstellar scattering increases at low galactic latitude.[40a]

[40a] A. C. S. Readhead and A. Hewish, *Nature (London)* **236**, 440 (1972).

(b)   Since $\theta_0 \propto \lambda^2$, the delay $Z\theta^2/2c$ increases markedly at low frequencies, and can become greater than the width of the pulse. This leads to a broadening of the pulse shape rather than scintillation, and at a sufficiently low frequency the broadening has a width greater than the pulse period. Then, the pulsar stops looking like one as its pulses merge together. These effects, which are most marked for the more distant pulsars, are discussed in Cronyn,[41] Lang,[42] Williamson,[43] and Sutton.[44]

[41] W. E. Cronyn, *Science* **168**, 1453 (1970).
[42] K. R. Lang, *Astrophys. J.* **164**, 249 (1971).
[43] I. P. Williamson, *Mon. Not. Roy. Astron. Soc.* **157**, 55 (1972).
[44] J. M. Sutton, *Mon. Not. Roy. Astron. Soc.* **155**, 51 (1971).

5. INTERFEROMETERS AND ARRAYS

5.1. Theory of Two-Element Interferometers*

5.1.1. Introduction

The basic observables in radio interferometry[1,2] are the correlated amplitude and relative phase of waves from a common source at two points on the wavefront. The vector between the two points on the wavefront is known as the interferometer baseline. Optical interferometers bring the two "signals" together using mirrors and superpose them onto a detector so that they either reinforce or cancel, depending on their relative phases. The interference of the signals in this manner produces interferometer "fringes." In radio interferometry, the "signals" at the ends of the baseline can be translated to a lower frequency, transmitted through cables, added and detected or cross correlated and Fourier transformed, or even recorded on magnetic tape for post real-time detection. The term "fringes" has been carried over from optical interferometry into radio interferometry where it refers to the correlated portion of the signals.

If the source of radiation is at infinity in the direction of the unit vector $\hat{\mathbf{i}}_s$, the signal arrives at a "remote" site late by an amount

$$\tau_g = -(\mathbf{D} \cdot \hat{\mathbf{i}}_s)/c, \qquad (5.1.1)$$

where the baseline $\mathbf{D}$ runs from the reference to the remote site as illustrated in Fig. 1. When both ends of the baseline are at the same height in a plane stratified atmosphere, the additional atmospheric delays to each site are equal so that c is the free space velocity of propagation and $\hat{\mathbf{i}}_s$ is the source direction† prior to entering the atmosphere. When the sites are widely separated or at different elevations, special atmospheric corrections have to

---

[1] R. N. Bracewell, *Proc. IRE* **46**, 97 (1958).
[2] R. N. Bracewell, *IRE Trans. Antennas Propag.* **2**, 59 (1961).

---

† More exactly, this is the apparent source direction after correction for aberration due to motion of the remote site.

---

* Chapter 5.1 is by A. E. E. Rogers.

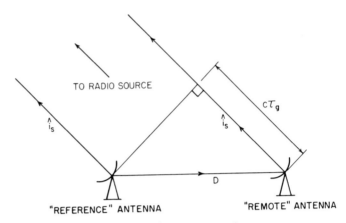

FIG. 1. Interferometer geometry.

be made to the above equation for different atmospheric paths to each site. In celestial coordinates.

$$\tau_g = -(D/c)[\sin \delta_B \sin \delta_S + \cos \delta_B \cos \delta_S \cos(L_S - L_B)], \qquad (5.1.2)$$

where L_S and δ_S are hour angle and declination of the source, respectively, and L_B and δ_B are hour angle and declination of the baseline, respectively. The "interferometer fringe phase" or the relative phase ϕ in radians of the signals received at each end of the baseline is

$$\phi = \omega \tau_g, \qquad (5.1.3)$$

where ω is the signal frequency in radians per second. Owing to the earth's rotation, the relative signal phase ϕ is changing with a rate F_R known as the fringe rate:

$$F_R = d\phi/dt \qquad (5.1.4)$$
$$= (d/dt)[\omega \tau_g(t)] \qquad (5.1.5)$$
$$= (\omega D/c) \cos \delta_B \cos \delta_S \sin(L_S - L_B)(dL_S/dt). \qquad (5.1.6)$$

The fringe rate results from the differential Doppler shift between the reference and remote sites.

The interferometer has angular resolution in the direction of the source given by the change of phase with source angle

$$d\phi/d\theta = -(\omega/c)D_T, \qquad (5.1.7)$$

where D_T is the component of the baseline normal to the source direction. The angle between lines of equal phase (modulo 2π) or "fringes" is λ/D_T, or twice Rayleigh's criterion for the resolution. The projection of the baseline on the plane of the incident wavefront has components in the direction of increasing right ascension and declination which are termed the u, v components, respectively:

$$u = (\omega/c)D \cos \delta_B \sin(L_S - L_B),\qquad (5.1.8)$$

$$v = (\omega/c)D[\sin \delta_B \cos \delta_S - \cos \delta_B \sin \delta_S \cos(L_S - L_B)],\qquad (5.1.9)$$

where u and v are in units of radians. As the Earth rotates, the projected baseline describes an ellipse in the uv-plane. This ellipse is centered at $u = 0$, $v = (\omega/c)D \sin \delta_B \cos \delta_S$, has eccentricity $\cos \delta_S$, and semimajor axis equal to the equatorial component of the baseline.

The interferometer power response is enveloped by the geometric mean of the beam patterns of two elements. In Section 5.1.5 it will be shown that the effective interferometer aperture is the geometric mean of the two apertures, and the effective system temperature is the geometric mean of the system temperatures.

The interferometer baseline is precisely defined as the vector between the phase centers (see Section 1.3.2) of the two antennas. However, since the antennas track the radio source, the ends of the baseline are effectively the intersection of antenna-mount axes since the dot product of the vectors from the axis intersection to the phase center with the source direction is a constant in time. Antennas with nonintersecting axes produce a changing baseline which can be accounted for with additional offset vectors.

5.1.2. Signal Analysis

For simplicity, it will be assumed that the signal comes from an unresolved point source of radio energy so that the signal amplitude $y(t)$ at the " remote " antenna is a delayed replica of the signal $x(t)$ at the " reference " antenna. By unresolved it is meant that the angular size is much less than the fringe spacing λ/D_T. The analysis for extended sources will be treated in Section 5.1.3. If the geometrical delay τ_g is defined as positive when the signal arrives at the remote antenna late, then

$$y(t) = x(t - \tau_g)\qquad (5.1.10)$$

and the double-sided Fourier transform of $y(t)$ is

$$y(\omega) = \int_{-\infty}^{+\infty} y(t)e^{-i\omega t}\, dt = x(\omega)e^{-i\omega\tau_g},\qquad (5.1.11)$$

where $x(\omega)$ is the transform of $x(t)$. The properties of Fourier integral transforms have been conveniently summarized by Zimmermann and Mason,[3]

[3] S. J. Mason and H. J. Zimmermann, " Electronic Circuits, Signals and Systems." Wiley, New York, 1960.

and by Bracewell.[4] If the signals within a band $\omega - (\Delta\omega/2)$ to $\omega + (\Delta\omega/2)$ are added and detected directly at radio frequency, the output A of a "square law" or power detector is

$$A = \int_{\omega-(\Delta\omega/2)}^{\omega+(\Delta\omega/2)} |x(\omega)(1 + e^{i\omega(\tau_g + \tau_e)})|^2 \, d\omega/2\pi \qquad (5.1.12)$$

$$= \int_{\omega-(\Delta\omega/2)}^{\omega+(\Delta\omega/2)} 2|x(\omega)|^2(1 + \cos \omega(\tau_g + \tau_e)) \, d\omega/2\pi \qquad (5.1.13)$$

$$= 2|x|^2\left[1 + \cos \omega(\tau_g + \tau_e)\left(\frac{\sin \pi B(\tau_g + \tau_e)}{\pi B(\tau_g + \tau_e)}\right)\right] \Delta\omega, \qquad (5.1.14)$$

where τ_e is some additional time offset introduced by unequal cable lengths to the detector. The detected power is modulated by 1 plus the cosine of fringe phase. This modulation produces complete nulls if the product of the predetection bandwidth B (Hz) and the time offset is much less than π.

In optical and radio interferometry, the signal paths are made nearly equal so that the time offset is small enough to produce fringes over a wide bandwidth. These fringes are often called "white" fringes. The simple adding interferometer is seldom employed in practice as it is difficult to combine the signal directly at radio frequency from two widely separated antennas. The fringes may be difficult to detect if the detector gain fluctuates. This problem can be eliminated if the radiometer is switched rapidly between the sum and difference of the signals, and the output is then synchronously detected. Also, fringes from an adding interferometer are only evident by observing the sinusoidal time variation in the detected power as the sources move through the peaks and nulls of the fringe pattern. Interferometric amplitude and phase can be measured, without the need to observe motion through the fringe pattern, by adding an additional channel and detector. This extra channel is known as a "quadrature" channel as it is obtained by shifting one of the arms of the interferometer by 90°. The left-hand portion of Fig. 2 shows various interferometer detector systems used directly at radio frequencies.

Usually, it is easier to translate the observing-frequency window down to a lower frequency by using a mixer and local oscillator (LO). In fact, many modern interferometers have many stages of mixing, going from the radio frequency (rf) to intermediate frequencies (i.f.), and finally to a band of frequencies ranging from almost dc to the desired rf bandwidth. This final frequency range is often termed "video" as it is often similar to that used in the transmission of a television image (a few Hz to a few MHz). If the reference signal is mixed with a frequency (in rad/sec) ω_x and phase ϕ_x, the remote

[4] R. N. Bracewell, "The Fourier Transform and Its Applications." McGraw-Hill, New York, 1965.

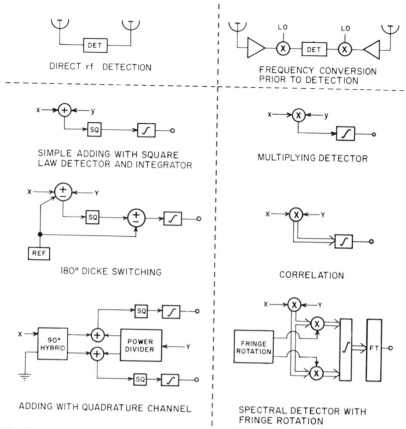

FIG. 2. Types of interferometer detectors. The left side shows those types commonly used directly at radio frequencies or intermediate frequencies, while the right side shows those used at intermediate or video frequencies. The double lines indicate multichannel signals.

signal with ω_y and ϕ_y, the frequency domain descriptions of the signals, are

$$X(\omega) = x(\omega + \omega_x)e^{-i\phi_x} + x(\omega - \omega_x)e^{i\phi_x}, \qquad \omega > 0 \qquad (5.1.15)$$

and

$$Y(\omega) = y(\omega + \omega_y)e^{-i\phi_y} + y(\omega - \omega_y)e^{i\phi_y}, \qquad \omega > 0, \qquad (5.1.16)$$

where the first term is the "upper" sideband, and the second the "lower" sideband. Multiplication of the spectral function $X(\omega)$ by the conjugate of $Y(\omega)$ gives a complex function S_{xy} which is the Fourier transform of the cross-correlation function $R_{xy}(\tau)$. S_{xy} is known as the cross-spectral function.

Because S_{xy} is the transform of a real function, the following relationship holds:

$$S_{xy}(\omega) = S_{xy}^*(-\omega). \tag{5.1.17}$$

The cross-spectral function contains all the information pertaining to interferometry and the spectrum of the radio source. The cross-spectral function $S_{xy}(\omega)$ for the upper sideband term is

$$S_{xy}^u = X(\omega)Y^*(\omega) \tag{5.1.18}$$

$$= x(\omega + \omega_x)e^{-i\phi_x}y^*(\omega + \omega_y)e^{+i\phi_y} \tag{5.1.19}$$

$$= x(\omega + \omega_x)x^*(\omega + \omega_y)e^{i(\phi_y - \phi_x)}e^{i(\omega + \omega_y)\tau_g} \tag{5.1.20}$$

$$= S_{xx}(\omega + \omega_x)e^{i(\phi_y - \phi_x)}\exp(i[(\omega + \omega_x)\tau_g + (\omega_y - \omega_x)t]), \tag{5.1.21}$$

where the last multiplicative factor is a quasi-time varying term if we assume that $S_{xy}(\omega)$ is derived over a portion of time during which the factor remains constant. The time variation can be removed if the fringe rate is canceled by the offset in local oscillators. In order to simplify the notation, the last equation can be rewritten

$$S_{xy}^u(\omega) = S_{xx}(\omega')e^{i\theta}\exp(i\omega^0\tau_g)e^{i\omega\tau_g}, \tag{5.1.22}$$

where

$$\theta = \phi_y - \phi_x + (\omega_y - \omega_x)t \tag{5.1.23}$$

is the local oscillator phase difference,

$$\omega' = \omega + \omega_x \tag{5.1.24}$$

is the radio frequency, and

$$\omega^0 = \omega_x \tag{5.1.25}$$

is the local oscillator frequency, while for the lower sideband

$$S_{xy}^l(\omega) = S_{xx}(\omega'')e^{-i\theta}e^{i\omega\tau_g}\exp(-i\omega^0\tau_g), \tag{5.1.26}$$

where

$$\omega'' = \omega - \omega_x. \tag{5.1.27}$$

The cross-spectral functions $S_{xx}(\omega')$ and $S_{xx}(\omega'')$ are the upper and lower sideband source spectra. The phase, $\tan^{-1}$ (imaginary part/real part), of the complex quantity S^u is the source fringe phase plus the instrumental phase. The fringe phase $\omega^0\tau_g$ is changing in time with the fringe rate F_R given in Eq. (5.1.6). The apparent fringe rate $F_R + (d\theta/dt)$ can be adjusted by offsetting the local oscillators or by multiplying the observed cross-spectral function by sine and cosine as illustrated in Fig. 2.

The factor $\exp(i\omega\tau_g)$ is the differential delay which appears as a frequency-dependent phase in the frequency domain. This term can be canceled by adding an equal delay in the opposite leg of the system. This added delay, known as delay compensation, is essential to all interferometers. Pictorially, the signal vector must not rotate significantly over the integrated bandwidth. If the vector rotates by $180°$, the "fringes" will be reduced by the factor $2/\pi$, while one rotation reduces the "fringes" to zero. In very long-baseline interferometry, using recorded-signal delay compensation is achieved by correct time alignment (to within a fraction of the inverse bandwidth) of the recorded tapes.

In systems with many frequency conversions, the effective local-oscillator frequency is the algebraic sum of the individual LO frequencies. Lower sideband conversions are added with a negative sign.

In single-sideband interferometer systems, the delay compensation at i.f. frequencies alters the fringe phase since adding delay at radio or intermediate frequencies has the same effect. However, in double-sideband systems, any changes in delay at intermediate frequencies leaves the fringe phase unaltered. The cross-spectral function for this case is

$$S^D_{xy}(\omega) = S_{xx} 2 \cos(\theta + \omega^0 \tau_g) e^{i\omega\tau_g}, \qquad (5.1.28)$$

which is obtained by adding Eqs. (5.1.22) and (5.1.26). The fringe phase is measured by performing a least-squares fit to the cosine factor in Eq. (5.1.28). Additional delay at intermediate frequencies alters the exponential term but leaves the fringe phase, estimated from the cosine term, unaltered.

The most complete method of interferometer detection is cross correlation, or the equivalent, in the frequency domain. Cross correlation automatically forms a "quadrature" channel and compensates for delay over the range of the cross-correlation delay shift.

The cross-correlation function

$$R_{xy} = \langle x(t)y(t - \tau)\rangle \qquad (5.1.29)$$

is most simply obtained from the previous analysis as the Fourier transform of the cross-spectral function $S_{xy}(\omega)$. For the upper sideband case

$$R_{xy}(\tau) = \int_{-\infty}^{+\infty} S_{xy}(\omega)e^{i\omega\tau} \, d\omega/2\pi \qquad (5.1.30)$$

$$= 2 \cos(\theta + \omega^0 \tau_g)B_1(\tau + \tau_g) - 2 \sin(\theta + \omega^0 \tau_g)B_2(\tau + \tau_g), \qquad (5.1.31)$$

where

$$B_1(\tau) = \int_0^\infty S_{xx}(\omega') \cos \omega\tau \, d\omega/2\pi \qquad (5.1.32)$$

and

$$B_2(\tau) = \int_0^\infty S_{xx}(\omega') \sin \omega\tau \, d\omega/2\pi. \qquad (5.1.33)$$

Figure 3 illustrates the cross-correlation function that would be obtained if a

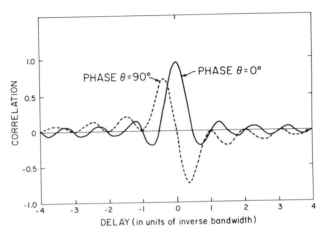

FIG. 3. The cross-correlation function obtained from correlation of correlated radio frequency signals after translation to video frequencies. Note how the correlation depends on θ, the relative phase of the local oscillators plus the fringe phase.

white-noise source is filtered by receivers with rectangular shaped bandpass of width B (Hz). This cross-correlation function is given by

$$R_{xy}(\tau) = 2\cos(\theta + \omega^0\tau_g + \pi B(\tau + \tau_g))\left(\frac{\sin \pi B(\tau + \tau_g)}{\pi B(\tau + \tau_g)}\right). \qquad (5.1.34)$$

This equation represents a cosine enveloped by a $(\sin x)/x$ function that has a delay width of approximately

$$\Delta\tau = 1/B. \qquad (5.1.35)$$

This is the same function that appears in Eq. (5.1.14). A multiplying interferometer (see Fig. 2) is a special case of a cross-correlation interferometer for which the cross correlation is performed for only one value of delay. The cross-correlation interferometer obtains both spectral and interferometric information as is most clearly illustrated in the frequency domain where the source spectrum and the interferometer phase are separable. When observing a continuum radio source for which no spectral information is desired, a one-sided transform of the cross-spectral function yields a complex delay function

$$D(\tau) = \int_0^\infty S_{xy}(\omega)e^{-i\omega\tau}\,d\omega/2\pi \qquad (5.1.36)$$

whose magnitude is the fringe amplitude and whose phase is the fringe phase when evaluated at

$$\tau = \tau_g \qquad (5.1.37)$$

since

$$D(\tau_g) = e^{i\theta}\exp(i\omega^0\tau_g)\int_0^\infty S_{xx}(\omega')\,d\omega/2\pi \qquad (5.1.38)$$

from Eq. (5.1.22). Alternately, the complex delay function can be obtained by convolving the cross-correlation function with

$$\int_0^\infty e^{-i\omega\tau}\,d\omega/2\pi \qquad (5.1.39)$$

whose real and imaginary parts are equal to the $\theta = 0°$ and $\theta = 90°$ curves of the cross-correlation function illustrated in Fig. 3.

Precise geodetic measurements made using very long-baseline interferometers have utilized the interferometer delay as an observable for small diameter radio sources. The delay can be measured with a precision of approximately $1/B$. However, in practice, the bandwidth that can be recorded and processed is severely limited so that narrowband samples taken over a wide bandwidth are required to improve the delay measurement precision. In order to make an optimum estimate of the delay we define a delay resolution or delay ambiguity function

$$D(\tau) = \int_0^\infty S_{xy}^u(\omega)e^{-i\omega\tau}\exp(-i\omega^0\tau)\,d\omega/2\pi. \qquad (5.1.40)$$

The best estimate of delay is found by selecting that value of τ which maximizes the magnitude of $D(\tau)$. For an interferometer sampling a wide bandwidth

$$D(\tau) = \sum_{k=1}^K D_k(\tau), \qquad (5.1.41)$$

where D_k are the delay functions for individual frequency windows. As an example, Figs. 4 and 5 show the delay structure of a multiwindow system using six frequency windows, each with 60 kHz bandwidth. A detailed error analysis shows that the delay measurement precision is the rms phase noise [given in Eq. (5.1.71)] divided by the rms frequency coverage in units of radians per second.

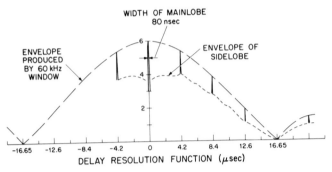

FIG. 4. Delay resolution (or ambiguity) function for six frequency windows spaced at 0, 0.24, 0.72, 1.68, 3.6, and 7.44 MHz; each channel is 60 kHz wide.

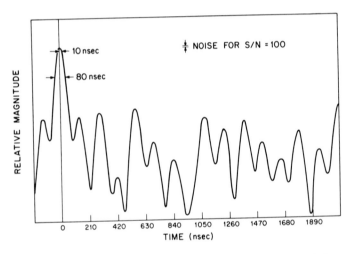

FIG. 5. Detailed sidelobe structure of the delay resolution function for windows spaced at 0, 1, 3, 7, 15, and 31 times 240 kHz.

5.1.3. Two-Dimensional Fourier Transform Relation between Brightness and Visibility

In the last section, the signal analysis was performed for a point radio source. An extended source consists of many independent radiators whose effect is the same as many point sources. The cross-spectral function for a multiple source is

$$S_{xy}^{u}(\omega) = \sum_{k=1}^{K} S_{k,xx}(\omega') \exp(i\omega^{0}\tau_{g,k})e^{i\theta}, \qquad (5.1.42)$$

which is known as the complex visibility function[5] when the instrumental phase is taken out and the magnitude is normalized to 1 for a baseline of zero length.

A two-element interferometer measures one component of the two-dimensional Fourier series description of source for each baseline. Stated mathematically, the complex visibility function $A(u, v)$ is the Fourier transform of the brightness function $B(x, y)$:

$$A(u, v) = \iint B(x, y)e^{-iux}e^{-ivy} \, dx \, dy, \tag{5.1.43}$$

where x and y are spatial coordinates in the direction parallel to increasing right ascension and declination, respectively; the brightness is normalized so that

$$\iint B(x, y) \, dx \, dy = 1; \tag{5.1.44}$$

and u and v are projected baseline components in the direction parallel to increasing right ascension and declination, respectively, in units of radians of phase per radian in the sky. Equation (5.1.43) physically represents the sum of independently radiating elements of source. Since the source brightness is a real function,

$$A(u, v) = A^*(-u, -v), \tag{5.1.45}$$

which states that if the projected baseline vector is reversed, the visibility is conjugated and no new information is obtained.

From Fourier transform theory, Eq. (5.1.43) can be inverted to obtain the source brightness from the visibility:

$$B(x, y) = (1/2\pi)^2 \iint A(u, v)e^{+iux}e^{+ivy} \, du \, dv. \tag{5.1.46}$$

This is the equation for "aperture" synthesis. If the visibility function $A(u, v)$ is measured over the entire uv-plane, the source distribution can be obtained. In fact, we have already seen that only half the uv-plane needs to be covered owing to the conjugate relation of Eq. (5.1.45). The coverage required can be further reduced by making use of the sampling theorem. If the source mapped is limited in extent to a width Δx in right ascension and Δy in declination, then the uv-plane need only be sampled at points on a grid with spacing $2\pi/\Delta x$ in u and $2\pi/\Delta y$ in v. (For further detail, see Section 5.2.3.) Further, the maximum baselines used can be limited if the source map is only required to a certain resolution. As a measure of the performance of an aperture-synthesis

---

[5] D. E. Gogg, G. H. MacDonald, R. G. Conway, and C. M. Wade, *Astron. J.* **74**, 1206 (1969).

interferometer, the equivalent beam pattern can be derived by taking all the points covered in the uv-plane with their conjugates and assigning a weight (usually unity) to each point, and then taking the transform. All points not covered are effectively zero so that the equivalent beam is

$$B(x, y) = (1/K) \sum_{k=1}^{K} W_K \cos(xu_K) \cos(yv_K),$$ (5.1.47)

where W_k is the weight assigned to the kth point located at (u_k, v_k). For example, if all points within a rectangle in the upper half of the uv-plane are sampled,

$$B(x, y) = \left[\frac{\sin(x\,\Delta u)}{x\,\Delta u}\right]\left[\frac{\sin(y\,\Delta v)}{y\,\Delta v}\right],$$ (5.1.48)

where $\pm\Delta u$ and $\pm\Delta v$ are the limits of the rectangle. If the uv-plane coverage is contained within the semicircle of radius R, the beam pattern for unit weighting is

$$B(x, y) = \frac{2J_1(R(x^2 + y^2)^{1/2})}{R(x^2 + y^2)^{1/2}},$$ (5.1.49)

where J_1 is the first-order Bessel function. If the weighting function is tapered with increasing radius in the uv-plane, the equivalent interferometer beam broadens and the sidelobes are reduced, analogous to the change in an antenna beam with tapered illumination of the parabolic surface. The analogy holds because the antenna beam pattern is the two-dimensional Fourier transform of the aperture illumination as shown in Chapter 1.3. In practice, the uv-plane coverage is seldom uniform. Holes in the uv-plane show up as sidelobes with the period and orientation of the hole.

Many attempts have been made to find a technique for interpolating the visibility function over regions in the uv-plane where no data are available. This is equivalent to an attempt to form a better equivalent beam pattern. Some progress can in fact be made using data-adaptive filtering techniques, which assign values in the missing regions of the uv-plane in an unbiased manner. In a sense, the Fourier-transformation algorithm gives a biased map since it assigns a value of zero to all regions where no data exist. While this map is consistent with the data, it may not be the most probable map. In fact, this map may contain negative values which have no physical reality. In addition, the map will have sidelobe structure similar to the equivalent beam pattern. While it is conceivable that the source may in fact look like the equivalent beam pattern, this is unlikely. A technique known as the "maximum entropy method" chooses a map which is one member of a family of possible maps, all consistent with the available interferometer data. The maximum entropy choice contains the minimum amount of information subject to the constraints imposed by the data, that is, the Fourier transform of the maximum entropy map agrees with the visibility function at points where measurements exist, while elsewhere it assigns values which maximize the entropy or equivalently minimize the extraneous content. At present, the maximum entropy method has no efficient algorithm for two-dimensional, nonequispaced data, and so has not yet been used to any great extent. Instead, an iterative procedure called " CLEAN " has been worked out by Hogbom[6] and has been widely used. This procedure is one in which the equivalent beam pattern centered at the maximum on the map is subtracted from the map successively. Finally, components subtracted are returned to the map with " clean " beam patterns. Figure 6 illustrates various methods of

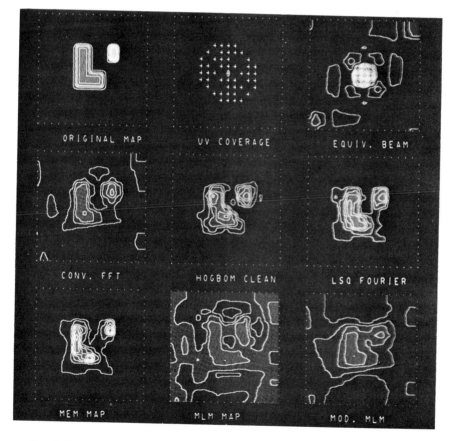

FIG. 6. Computer test of various methods of inverting data in the uv-plane to form a map free of sidelobe effects. The figure is one sample from a set of computer tests on many different sources and uv sample sets.

[6] J. A. Hogbom, *Astron. Astrophys. Suppl.* **15**, 417 (1974).

interpolating the visibility function using noiseless computer-generated data. The conventional Fourier transform, the Hogbom CLEAN, the maximum entropy method[7] (MEM), as well as least-squares Fourier series interpolation (LSQ Fourier) and maximum likelihood methods[8] were tested on artificial sources (in this case, an L-shaped object adjacent to a point source) with incomplete uv-sampling.

5.1.4. Polarization Measurements with Interferometers

When a single-antenna and power-measuring radiometer is used to measure polarization, a combination of circular and linear polarization is required for a complete analysis.[9] More specifically, a complete set of polarization parameters can be obtained by measuring Stokes[10] parameters

$$S_0 = I(0°) + I(90°),   \tag{5.1.50}$$

$$S_1 = I(0°) - I(90°),   \tag{5.1.51}$$

$$S_2 = I(45°) - I(135°),   \tag{5.1.52}$$

$$S_3 = I(R) - I(L),   \tag{5.1.53}$$

where $I(\phi)$ is the intensity for linear polarization at a position angle ϕ, and $I(R)$ and $I(L)$ are the intensities for right and left circular polarization. Alternately, if two radiometers are available on a single antenna, one for each of two orthogonal polarizations, the polarization parameters can be measured from the autocorrelation and cross-correlation functions or from their transforms, the cross-spectral functions. In practice, the orthogonal-polarization pairs would be either left and right circular or two orthogonal linear polarizations, say, horizontal and vertical. In this case

$$S_0(\omega) = S_{HH}(\omega) + S_{VV}(\omega) = S_{RR}(\omega) + S_{LL}(\omega),   \tag{5.1.54}$$

$$S_1(\omega) = S_{HH}(\omega) - S_{VV}(\omega) = 2 \text{ Re } S_{RL}(\omega),   \tag{5.1.55}$$

$$S_2(\omega) = 2 \text{ Re } S_{HV}(\omega) = 2 \text{ Im } S_{RL}(\omega),   \tag{5.1.56}$$

$$S_3(\omega) = 2 \text{ Im } S_{HV}(\omega) = S_{RR}(\omega) - S_{LL}(\omega),   \tag{5.1.57}$$

where $S_{RL}(\omega)$ and $S_{HV}(\omega)$ are the Fourier transforms of cross correlation between right and left circular and horizontal and vertical polarizations, respectively. The four correlation parameters are elements of the polarization

[7] J. G. Ables, *Astron. Astrophys. Suppl.* **15**, 383 (1974).
[8] G. D. Papodopoulos, *IEEE Trans. Antennas Propng.* **AP-23**, 45 (1975).
[9] M. L. Meeks, J. A. Ball, J. C. Carter, and R. P. Ingalls, *Science* **153**, 978 (1966).
[10] M. Born and E. Wolf, "Principles of Optics." Macmillan, New York, 1964.

coherency matrix.[10] The relations above are easily derived using the transformations

$$\sqrt{2}R(\omega) = H(\omega) + iV(\omega), \qquad (5.1.58)$$

$$\sqrt{2}L(\omega) = H(\omega) - iV(\omega), \qquad (5.1.59)$$

where $R(\omega)$, $L(\omega)$, $H(\omega)$, and $V(\omega)$ are the frequency-domain descriptions of the signals in right, left, horizontal, and vertical polarizations, respectively.

Interferometric polarimetry is very simply accomplished through the correlation approach because the interferometer already consists of one radiometer in each antenna, the outputs of which can be correlated. Measurements with both antennas of the interferometer having right circular polarization, both having left, and one having left, one having right ("cross" polarizations) give, for a point source,

$$S_{RR}(\omega)A(\omega) \qquad (5.1.60)$$

$$S_{LL}(\omega)A(\omega) \qquad (5.1.61)$$

$$S_{RL}(\omega)A(\omega) \qquad (5.1.62)$$

$$S_{LR}(\omega)A(\omega) = S_{RL}^{*}(\omega)A(\omega), \qquad (5.1.63)$$

and similar results for linear polarizations. Measurements of both crossed polarizations, right versus left and left versus right, are necessary to separate the phase of $A(\omega)$ or fringe phase from the phase of $S_{RL}(\omega)$ which is nonzero since it is a complex quantity. A complete map of the polarization parameters for an extended source requires measurement of four correlation functions over the uv-plane. If each antenna is equipped with two radiometers, one connected to each orthogonal polarization, then the four polarization parameters of the visibility function can be determined simultaneously. This requires four simultaneous cross correlations. If only two radiometers are available, then the polarization parameters can be obtained sequentially by switching the polarization entering each radiometer. The interferometer observing in cross-circular mode makes an exceedingly sensitive detector of linear polarization[11] and provides a simple means of measuring the position angle given by

$$\tan(2\phi) = S_2/S_1 = \operatorname{Im} S_{RL}/\operatorname{Re} S_{RL}. \qquad (5.1.64)$$

5.1.5. Signal-to-Noise Ratio Analysis for Interferometers

The signal and noise wave forms at the output of each interferometer receiver can be represented by the sum of two independent noise voltages with Gaussian statistics. The antenna temperatures are T_A and T_A' and the

[11] R. G. Conway and P. P. Kronberg, *Mon. Not. Roy. Astron. Soc.* **142**, 11 (1969).

system temperatures are T_s and $T_s{'}$. The time functions $x(t)$ and $y(t)$ from the two receivers can be written as

$$x(t) = (T_A)^{1/2}s(t) + (T_s)^{1/2}n(t),$$
$$y(t) = (T_A{'})^{1/2}s'(t) + (T_s{'})^{1/2}n'(t),$$
(5.1.65)

where the normalized time functions $s(t)$, $s'(t)$, $n(t)$, and $n'(t)$ can each be represented over the time interval T by a Fourier series expansion that has BT independent components. In the following analysis it is assumed the source is unresolved. The complex cross power (the cross-spectral function integrated over the bandwidth B) S_{xy} is the sum of terms

$$S_{xy} = (T_A T_A{'})^{1/2}S_{ss'} + (T_s T_s{'})^{1/2}S_{nn'}$$
(5.1.66)

$$+ (T_A T_s{'})^{1/2}S_{sn'} + (T_A{'}T_s)^{1/2}S_{s'n}$$
(5.1.67)

which are uncorrelated and hence statistically independent.[12] The proof that these components are uncorrelated uses the expansion of the expectation of the product of four random variables; this expansion was derived by Davenport and Root.[12] If each component of the Fourier series has an rms magnitude of unity, the mean and variance of the magnitude of $S_{ss'}$ are equal to BT. The cross powers between uncorrelated signals $S_{sn'}$, $S_{s'n}$, and $S_{nn'}$ have a magnitude with variance of BT and have random phase with respect to $S_{ss'}$.

The vector representation of the random variable S_{xy}/BT shown in Fig. 7 is useful. If the real and imaginary parts are the components of a vector, then $(T_A T_A{'})^{1/2}(S_{ss'}/BT)$ can be considered the "signal vector" since its magnitude is $(T_A T_A{'})^{1/2}$ and its phase is the fringe phase. $(T_A T_A{'})^{1/2}$ is related to the antenna apertures by

$$(T_A T_A{'})^{1/2} = 10^{-26}F(AA')^{1/2}/2k,$$
(5.1.68)

where F is the source flux, k Boltzmann's constant, and $(AA')^{1/2}$ the effective interferometer aperture.

Added to this signal vector is a noise vector N_1 with zero mean (and in the same direction as the signal vector) which is due to the power fluctuations in the signal. The other terms produce a noise vector N_2 with uniformly distributed random phase θ and mean square magnitude

$$(T_s T_s{'} + T_A T_s{'} + T_A{'} T_s)/BT.$$
(5.1.69)

The probability distribution of the magnitude of N_2 is a Rayleigh distribution (see Davenport and Root[12]) shown in Fig. 8. There are two ways of

[12] W. B. Davenport and R. L. Root, "An Introduction to the Theory of Random Signals and Noise." McGraw-Hill, New York, 1958.

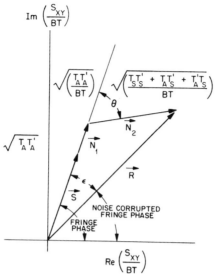

FIG. 7. Vector diagram of signal **S** plus its self noise N_1 plus system noise N_2. N_1 is Gaussianly distributed in the direction of **S**, whereas N_2 is uniformly distributed in angle θ and has a Rayleigh distribution in length.

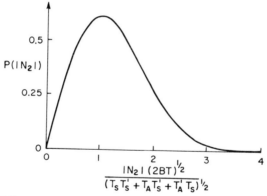

FIG. 8. Probability distribution of the noise vector $|N_2|$. This distribution is commonly known as the Rayleigh distribution.

expressing the effect of the noise. For large signal-to-noise ratios the phase error ε is a random variable (approximately Gaussian) with zero mean and rms from Fig. 7:

$$\varepsilon_{rms} = \left(\frac{T_s T_s' + T_A T_s' + T_A' T_s}{BT T_A T_A'}\right)^{1/2} (\sin^2 \theta)^{1/2} \qquad (5.1.70)$$

$$= \left(\frac{T_s T_s' + T_A T_s' + T_A' T_s}{2BT T_A T_A'}\right)^{1/2} \text{[rad]}. \qquad (5.1.71)$$

The rms error in fringe amplitude using

$$\mathbf{R} = \mathbf{S} + \mathbf{N}_1 + \mathbf{N}_2, \tag{5.1.72}$$

$$\mathbf{R} \cdot \mathbf{R} = (\mathbf{S} + \mathbf{N}_1) \cdot (\mathbf{S} + \mathbf{N}_1) + \mathbf{N}_2 \cdot \mathbf{N}_2 \tag{5.1.73}$$

is

$$\Delta A_{\mathrm{rms}} = (\mathbf{R} \cdot \mathbf{R} - \mathbf{S} \cdot \mathbf{S})^{1/2} \tag{5.1.74}$$

$$= [(T_{\mathrm{A}} + T_{\mathrm{s}})(T_{\mathrm{A}}' + T_{\mathrm{s}}')/BT]^{1/2} \quad [^\circ\mathrm{K}]. \tag{5.1.75}$$

The variance of the fringe amplitude depends on signal-to-noise ratio and is given by Eqs. (5.5.29) and (5.5.35) for the strong and weak cases, respectively. Alternatively, one might consider the noise in each component of S_{xy}/BT (if we assume the signal to be equally split between components) which is Gaussian with rms value

$$(\Delta T)_{\mathrm{rms}} = \left(\frac{(T_{\mathrm{A}} + T_{\mathrm{s}})(T_{\mathrm{A}}' + T_{\mathrm{s}}') + T_{\mathrm{A}} T_{\mathrm{A}}'}{2BT} \right)^{1/2} \quad [^\circ\mathrm{K}] \tag{5.1.76}$$

for each component. This last expression enables us to make a direct comparison between a "correlation" and total power receiver. If the signal power is equally split between the two receivers and the outputs, cross multiplied, and integrated to obtain the signal power, one component of the complex cross power is measured, and the noise-to-signal ratio is

$$\frac{(\Delta T)_{\mathrm{rms}}}{T} = \left[\frac{[(T_{\mathrm{A}}/2) + T_{\mathrm{s}}][(T_{\mathrm{A}}/2) + T_{\mathrm{s}}'] + T_{\mathrm{A}}^2/4}{(T_{\mathrm{A}}^2/4)2BT} \right]^{1/2}, \tag{5.1.77}$$

which is a factor of $\sqrt{2}$ larger (poorer signal-to-noise ratio) than a total-power radiometer. When the antenna temperatures are small, the major part of the noise originates from $\mathbf{N}_2$. The rms magnitude of $\mathbf{N}_2$, $(T_{\mathrm{s}} T_{\mathrm{s}}')^{1/2}/(BT)^{1/2}$, acts as a "noise threshold" since fringe phases will be almost completely random until the signal level exceeds this noise threshold. In this respect, the noise threshold may be considered as the limit of sensitivity. The theory can easily be extended from a point source to a distributed radio source by simply noting that the antenna temperature that contributes to the "signal" is the correlated portion, whereas the uncorrelated portion effectively contributes to the system temperature.

The analysis above assumes the optimum interferometer signal processing. It can be shown that correlation or its equivalent performs the maximum likelihood estimate of fringe amplitude and phase.[13,14] The best obtainable

[13] A. E. E. Rogers, *Radio Sci.* **5**, 1239 (1970).
[14] J. M. Wozencraft and I. M. Jacobs, "Principles of Communication Engineering." Wiley, New York, 1965.

signal-to-noise ratio for the weak signal case is

$$\frac{S}{N} = \frac{|S|}{(|N|^2)^{1/2}} = \frac{(T_A T_A')^{1/2}(BT)^{1/2}}{(T_s T_s')^{1/2}}, \qquad (5.1.78)$$

while other simpler processing techniques yield the poorer signal-to-noise ratios reflected in Table I. Simplifying factors such as clipping and sampling also reduce the signal-to-noise ratio. The one-bit correlation scheme now frequently used in VLBI and spectral-line radio astronomy results in a reduction by a factor of about $\pi/2$.

TABLE I.   Comparison of the Signal-to-Noise Ratios for Various Interferometer Detection Schemes

Interferometer detector type (SSB case, except where noted)	Normalized output in the frequency domain prior to integration	Signal-to-noise ratio relative to ideal case	Remarks
Correlation with adequate delay range	$e^{i\omega\tau_g}$	1	Delay range has to be at least $1/B$ (see Fig. 3); increased delay range increases spectral resolution.
Fourier transform of each video signal and conjugate cross multiplication	$e^{i\omega\tau_g}$	1	Equivalent to correlation related via FT.
Delay compensated multiplying interferometer with quadrature component	$\cos \omega\tau_g, \sin \omega\tau_g$	1	Same as above (for continuum sources), except no spectral information is obtained.
As above, without quadrature component	$\cos \omega\tau_g$	$1/\sqrt{2}$	Least squares fringe fitting required to obtain fringe phase.
Adding	$1 + \cos \omega\tau_g$	$1/\sqrt{2}$	Sensitive to gain fluctuations.
Simultaneous add and subtract	$\cos \omega\tau_g$	$1/\sqrt{2}$	Not sensitive to gain fluctuations.
Adding, with 180° Dicke switch	$\cos \omega\tau_g$	$1/2$	Additional signal-to-noise loss incurred by Dicke switching.
Double sideband correlation with adequate delay range or delay compensation	$\cos \omega\tau_g$	$1/\sqrt{2}$	Fringe fitting required; rf bandwidth, double SSB case.

5.2. Connected-Element Interferometry*

5.2.1. Introduction

This chapter deals with two applications of phase-measuring interferometers: the accurate measurement of position (astrometry), and the determination of source structure—in particular, mapping observations. In both cases, the determination of the difference in phase between incoming signals (the interferometer phase) is crucial, and we will begin by discussing aspects of phase measurement without regard for which application is being considered.

Essential features for the interpretation of measurements of interferometer phase are a known baseline and stable electrical lengths for the various local oscillator and signal paths. The effective baseline may be varied, either by tracking the source over a range of hour angles so that the orientation and projected length of the baseline change, or by using movable elements in the interferometer. Both methods may be employed in the same instrument.

In a tracking interferometer, the difference in electrical paths from the source to the two antennas changes continuously. There are two important consequences for the receiving and recording arrangements:

(a) Extra delay cable must be inserted in one of the signal paths (conveniently in the intermediate frequency section) so that the overall paths are nearly equal (this is the "white light fringe" condition). There will be a significant loss in the observed signal if the path difference remaining after the correction is as large as c/B, where c is the velocity of light and B the bandwidth. The intermediate frequency is normally several times higher than the bandwidth (e.g., 45 MHz when $B = 10$ MHz), and the cables may be switched in units of $\lambda_{i.f.}$ without altering the observed phase. When observing a region of sky of large angular extent, a more restrictive condition may apply. Since the path difference varies across the area of sky, the compensation will be in error at some points (Fig. 1). Signals may be received from within an angle $\sim \lambda/d$, where λ is the observing wavelength and d the size of the individual elements. The maximum variation in path difference across the primary beam of the antennas is therefore about $D\lambda/d$, and the resulting limitation in fractional bandwidth is[1]

$$B/f \lesssim d/D.$$

[1] B. Elsmore, S. Kenderdine, and M. Ryle, *Mon. Not. Roy. Astron. Soc.* **134**, 87 (1966).

* Chapter 5.2 is by Guy Pooley.

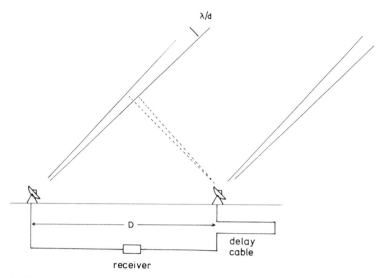

Fɪɢ. 1. The variation in path difference across the field of view of the interferometer.

Here f is the observing frequency and D is the maximum baseline used.

(b)   Since the path difference usually changes rapidly during tracking, so does the phase of the interferometer output. The high frequencies (perhaps several Hz) in this signal may be conveniently recorded if a " phase rotation " (or " fringe stopping ") system is used to insert a variable phase delay which (almost) removes the phase change introduced by the rotation of the Earth. This system is possible when the primary antennas are directive (as is usually the case), since this means that only a small range of frequencies (perhaps 0.01 Hz) is present in the interferometer output. The final sampling rate is then determined by this range.

In most interferometers operating at frequencies above about 1 GHz, the phase stability of the local oscillator system is crucial to the performance of the telescope. The signal is usually transmitted from the central local oscillator to the antennas via buried cables, whose temperature remains nearly constant for many days. In order to reduce power losses in these cables, a subharmonic of the final frequency may be used; a frequency multiplier at each antenna is then used to derive the local oscillator signal. The amplifiers and multipliers in this system must be designed to maintain a constant phase output.

The observing frequency itself must be known accurately in order to express the length of the baseline in terms of the observing wavelength. The frequency is determined by the sum (or difference) of the local oscillator and intermediate frequencies. While the first may be measured with high precision, it seems necessary at first sight to measure the frequency response of the intermediate frequency amplifiers, cables, and filters very carefully. Elsmore and Mackay[2] show that this is not the case, but that the effective intermediate frequency is that at which the lengths of the delay cables are integral numbers of wavelengths. The true observing frequency may therefore be accurately defined, provided that the electrical lengths of the delay cables are stable.

If the interferometer axis is horizontal, then the presence of a plane stratified atmosphere or ionosphere above the telescope does not affect the measured interferometer phase or, hence, the derived source position. (The free-space wavelength must still be used in any calculations of the electrical path difference; this then allows exactly for any change in direction of the plane wavefront.) Corrections must be applied for any departure from this idealized condition, which can arise in three ways:

(i)   Antennas not at the same height. This is so for the Green Bank interferometer of the National Radio Astronomy Observatory (NRAO), which incorporates an automatic measurement of the refractivity of the air and allowance for the change in phase.

(ii)   Curvature of the atmosphere and/or ionosphere or gradients in the electron density of the ionosphere. The spherical component of atmospheric refraction becomes important for angles of elevation below about 20°; for this reason, the NRAO astrometric program included no measurements at lower elevations. The observations of sources at low declinations (and hence low elevations) made with the One-Mile Telescope at Cambridge University are corrected for atmospheric refraction using a model atmosphere to calculate the phase changes.

Ionospheric refraction becomes rapidly less important as the observing frequency increases (since the departure of the refractive index from unity varies as λ^2). For the earlier observations at 1.4 GHz made at Cambridge, the errors introduced by ionospheric gradients were usually less than 0.3 arc sec (ref 2) (see Chapter 2.1). For higher observing frequencies, refraction in the ionosphere can normally be neglected.

(iii)   Tropospheric irregularities, which are important under some conditions (see Chapter 2.5). Hinder,[3] for example, has studied conditions in England at an observing frequency of 5 GHz. He showed that under winter nighttime conditions, the phase variations introduced by the troposphere were very small ($<1°$). In warmer daytime weather, atmospheric conditions may occasionally introduce very large phase deviations ($>100°$ at 5 GHz). The scale size of the most important irregularities is about 700 m, corresponding to time scales of several minutes as the irregularities blow across

[2] B. Elsmore and C. D. Mackay, *Mon. Not. Roy. Astron. Soc.* **146**, 361 (1969).
[3] R. A. Hinder, *J. Atmos. Terr. Phys.* **34**, 1171 (1972).

the interferometer (Fig. 2). Subsequent observations with the 5-km telescope[4] have shown that irregularities of larger scale sizes (several km) also occur throughout the year.

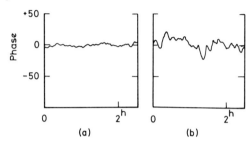

FIG. 2. The disturbing effects of tropospheric irregularities under (a) calm and (b) disturbed conditions.

If two correlation receivers are used, with an extra $\lambda/4$ delay in one signal path to one receiver, the correlated signal can be measured uniquely, and is usually displayed as amplitude and phase. The ways in which this information may be used for astrometry and source mapping are described further in Sections 5.2.2 and 5.2.3.

5.2.2. Astrometry

Radio methods of measuring astronomical positions now rival optical methods in their precision. Radio and optical methods involve somewhat different principles and different corrections for atmospheric refraction. A close comparison of optical and radio measurements is therefore very useful. Reviews of this area are to be found in ref. 5.

Source positions have now been measured with an accuracy of about 1 arc sec or better by workers at the Mullard Radio Astronomy Observatory, University of Cambridge (UK), the National Radio Astronomy Observatory (NRAO), Green Bank (USA), and the Royal Radar Establishment (RRE), Malvern (UK). The methods used by these three groups differ considerably in the techniques for determining the parameters of their baselines, but their results are in substantial agreement. We will describe first the principles involved and then the individual telescopes and their modes of operation.

The basis of most accurate determinations of radio-source positions is the precise establishment of an interferometer baseline. The known geometry of the telescope and the measured phase of the correlated signal are then suf-

[4] P. J. Hargrave and L. J. Shaw (private communication).
[5] W. Gliese, C. A. Murray, and R. H. Tucker (eds.), in " New Problems in Astrometry." Reidel, Dordrecht, Netherlands, 1974.

ficient to determine the source positions. (In the whole of this section on astrometry, we are concerned with sources which are not significantly resolved by the interferometer. The amplitude of the correlated signal is then independent of the interferometer baseline, and is not used in the analysis.)

For any interferometer observing an unresolved source over a range of hour angle H, the observed phase Φ will vary as

$$\Phi(H) = \phi_1 \cos H + \phi_2 \sin H + \phi_3. \qquad (5.2.1)$$

The basic measured quantities are ϕ_1, ϕ_2, and ϕ_3. Once the baseline is known, the source position may be derived. The One-Mile (1.6-km) and 5-km Telescopes at Cambridge and the Green Bank three-element interferometer are each operated in such a way that the source positions measured do not depend on previous optical measurements, so far as is possible. (The reservations for each particular telescope will be discussed below.) The RRE interferometer measures positions relative to a number of measured optical positions of radio sources. The differences in technique are best described by discussing each telescope in turn.

5.2.2.1. The One-Mile Telescope at Cambridge. For an East–West (E–W) telescope, the variation of phase Φ with hour angle takes on a particularly simple form:

$$\Phi(H) = (2\pi D/\lambda) \cos \delta \sin H + \phi_3, \qquad (5.2.2)$$

where D is the separation of the antennas, λ the observing wavelength, and δ the declination of the source. The quantity ϕ_3 is an instrumental constant giving no potential information. The One-Mile Telescope uses three equatorially mounted paraboloids, each 18 m in diameter on an E–W line. They are used as two independent interferometers with spacings which can be varied by moving one element on a rail track.[1] The maximum separation is 1500 m. It has been operated[2] at 0.4 and 1.4 GHz, and[6] at 2.7 and 5.0 GHz.

It can be seen from Eq. (5.2.2) that the amplitude of the $\Phi(H)$ curve yields the product $D(\cos \delta)/\lambda$, while its phase yields the right ascension, provided the sidereal time is known. Declination determination therefore requires a precisely known separation D and observing wavelength λ. The first of these was determined by Froome and Bradsell (private communication) of the UK National Physical Laboratory, using the Mekometer III laser distance-measuring system. The lengths are thought to be accurate to ± 1.5 mm.

The declination of a source may therefore be determined (with an ideal E–W telescope) from length and frequency measurements alone, and in this way the measurement is independent of any previous determination.

In order to measure right ascension, the hour angle of the source is deter-

[6] J. W. Smith, *Nature (London)* **232**, 150 (1971).

mined at a known sidereal time [through Eq. (5.2.2)]. This requires a precise clock which, in practice, is calibrated via radio transmissions of standard time signals, and ultimately from optical astrometric data on the rotation of the earth.

At this stage, it is necessary to consider the effects of small errors in the baseline. It is convenient to measure the departures from a true E–W line in terms of displacements in, and normal to, the equatorial plane.

An error in the length of the baseline yields systematically wrong declinations; if the length is known to ± 1.5 mm in 1.5 km, the systematic error will not exceed about 0.2 arc sec in declination.

An error in the azimuth angle of the baseline (as projected on to the equatorial plane) is a change in the effective longitude of the telescope. In the Cambridge telescope, the azimuth angle of the baseline was determined by theodolite surveying, but the principal difficulty in this technique is to locate a suitable point of reference on the antenna structures themselves. The azimuth determined in this way led to a systematic difference between the derived right ascensions of compact radio sources. It was therefore decided to adopt a value for the azimuth angle which reduced the mean error to zero. The zero point of the right-ascension scale is linked to optical measurements of the positions of radio sources, but the accuracy of the relative right ascensions depends only on the stability of the timing system.

If the baseline is not parallel to the equatorial plane, the value of the phase ϕ_3 (which remains constant during an observation of one source) varies with the declination of the source. When the angle between the baseline and the equatorial plane is small, this variation adds no useful information.

Further departures from an ideal telescope will arise if the physical structures of the individual antennas are not identical. In the case of the telescopes at Green Bank and Cambridge, the important differences are in the directions of the polar axes and in the distances between the polar and declination axes. (If the two axes intersect, these errors do not arise.)

The parameters for the individual antenna structures in the Cambridge telescope have been determined directly (by star photography for the directions of the polar axes, and by direct measurement for the interaxis distances). By contrast, the corresponding parameters of the NRAO interferometer at Green Bank are derived indirectly, as described later.

The observing technique used with the One-Mile Telescope normally involves tracking the source over a period of 12^h, and the data are presented as the amplitude of the correlated signal and its phase relative to that which would be expected if the source were in some assumed position. Figure 3 shows the amplitude and phase plots for one source, using four different assumed positions.[2]

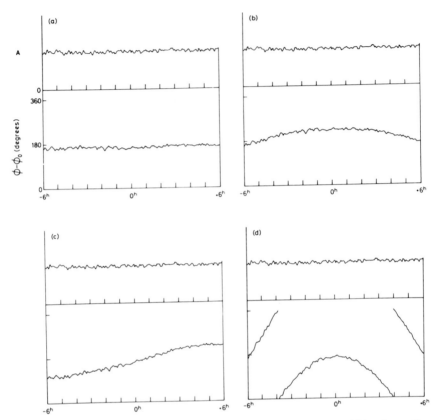

FIG. 3. The amplitude and phase of the interferometer signal during a 12-hr observation with the One-Mile Telescope at 1.4 GHz. The phase is plotted relative to that expected for a source at each of four assumed positions. The assumed positions are: (a) at the true position; (b) 5 arc sec east of the true position; (c) 5 arc sec north of the true position; and (d) 1 lobe (14 arc sec) east of the true position.

5.2.2.2. Green Bank Three-Element Interferometer.

The interferometer[7] at the National Radio Astronomy Observatory (NRAO), Green Bank, employs three 26-m equatorially mounted paraboloidal antennas on a 2700-m baseline. Each pair of antennas is connected as an interferometer, giving three element spacings. The baseline is 28° away from E–W in azimuth, a fact which is important in analysis of their results. Wade[8] describes astrometric observations made at 2.7 GHz, in which the baseline of the instrument

[7] D. E. Hogg, G. H. Macdonald, R. G. Conway, and C. M. Wade, *Astron. J.* **74**, 1206 (1969).

[8] C. M. Wade, *Astrophys. J.* **162**, 381 (1970).

was determined by an entirely different technique from that used for the Cambridge instrument. The instrument has subsequently been operated at 2.7 and 8 GHz simultaneously. More recently, a fourth element has been added 35 km away.

The observations to determine the three constants ϕ_1, ϕ_2, and ϕ_3 in Eq. (5.2.1) are made in three observing periods for each source, each about 20^m in length and covering a wide range of hour angles. For a telescope which does not lie in a plane parallel to the equator, the value of ϕ_3 gives an estimate of the declination of the source which is independent of the estimate given by ϕ_1 and ϕ_2. The method actually used is to find a self-consistent baseline with which the independent values of declination agree for sources at different declinations. This process requires observations of at least four sources (in practice about 20 are used) having declinations scattered over a wide range. The analysis is described further by Wade.[8]

Two of the baseline parameters for each spacing which are not derived by this procedure are (1) a zero point in the right ascension scale and (2) a term relating to the polar axes of the aerials. These are derived by comparison of the measured positions with optically derived positions[9]; it is assumed that the mean difference between radio and optical right ascensions is zero (as for the One-Mile Telescope), and that this difference does not vary systematically with declination.

Both the Cambridge and Green Bank telescopes give measurements of declination that are independent of any previous astronomical determinations.

5.2.2.3. The Royal-Radar-Establishment Interferometer. This telescope employs two 26-m paraboloidal antennas on a baseline whose length and orientation may be varied. The analysis of the observations depends more heavily on the measured optical positions of identified radio sources than in either of the two telescopes described previously.

In a series of observations (Adgie et al., unpublished), the interferometer was used at 2.7 GHz with a North–South (N–S) baseline of 640 m (5800λ). Sources were observed at two hour angles, near $\pm 4^h$. The two fringe patterns were then approximately at right angles for sources over a wide range of declinations. In these observations, however, the telescope was calibrated by frequent observations of 36 optically identified sources, using their optical positions directly in the analysis.

Wade et al.[10] have compared the measurements of the positions of 28 sources which were common to the RRE and Green Bank lists. The standard error for the differences in each coordinate was 0.5 arc sec.

5.2.2.4. The 5-km Telescope at Cambridge. This instrument, completed in

[9] J Kristian and A. Sandage, Astrophys. J. 162, 391 (1970).
[10] C. M. Wade, H. Gent, R. L. Adgie, and J. H. Crowther, Nature (London) 228, 146 (1970).

1972, is described in more detail by Ryle.[11] It comprises eight 13-m antennas on a baseline (formerly used as a railway) which runs almost E–W; four of the antennas are fixed at separations of 1.2 km, and the remaining four may be moved along a length of rail extending for a further 1.2 km. Sixteen independent interferometer spacings (one fixed and one moving antenna) are thus observed simultaneously. The initial observing frequency is 5 GHz. The observing technique and the principles involved in the data analysis are very similar to those described for the One-Mile Telescope and will not be discussed in detail, but some points of particular interest are noted.

The instrument was surveyed by the UK Ordnance Survey, who established the astronomical coordinates of the 5-km baseline with an accuracy of about 0.1 arc sec, and the location of each of the fixed telescopes and mounting points for the moving telescopes with a precision of a few millimeters. The accuracy of these figures is confirmed by observations of the star β Persei (Algol), which has recently been found to be an erratic radio emitter.[12] On rare occasions, the flux density of the star is sufficient for astrometric use with the telescope, and the possibility of a close link between radio astrometric measurements and fundamental star catalogs is clearly of great importance[5].

First results of astrometric measurements[13,14] with the instrument indicate that a precision of 0.02 arc sec in the positions of compact radio sources may be reached. A new method of determining the length of the baseline with an accuracy of 0.25 mm from the observations of sources near the equator has been used.

5.2.3. Mapping of Radio Sources

Interferometers provide the most convenient means of studying the angular structure of radio sources on scales less than a few minutes of arc. An interferometer of separation D will give information about those components of radio sources whose angular scales are comparable with the fringe separation λ/D. If the interferometer phase is not measured, or if only a few separate baselines are available, the interpretation of the data is usually limited to the construction of possible models for the distribution of radio emission. When the amplitude and phase of the interferometer signal is measured over a systematic range of baselines, it becomes possible to use "aperture synthesis" techniques, in which we construct an effective aperture comparable in size with the largest interferometer spacing used.

We will consider first the case of a "one-dimensional" synthesis, involving

[11] M. Ryle, *Nature (London)* **239**, 435 (1972).
[12] R. M. Hjellming, C. M. Wade, and E. Webster, *Nature (London)* **236**, 43 (1972).
[13] M. Ryle and B. Elsmore, *Mon. Not. Roy. Astron. Soc.* **164**, 223 (1973).
[14] B. Elsmore and M. Ryle, *Mon. Not. Roy. Astron. Soc.* **174**, 411 (1976).

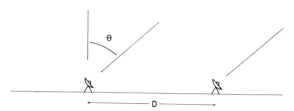

FIG. 4. The basic interferometer configuration.

a baseline of fixed orientation and variable length (Fig. 4). The signal from a small patch of sky, subtending an angle $d\theta$, is proportional to

$$A(D, \theta) \, d\theta = B(\theta) \, d\theta \, \exp[2\pi i \, D(\sin \theta)/\lambda].$$

(We include the response of the primary antennas in the apparent brightness $B(\theta)$.) The total signal is

$$A(D) = \int B(\theta) \, \exp[2\pi i \, D(\sin \theta)/\lambda] \, d\theta. \qquad (5.2.3)$$

If we write $C(\sin \theta) \equiv B(\theta)/\cos \theta$, it is clear that $A(D)$ and $C(\sin \theta)$ are related as a Fourier pair.

The function $A(D)$, the interferometer response, is measured over a range of values of (projected) baseline. The data may then be inverted to produce a map of the area of sky within the primary response of the antennas.

The data are conveniently measured at equal intervals of projected spacing: separations a, $2a$, $3a$, ..., na. Observations at spacings $-a$, $-2a$, etc. are unnecessary, since a negative spacing $(-D)$ may be obtained by interchanging the antenna connections, or more simply, by taking the complex conjugate of $A(D)$. The function $A(D)$ has been sampled only at this array of points, so the map of the sky we obtain from an inversion of Eq. (5.2.3) represents the apparent sky brightness $B(\theta)$ convolved with the transform of the sampling pattern. We may call this transform the "synthesized beam" of the telescope.

The size and shape of the principal response depend on the area of the aperture which has been sampled (the value of na in particular), and the weighting given to the data before transformation. As an alternative to weighting the data before the transformation, a uniform weighting may be used and the final response obtained later by a numerical convolution of the maps. The spatial frequencies (corresponding to the projected interferometer spacing measured in wavelengths) lie in the range $-na/\lambda$ to $+na/\lambda$. If the data are given equal weight in the transformation, the response to an unresolved source at $\theta = 0$ varies as

$$\sin(2\pi na\theta/\lambda)/(2\pi na\theta/\lambda)$$

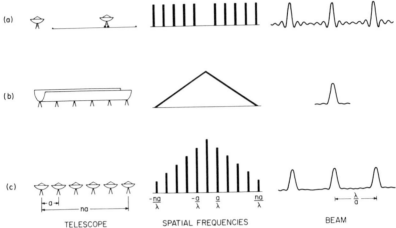

FIG. 5. Three telescopes, each of length na, showing the Fourier components measured and the corresponding reception patterns: (a) synthesis telescope; (b) single-element, total power telescope; (c) phased array.

(see Fig. 5a). This response does not represent that of any real aperture, of course, since it is negative in some regions. It is the same as the voltage response of an aperture of length $2na$.

There will also be grating-sidelobe responses, analogous to those formed by an optical diffraction grating. These occur at angles of $m\lambda/a$ from the main response, where m is an integer. For a one-dimensional aperture such as this, the grating responses all have the same amplitude as the main response.

These sidelobe responses do not introduce any ambiguity, provided that the angle λ/a is larger than the diameter of the source, or that it is larger than the primary reception pattern of the antennas. When observing bright, isolated sources, the first criterion is appropriate, and the spacing increment can be chosen according to the angular size of the source. Survey observations, on the other hand, require a complete mapping of the region in the primary beam of the antennas. Using elements of size d, the main response is about λ/d in extent, and the condition $\lambda/a > \lambda/d$ reduces to the physically reasonable one that the spacing increment a should be less than the size of the individual elements. In practice, a value $a \simeq 0.7d$ might be used.

It is useful here to compare the response of a synthesis array, length na, with those of two related antenna systems: a single-element telescope of length na (Fig. 5b), and a phased array of $n + 1$ elements with separation a.

The response of the single-element telescope (if the aperture is uniformly illuminated) varies as

$$[\sin(n\pi a\theta/\lambda)/(n\pi a\theta/\lambda)]^2.$$

The Fourier components in this response have a triangular weighting, with a maximum at zero and falling to zero at $\pm na/\lambda$. (A slightly longer instrument is required to measure the Fourier terms at these points with a finite weighting.)

The grating array, in which all the elements are connected to one receiver, may be phased to point in any chosen direction (this action is performed by numerical methods in the synthesis telescope). The Fourier terms (see Fig. 5c) are just those measured by the synthesis telescope, but with a different weighting. The response is similar to that of the single-element telescope, but with grating responses at angles of $\pm m\lambda/a$ from the main response.

All the telescopes measure spatial frequencies in the range $\pm na/\lambda$, and, apart from the sampling at intervals of a/λ by the synthesis and grating systems, the maps produced are equivalent; a numerical convolution could be used to change the weighting of the spatial frequencies from one to the other.

A variety of instruments at Cambridge have been used in one-dimensional synthesis observations. Most of these have used "T" systems, in which the resolution of the instrument in the E–W direction is determined by a cylindrical paraboloid, and the N–S resolution is achieved by a smaller movable element on an N–S rail (see Fig. 6a). Surveys of the whole sky have been

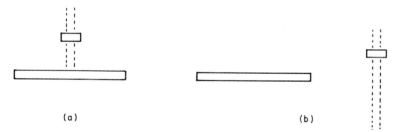

FIG. 6. Configurations used for one-dimensional synthesis at Cambridge: (a) "T" antenna; (b) synthesized interferometer.

made with instruments of this type, at frequencies of 178 MHz,[15] 81.5 MHz,[16] and 38 MHz.[17,18] Each of the surveys is made up of observations at a series of different declinations, using the telescope as a transit instrument.

Since the whole of the sky within the primary beam must be mapped when making survey observations, the spacing increment a must be less than the width d of the elements. It is therefore not possible to observe the smallest

[15] J. L. Caswell, J. H. Crowther, and D. J. Holden, *Mem. Roy. Astron. Soc.* **72**, 1 (1967).
[16] M. A. Smith, Ph.D. Thesis, Univ. of Cambridge (1970).
[17] J. H. Blythe, *Mon. Not. Roy. Astron. Soc.* **117**, 652 (1957).
[18] P. J. S. Williams, S. Kenderine, and J. E. Baldwin, *Mem. Roy. Astron. Soc.* **70**, 53 (1966).

projected spacing in the sequence; a technique described by Williams *et al.*[18] can be used to estimate the results for this spacing. Total power observations with the long element alone, and observations with the smallest N–S interferometer spacing that is possible, include Fourier terms of spatial frequencies $\sim d/\lambda$ and in the range 0–2 d/λ, respectively. A combination of these two can be used to give the best estimate of the missing "low order" term. The efficiency of this process is illustrated by the maps of Caswell *et al.*[15] where the successive declination strips match up well.

A slightly different use of one-dimensional synthesis is to produce a synthesized interferometer, as was done for the 4C source survey.[19,20] The arrangement of antennas is shown in Fig. 6b. It is important to note that the moving element positions north and south of the line of the fixed element are no longer equivalent, and both must be observed in order to synthesize an E–W instrument.

Two-dimensional synthesis involves straightforward extensions of the principles discussed above. The first observations using a two-dimensional aperture were made at Cambridge by O'Brien,[21] who observed the sun at several frequencies by moving the elements of his interferometer on the surface of the earth to all the required relative positions.

A more convenient method of building two-dimensional synthesis telescopes is to use the rotation of the Earth to vary the orientation of the baseline, so that one element appears to move around the other when seen from the direction of the source. The baseline separation is varied by moving one element along a straight line, or by having an array which can measure several spacings at once. Ryle and Neville[22] first used Earth rotation synthesis in a survey of the North celestial pole at 178 MHz.

If we observe a source with an interferometer for a range of hour angles, the path described by one element relative to the other lies on a circle. This circle lies in a plane at $z = z_0$ ($z = 0$ being the position of the reference antenna and the z-axis pointing in the direction of the celestial pole: Fig. 7). When observations are made for 12^h, the segment is a semicircle, and the corresponding part obtained by taking the complex conjugate of the signal (equivalent to interchanging the antennas) is a semicircle at $z = -z_0$. Only in the special case of an E–W telescope do the semicircles lie in the same plane and form a complete circle.

When observations are made with a series of spacings along the same line, the semicircles lie on the surface of a cone. The effective aperture in the synthesis is the projection of this array of semicircles as seen from the direc-

[19] J. D. H. Pilkington and P. F. Scott, *Mem. Roy. Astron. Soc.* **69**, 183 (1965).
[20] J. F. R. Gower, P. F. Scott, and D. Wills, *Mem. Roy. Astron. Soc.* **71**, 49 (1967).
[21] P. A. O'Brien, *Mon. Not. Roy. Astron. Soc.* **113**, 597 (1953).
[22] M. Ryle and A. C. Neville, *Mon. Not. Roy. Astron. Soc.* **125**, 39 (1962).

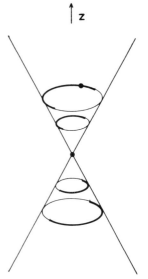

FIG. 7. The relative path of one interferometer element around the other during a 12-hr observation, as seen from the direction of the source.

tion of the source, and an adequate mapping of the source requires an adequate coverage of the aperture plane by these projected lines. If the axis of the telescope is not E–W, there may be substantial areas of the aperture plane which are not accessible, and some terms in the transform will be missing. This makes the interpretation of the data much more difficult.

An E–W interferometer does not suffer from this disadvantage since the circular tracks in the aperture plane merely distort into ellipses when viewed from the direction of a source away from the celestial pole. The corresponding synthesized beam is, of course, elliptical, with its major axis in the N–S direction, and larger by a factor csc δ than the minor axis. A disadvantage of the E–W instrument is that its resolution in declination becomes poor near the equator.

The grating responses from a telescope making a series of circular tracks in the aperture plane (of radii $a, 2a, \ldots$) are circles of radii $m\lambda/a$. When observing a source away from the pole, these become ellipses, larger by a factor of csc δ in declination. The grating responses from a telescope which is not E–W are more complicated, and will depend in detail on the axis of the telescope and the declination of the source.

The synthesis computation itself requires particular consideration when the interferometer axis is not E–W. A map can always be made by combining the signals at each point on the aperture with the appropriate phase delays for each point on the sky in turn. This process usually takes two steps. First, all the interferometer records are phased so that the beam points at a chosen reference point in the map. Second, additional phase distributions are applied to the data in order to point the beam at successive points away from the map center. In many cases, the phase delays concerned are, to first order, linear in distance along some plane onto which the sky may be projected. This part of the analysis is then a Fourier transform. In the particular case of an E–W interferometer, the aperture is always a plane, and the reduction can always be done by a Fourier transform. If the aperture is not a plane, then the linearity of phase difference with distance breaks down for large angles, and a Fourier transform can only be used for a small range of angles in the sky.

The angular sizes of sources so far observed with the NRAO telescope are not so large that the analysis is restricted by this effect. Restrictions become more severe for larger telescopes, shorter wavelengths, or sources of larger angular size; a separate Fourier transform will be required for each of a number of phase centers across the map.

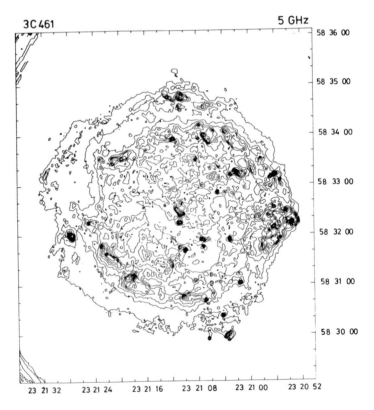

FIG. 8. A map of the supernova remnant Cassiopeia A, observed at 5 GHz with the Cambridge 5-km Telescope.

As an example of the use of aperture synthesis, Fig. 8 shows a contour map of Cassiopeia A, observed at 5 GHz with the Cambridge 5-km Telescope.[23] The resolution is 2 × 2.3 arc sec.

The ultimate size of a telescope which might be built using aperture synthesis techniques is, now, not limited by the technical difficulties of constructing the antenna and electrical systems, but by the disturbing effects of atmospheric and ionospheric refraction (see Chapters 2.1 and 2.5). Hinder and Ryle[24] and Hargrave and Shaw[4] have considered the restrictions imposed both by irregularities and by systematic refraction imposed by curvature of the Earth. The systematic effects could, in principle, be greatly reduced by determining the properties of the atmosphere along a ray path; irregularities would be much more difficult to remove. Hinder and Ryle[24] conclude that a telescope of size ∼ 10 km, operating in temperate latitudes, might achieve a limiting resolution of a few tenths of a second of arc.

[23] A. R. Bell, S. F. Gull, and S. Kenderdine, *Nature (London)* **257**, 463 (1975).
[24] R. A. Hinder and M. Ryle, *Mon. Not. Roy. Astron. Soc.* **154**, 229 (1971).

5.3. Very Long Baseline Interferometer Systems*

5.3.1. Introduction

The principles and operation of the recording systems used in very long baseline interferometry (VLBI)— also known as tape-recorder interferometry —are described here. In VLBI, the undetected signals received at remote, unconnected radio telescopes are stored in such a way that they can be recovered at a later time and processed in a fashion similar to conventional interferometry. Several essential components are required to make two remote telescopes into an interferometer:

(1) A stable local-oscillator system for signal frequency conversion.
(2) A video converter for conversion of the i.f. signal to a video band.
(3) A means of synchronizing the clocks at the stations.
(4) A device for storing the received signal and precise timing information.
(5) A playback system.

These basic parts, shown in the block diagram in Fig. 1, will be considered briefly in the next section. This block diagram is hopefully general enough to describe all VLBI systems in use today and in the foreseeable future. Three specific VLBI systems will be discussed in Section 5.3.3:

(1) The National Radio Observatory† (NRAO) Mark I digital, IBM‡-compatible system.
(2) The NRAO Mark II digital system.
(3) The Canadian analog system.

5.3.2. Basic Parts

The antennas and front-end amplifiers used in other applications of radio astronomy are generally adequate for VLBI measurements. There are no special tracking requirements for the antennas. However, exceptional care must be taken in converting the rf signal to an i.f. band in order not to

---

† NRAO is operated by Associated Universities, Inc., under contract with the National Science Foundation.

‡ International Business Machines Data Processing Division, 1133 Westchester Avenue, White Plains, New York 10604.

---

* Chapter 5.3. is by J. M. Moran.

174

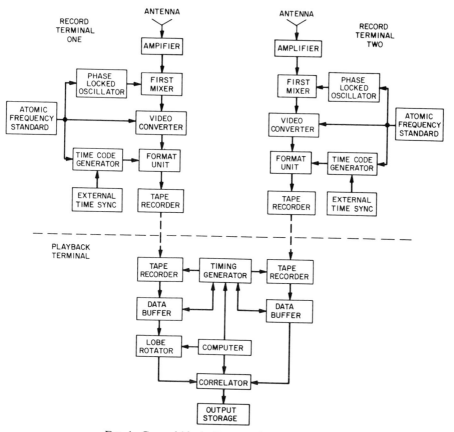

FIG. 1. General block diagram of a VLBI system.

destroy the coherence by introducing phase noise into the signal. Hence, whereas the local oscillators in single-telescope continuum studies can be free running and, in spectral-line studies, need only be frequency locked, those in VLBI applications must be phase locked to high-performance frequency standards. (See Chapter 5.4.) The requirement on the phase stability of the local oscillator can be stated by defining a coherence function $A(T)$, which describes the reduction in fringe visibility with integration time T due to phase noise, by

$$A(T) = \left| \left\langle 1/T \int_0^T e^{i\phi(t)} \, dt \right\rangle \right|, \qquad (5.3.1)$$

where ϕ is the difference in the local-oscillator phases, t the time, and $|\;|$ and $\langle \rangle$ the absolute value and the expectation, respectively. Linear phase drift with time i.e., (frequency offset) is ignored because it simply represents a fringe-frequency error that can be removed during data processing. Let us assume, therefore, that ϕ is a Gaussian random variable with zero mean and variance $\langle \phi^2 \rangle_T$. Use of the relation

$$\langle e^{i\phi} \rangle = \exp(-\langle \phi^2 \rangle/2) \tag{5.3.2}$$

enables us to write

$$A(T) \sim \exp(-\langle \phi^2 \rangle_T/2). \tag{5.3.3}$$

Consider the phase noise due to the frequency standard alone. Frequency standards are usually described by their fractional frequency stability $(\Delta f/f)_T$, which is readily determined in the laboratory from measurements of the phase difference between two standards over time intervals T. Since $\phi = 2\pi fT$,

$$\langle \phi^2 \rangle_T = [(\Delta f/f)_T \, 2\pi fT]^2. \tag{5.3.4}$$

The fractional frequency stability of a pair of hydrogen-maser oscillators is about[1]

$$(\Delta f/f)_T = \begin{cases} 10^{-12}T^{-1}, & 10^{-1} < T < 10^2, \\ 10^{-14}, & 10^2 < T < 10^4. \end{cases} \tag{5.3.5}$$

The short-term stability is governed by white noise inherent to the frequency standard, and the long-term stability by environmental fluctuations and other poorly understood physical changes (see Chapter 5.4). The coherence time T_c of the interferometer is the time for which $A(T) = 0.5$; i.e., $\langle \phi^2 \rangle^{1/2} = 68°$. Combining Eqs. (5.3.4) and (5.3.5) with the definition of coherence time, we obtain

$$T_c = \begin{cases} 1.9 \times 10^4/f, & 0 < f < 190 \quad [\text{GHz}], \\ 0, & f > 190 \quad [\text{GHz}], \end{cases} \tag{5.3.6}$$

where f is the frequency in gigahertz. Because $\langle \phi^2 \rangle$ is independent of T for times shorter than 100 sec, there is a maximum frequency, 190 GHz, above which the phase noise is too great (no matter how short the integration time) to maintain coherence; i.e., $A \geq 0.5$. The coherence time will exceed 1 day, the maximum requirement for any VLBI experiment, at frequencies below 220 MHz. The coherence function $A(T)$ is shown in Fig. 2 for a number of frequencies.

    The interferometer will have poorer phase stability than the frequency standards. First, the atmosphere and ionosphere introduce phase noise. Second, the local oscillators have poorer phase stability than the frequency standards to which they are slaved. Frequency standards usually provide a low-frequency signal, typically 5 or 100 MHz. The local-oscillator signal is derived through multiplier chains or phase-locked oscillators. Signals from

[1] M. W. Levine and R. F. C. Vessot, *Radio Sci.* **5**, 1287 (1970).

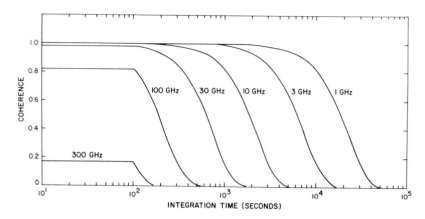

FIG. 2. The coherence of an interferometer at various frequencies versus integration time based on the stability of two hydrogen-maser frequency standards [M. W. Levine and R. F. C. Vessot, *Radio Sci.* 5, 1287 (1970)]. Actual performance will be somewhat worse because of atmospheric and local-oscillator phase noise.

commercially available frequency synthesizers are generally not multiplied because their phase noise is considerable. When frequencies that are not simple harmonics of 5 or 100 MHz are required, as in spectral-line work, they can be obtained by using synthesizers to provide either an offsetting frequency in a phase-lock loop or a second-stage local-oscillator signal.

The stable local-oscillator signals are used to convert the r.f. signal of bandwidth B to a band from 0 to B Hz, where it can be efficiently recorded. Conversion to such a low frequency is generally not required for single-antenna observations in continuum or spectral-line work. For example, a typical filter-bank receiver might have its filters and detectors at 10 MHz. The final conversion of the signal, from an i.f. to a video band from 0 to B Hz, is performed in a video converter. The problem in doing this conversion with the conventional filter-mixer system, and maintaining image rejection for single sideband, is that the smallest ratio of bandwidth to center frequency of typical lumped element realizable filters is about 0.1. Hence, if the i.f. is at 30 MHz and bandwidths of 2 MHz and 2 kHz are desired, then an additional mixer stage is required for the 2-kHz bandwidth. Because of the lack of versatility of this approach, the technique of single-sideband conversion was adapted for VLBI application. A block diagram of a single-sideband converter is shown in Fig. 3. The i.f. signal is mixed, without filtering, with quadrature local-oscillator signals at frequency f_0 in separate mixers. One of the two output signals is shifted in phase by 90°, and both are then added or subtracted. The resulting signal is passed through a low-pass filter of bandwidth B. The sum signal is the converted signal from f_0 to $f_0 + B$ (upper sideband),

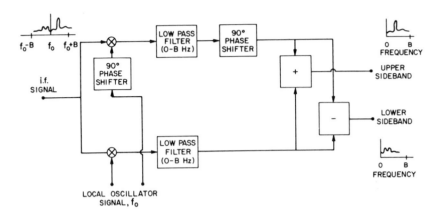

FIG. 3. Single-sideband video converter. This device will convert a signal band either from $f_0 - B$ to f_0 or from f_0 to $f_0 + B$ to a video band from 0 to B. Variations of this form are possible. The low-pass filter can be placed after the signals are combined. The 90° phase shifter in one signal path can be replaced with 45° phase shifters in both signal paths.

and the difference signal is the converted signal from f_0 to $f_0 - B$ (lower sideband). A set of video converters was built by NRAO for their Mark II VLBI terminals (the broadband phase shifter operating between 200 Hz and 50 MHz was developed by A. E. E. Rogers), with bandwidths of 2000, 1000, 500, 250, 62.5, 31.25, and 15.625 kHz. Any other filter can be externally connected. Butterworth filters, because of their flat phase response, are generally used.

In order for interference fringes to be observed, the time relationship between the two recorded signals must be known to within the reciprocal of the signal bandwidth. If the uncertaintly is greater than this, the correct time delay can be found by shifting the two data streams with respect to one another until fringes are detected. The *a priori* timing should be accurate to at least $100/B$, which is 50 μsec with the NRAO Mark II system. This accuracy is about four orders of magnitude greater than required for pointing filled-aperture radio telescopes.

The most straightforward way to synchronize the interferometer is to carry to all the participating stations a clock with a second tick that can be delayed in steps of 1 μsec. The second tick generated by the recording terminal is simply brought into alignment with that of the clock. The best type of clock for this purpose is the cesium-beam clock, which has a long-term stability of better than 10^{-13} (1-μsec error per 4 mo). Its frequency is independent of the magnetic field and other environmental influences such as temperature and pressure, and it therefore is referred to as a " primary standard." The United

States Naval Observatory (USNO) routinely carries cesium clocks to observatories around the world to synchronize them to within 1 μsec. A rubidium clock, which is susceptible to rate changes, can be used to synchronize two stations to 1 μsec if the comparison is done in the same day. A crystal clock is accurate to about 10 μsec/day.

Stations can also be synchronized by monitoring the broadcast time signals. Transmission from high-frequency stations such as WWV (2.5, 5, 10, 15, 20, and 25 MHz) in Fort Collins, Colorado, CHU (3.33, 7.34, and 14.670 MHz) in Ottawa, Canada, and MSF (2.5, 5, and 10 MHz) in Rugby, England, is accurate only to a few milliseconds, owing to uncertainties in the propagation path which involve reflections from ionospheric layers. These signals are useful for coarse timing and for discovering gross timing errors.

Much higher accuracy can be obtained from Loran C (LOng-RAnge Navigation) signals. Loran C is a navigation system that was developed during World War II, and has become an important tool in synchronizing radio observatories. The system is set up on eight chains operated by the US Coast Guard with assistance from USNO. Each chain (see Table I) has a master station and up to four slave stations. The transmission frequency is 100 kHz. Eight pulses, 300 μsec long (30 cycles of the 100-kHz carrier) and separated by 1000 μsec, are transmitted every repetition period. The slave station transmits the same pattern, delayed by a fixed time. The repetition rates vary between 0.05 and 0.10 sec among the chains. The ground wave can be received up to 1500 miles over land and 3000 miles over water. The propagation time is stable to about 0.1 μsec. The positive-going zero crossing at the end of the third cycle of the first transmitted pulse from the master station is coincident with a Universal Coordinated Time (UTC) second every 993 sec (East Coast chain). A listing of these times is called a TOC table. The reference epoch is 0 hours, January 1, 1958. The absolute timing accuracy in the field is probably about 10 μsec because of uncertainties in the antenna and receiver delay, and in the identification of the correct zero crossing in the signal. The use of transmissions for synchronization requires a pulse generator, which produces a sync signal with a period equal to the chain repetition rate, and a receiver, which can be as simple as a long wire and an oscilloscope. At observatories used frquently in VLBI operations, time is kept continuously by careful monitoring of Loran C, plus occasional visits of USNO's traveling clock, which provides absolute synchronization. If a phase-tracking Loran receiver is used, time can be kept to an accuracy of 0.1 μsec. More details of the Loran C system are given by Shapiro and Fisher[2] and by Potts and Wieder.[3]

The nonuniform rotation rate of the Earth gives rise to a number of prob-

[2] L. D. Shapiro and D. O. Fisher, *Radio Sci.* **5**, 1233 (1970).
[3] C. E. Potts and B. Wieder, *Proc. IEEE* **60**, 530 (1973).

TABLE I. Loran C Stations Now in Operation[a]

| Chain[b] | Period (μsec) | Coincidence (sec) | Master M | W | Slaves[c] | | | |
					X	Y	Z
East Coast	99,300	993	Carolina Beach, North Carolina	Jupiter, Florida 13,695.48	Cape Race, Newfoundland 36,389.56	Nantucket Island, Massachusetts 52,541.27	Dana, Indiana 68,560.68
Central Pacific (Hawaiian)	49,900	499	Johnston Island		Upolo Point, Hawaii 15,972.44	Kure, Midway Islands 34,253.02	
Mediterranean	79,900	799	Simeri Crichi, Italy		Lampedusa, Italy 12,757.12	Targabarun, Turkey 32,273.28	Estartit, Spain 50,999.68
North Atlantic	79,300	793	Angissoq, Greenland	Sandur, Iceland 15,068.10	Ejde, Faeroe Islands 27,803.80		Cape Race, Newfoundland 48,212.80

North Pacific (Alaskan)	59,300	593	St. Paul, Pribiloff Islands		Attu, Aleutian Islands 14,875.30	Port Clarence, Alaska 31,069.07	Sitkinak, Alaska 45,284.39
Norwegian Sea	79,700	797	Ejde, Faeroe Islands	Sylt, Germany 30,065.69	Bo, Norway 15,048.16	Sandur, Iceland 48,944.47	Jan Mayen, Norway 63,216.20
Northwest Pacific	99,700	997	Iwo Jima, Bonin Island	Marcus Island 15,283.94	Hokkaido, Japan 36,684.70	Gesashi, Okinawa 59,463.34	Yap, Carolina Islands 80,746.78
Southeast Asia	59,700	597	Sattahip, Thailand		Lampang, Thailand 13,182.87	Con Son, South Vietnam 29,522.16	Tan My, South Vietnam 43,807.30

[a] Data from C. E. Potts and B. Wieder, *Proc. IEEE* **60**, 530 (1972).

[b] As of May 1972, the East Coast, Norwegian Sea, Central Pacific, and Northwest Pacific chains are time synchronized with UTC (USNO) to within ±15 μsec. The Mediterranean and North Altantic chains are time monitored, and corrections to UTC are published. The North Pacific and Southeast Asia chains are unsynchronized.

[c] Emission delays (in μsec) are listed below station name.

181

lems in defining time scales. The International Radio Consultative Committee (CCIR) attempts to coordinate radio transmissions of time and frequency in a way that satisfies the needs of both laboratory scientists, who require atomic time (AT), and astronomers, navigators, and surveyors, who require universal time (UT). The second of atomic time is defined as the duration of 9, 192, 631, 770 cycles of radiation from the $(4, 0) \leftrightarrow (3, 0)$ hyperfine transition of the ground state of the ^{133}Cs atom, which equals $1/86,400$ of a mean solar day in 1900. Universal time, or mean solar time at Greenwich, is derived mainly from measurements of star positions by photographic zenith tubes, and is denoted as UT0. When corrected for polar variations ($\lesssim 30$ msec), UT0 is denoted UT1, and when also corrected for seasonal variations, it is denoted UT2. The Earth's rotation rate has decreased since 1900, and the day is now about 3 msec longer. The second of UTC was therefore offset from the second of atomic time by the factor 3×10^{-8} before January 1, 1973. However, the new system, in effect since that date, redefines the second of UTC as being equal to that of AT. Whenever necessary, about once a year, a 1-sec jump is made to keep UTC approximately aligned with UT. UT1, which is required for VLBI measurements, can therefore differ from UTC by up to about 0.5 sec. The exact difference is published periodically by USNO. Failure to correct for this difference can cause a fringe-rate offset of up to 1 Hz on a 4000-km baseline at a wavelength of 1 cm. The subtleties of time and frequency measurement, definition, and dissemination are discussed by Smith.[4,5]

The received signals, converted to a video band, must be stored at each station, along with precise time information, so they that can be recovered later. The signal can be represented by samples taken at a rate equal to the reciprocal of twice the bandwidth, $1/2B$. In digital recording, each sample is usually represented by one bit, which gives the sign of signal voltage. Because the signals and the receiver noise voltages are Gaussian random variables, they can be characterized entirely by their second moments, or second-order correlation functions. The Van Vleck equation,[6] which gives the relationship between the autocorrelation functions of a Gaussian signal and its associated clipped signal, can be extended to the cross-correlation function since the two received signals and their associated receiver noise voltages can be considered as a joint Gaussian random process. If the voltages at each station are $v_1(t)$ and $v_2(t)$, then the desired correlation function is

$$R_{12}(\tau) = \langle v_1(t) \, v_2(t + \tau) \rangle. \tag{5.3.7}$$

[4] H. M. Smith, *Nature (London)* **221**, 221 (1969).
[5] H. M. Smith, *Proc. IEEE* **60**, 479 (1973).
[6] J. H. Van Vleck and D. Middleton, *Proc. IEEE* **54**, 2 (1966).

The one-bit representation is

$$x_i = \begin{cases} 1, & v_i > 0, \\ -1, & v_i < 0, \end{cases} \qquad (5.3.8)$$

where $i = 1, 2$, and the associated correlation function is

$$\rho_{12}(\tau) = \langle x_1(t)x_2(t + \tau) \rangle. \qquad (5.3.9)$$

The Van Vleck relation gives

$$R_{12}(\tau) = [(\overline{T}_{A_1} + T_{R_1})(\overline{T}_{A_2} + T_{R_2})]^{1/2} \sin[(\pi/2)\rho_{12}(\tau)], \qquad (5.3.10)$$

where $\overline{T}_{A_1}$, $\overline{T}_{A_2}$, T_{R_1}, and T_{R_2}, are the average antenna and receiver temperatures, respectively, at stations 1 and 2. Hence, the cross-correlation function of the true signals can be recovered from that of the clipped signals. The normalization factor $[(\overline{T}_{A_1} + T_{R_1})(\overline{T}_{A_2} + T_{R_2})]^{1/2}$ is lost, but can be measured separately. The signal-to-noise ratio is degraded by a factor of $\pi/2$.

The sensitivity of the interferometer is proportional to the square root of the number of data samples processed. However, the bandwidth capability, or recording rate, of the system is important because the coherence time places a limit on the time available to acquire a given number of data samples. Magnetic tape is the best medium for data storage. A standard television tape can be recorded at a rate of 8 megasamples/sec, with an information density of about 1800 samples/mm, at a tape cost of approximately 10^{-3} dollars/megabit (Mbit). This information density is about the same as on an astronomical photograph. The storage rate can be increased by parallel recording of several signal bands. Instrumentation tape recorders, for example, can typically record at a 3-Mbit rate on up to 14 channels simultaneously. At bandwidths of 100 MHz and greater, laser beam recorders are being developed where the information density is about 5000 samples/mm². However, this technique is very expensive and impractical at present.

The limitation on the density of data storage on magnetic tape is governed by the size of the recording head gap and the associated electromechanical problems. The achievable bandwidth, or data-recording rate, depends also on the timing accuracy, since every data sample must have a time associated with it. In the NRAO Mark I system, the data samples are recorded in 0.2-sec records, and the timing is recovered by counting data samples within each record. In the NRAO Mark II and Canadian systems, there is a hierarchy of time markers. The coarse time (hours, minutes, and seconds) is written on an auxiliary audio track. The basic fine-time mark is the frame count or frame sync, which occurs every 1/60 sec (16.67 msec). In the digital system, a sync word is included in the data stream about every 512 μsec, while in the analog system, a sync pulse (called the line sync in television terminology) is included every 64 μsec. The data bits between the sync words can be counted in the digital system, whereas mechanical uniformity in the tape motion is relied on in the analog system.

After an observing period, the tape recordings are brought together, and the data are recovered and processed by the playback system. The main functions of the playback system are to (1) buffer the data from the tape recorders, (2) compensate for the differential delay and Doppler frequency shift between the two stations, (3) cross correlate the data streams, and (4) accumulate and store the cross-correlation functions for the desired integration time. Further processing is done in a general-purpose computer. The processor must have a buffering capability in order to produce data streams having a uniform time base in spite of the mechanical jitter of the tape recorders. This uniform time base is achieved by using the timing information recorded with the data. The delay compensation is accomplished simply, by shifting one data stream with respect to the other. The differential frequency compensation, called lobe rotation, can be done either before or after correlation, as shown in Fig. 4. When the lobe rotator comes after the correlator, it functions as a synchronous detector, multiplying the correlation functions by $\exp[i\Phi_0(t)]$, where $\Phi_0(t)$ is the expected apparent fringe phase (natural fringe phase plus local phase). In digital systems, this multiplication must occur at intervals short with respect to the apparent fringe period. Because this period can be very short, it is generally preferable to place the lobe rotator before the correlator, where it functions as a single-sideband mixer. In the case of the digital, single-sideband, clipped system, the lobe rotator multiplies one data stream by a digital approximation of $\sin[\Phi_0(t)]$ and $\cos[\Phi_0(t)]$, thereby producing two data streams, each of which is cross correlated with the data stream from the other station (see Fig. 4). The disadvantage of this configuration is that twice the number of correlator channels is required. The simplest digital approximation to the fringe rotation signal is a two-level sine wave or square wave. The lobe rotator merely changes the "sign" of the data bits during half of the fringe period. Both the NRAO Mark II and the Haystack Mark I processors use a three-level approximation for the fringe rotation signal, with the values 1, 0, and -1, as shown in Fig. 4. When the value is 0, the correlation of the corresponding data bit is inhibited. Some fringe power is lost because of the harmonic content in the waveform of the digital approximation. The loss in fringe power is about 11 % for the two-level approximation, and 4 % for the three-level approximation.

The data streams are cross correlated either in digital circuitry, by a digital processor that uses shift registers and accumulators, or in analog circuitry, by delay lines and multipliers. The correlation functions are generally written on tape, for permanent storage and further analysis, at intervals of about 1 sec. The processor thereby reduces the data by factors from 10^3 to 10^5. For example, the NRAO Mark II processor takes two 4-Mbit/sec data streams

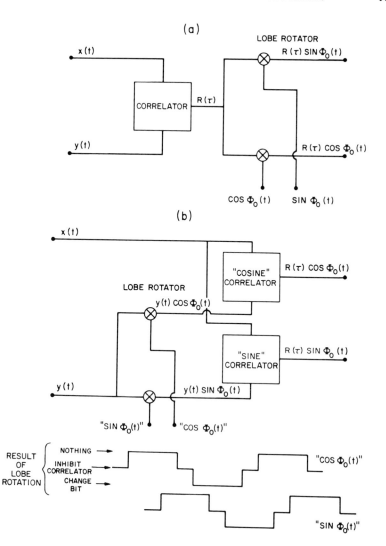

FIG. 4. Correlator and lobe rotator. When the signal processing is done in the computer, it is generally faster to cross correlate first and then compensate for the fringe rate $d\Phi_0(t)/dt$, as in (a). In hardware, it is more expedient to interchange the operations, as in (b), by quadrature mixing one of the data streams with a local-oscillator signal having a frequency equal to the expected fringe rate. When the data are clipped, the three-level square wave is a good approximation to a sine wave, and 96% of the fringe power is retained.

and outputs five 95-point complex correlation functions (12 bits per point) every second, giving a data-compression factor of about 10^3.

5.3.3. Specific VLBI Systems

At least six different VLBI recording systems were in operation in January 1973. Three have been selected for discussion here because they represent diverse approaches to the recording problem, and because these particular systems are in wide use in the astronomical community. The information densities and the data-storage costs for these systems are compared in Table II.

5.3.3.1. IBM-Compatible, Digital-Recording System. The first IBM-compatible recording system was built in 1966 by NRAO, and is referred to as the NRAO Mark I VLBI system.[7] It is "IBM-compatible" in that the data are written on half-inch computer tapes in a format that can be read on any general-purpose computer. Seven portable recording terminals are available, plus a fixed terminal at the Haystack Observatory in Westford, Massachusetts, which utilizes a direct core-access channel to a CDC 3300 computer.

A conceptually simple system, it embodies a very conservative design in the sense that the density of data on computer tapes is quite low, in order to ensure that the error rate is essentially zero. An error rate of a few percent is quite acceptable for VLBI recording. The first tape recorders employed in the system were Ampex† model TM-12 units, which record on seven tracks at 315 bits/cm [800 bits per inch (bpi)] and 59.1 cm/sec [150 inches per second (ips)] (there are 6 × 800, or 4800, bpi of tape; a parity bit is written on the seventh track). The data are clipped and recorded at 720,000 bits per second (bps). The tape format is shown in Fig. 5. The widest bandwidth is 360 kHz; narrower bandwidths are achievable by reducing the signal bandwidth before recording and then discarding the redundant data bits during processing. The data are blocked into records 0.2 sec long, each containing 140,400 data bits. While the record gap is being written, 3600 bits (2.5% of the data) are lost. There is no explicit timing information on the tape. The first bit recorded is sampled 310,000 μsec after the nominal start time, and the time of each subsequent data bit is determined by counting records from the beginning of the tape and counting bits within the record.

The timing signals are all derived from a 5-MHz reference signal. The sampling signal is derived, for example, by dividing the reference signal by

---

[7] C. Bare, B. G. Clark, K. I. Kellermann, M. H. Cohen, and D. L. Jauncey, *Science* **157**, 189 (1967).

---

† Ampex Corporation, 401 Broadway, Redwood City, California 94063.

TABLE II. Comparison of Information Storage Capabilities of Three VLBI Systems

System	Recording	Tape recorder	Sample rate (10^6 samples /sec)	Tape cost (dollars)	Date density (samples/mm²)	Tape cost (dollars/10^6 samples)
MRAO Mark I	IBM-compatible, digital	Ampex TM-12	0.720	20 (3 min)	15	0.15
NRAO Mark II	Digital	Ampex VR660C	4	45 (60 min)	830	0.0017
Canadian	Analog	IVC 800-900	8	35 (60 min)	1800	0.00012

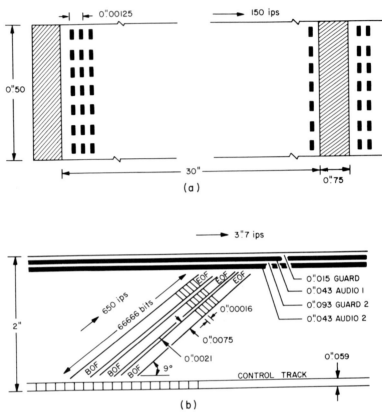

FIG. 5. Magnetic-tape data-storage formats for (a) the NRAO Mark I and (b) the NRAO Mark II systems.

$6\frac{17}{18}$. The 5-MHz signal is also divided down to 1 and 0.1 Hz and other frequencies. The 1-Hz pulse is aligned with an external second tick from a Loran C receiver or traveling clock. The 0.1-Hz pulse starts the tape recorder.

The recorded data can be processed by any general-purpose computer with the appropriate software. The processing is done record by record since there is generally not enough core space to read in all the data at once. The most time-consuming part of the data processing is the calculation of the cross-correlation functions. An assembly-language subroutine does the correlation one word at a time, using the "Exclusive–Or" instruction. On the CDC 3300 and the IBM 360 model 50, the correlation takes about 1.5 μsec per one-bit multiply, even though many bits are "multiplied" at one time. Hence, a seven-delay correlation function of 1 sec of data requires 15 sec to compute.

A digital correlator constructed at the Haystack Observatory was connected to the CDC 3300 computer† as a peripheral device to perform the cross correlation and fringe rotation on the data. The unit has two 16-delay correlators. The lobe rotation is done by quadrature mixing of one data stream, as shown in Fig. 4b. It is accessed via a Fortran subroutine call with the following arguments: the address of the first word in data sequence A to be correlated, the same for data sequence B, the number of words to be correlated, the bit number in the first word of B to be correlated with the first bit of A, the operating mode (cross-correlation, autocorrelation, etc.), the fringe rotation frequency, the initial fringe rotation phase, the answer array, and a status word. The possible fringe rotation frequencies are quantized between 0 and 45 kHz in steps of 0.043 Hz. The processor requires about 36 msec to calculate the 16-point complex correlation function of one record of data (0.2 sec of data, or 140,000 bits). When correlation functions with more delays are desired, as in spectral-line observations, the subroutine is called again, with a new starting location in one of the data streams so as to insert the correct delay offset.

5.3.3.2. NRAO Mark II System. The NRAO Mark II system,[8] in operation since March 1971, now consists of 10 recording terminals and one playback terminal located at NRAO, Charlottesville, Virginia.‡ The received signals are clipped and recorded digitally at a 4-Mbit rate on 2-in.-wide magnetic tape. The recorders—color-television tape recorders, Ampex model VR660C —have a time-base stability of about 100 μsec. Precise control of the timing is accomplished by encoding the data with a self-clocking code, and inserting a special sync word in the data every 512 μsec. The playback system automatically brings the data recorded at two terminals into time synchronization, removes the geometric delay and fringe rate, and calculates a 95-point cross-correlation function, which is written onto magnetic tape every 0.2 sec. Subsequent processing is performed on a general-purpose computer.

The VR660C is a helical-scan tape recorder. The tape is wound around a slotted drum in a one-turn helix. Two record heads, mounted on a disk, contact the tape through the slot. The disk spins at 30 rps, so that the head-to-tape speed is 650 ips, while the tape speed is only 3.7 ips. The recording signal is switched between heads every 1/60 sec, during which time a head makes one diagonal sweep across the tape. At the 4-Mbit data rate, the linear density is 6000 bpi. The space allotted for each bit is 0.0016 in., and the record head gap

---

[8] B. G. Clark, *Proc. IEEE* **61**, 1242 (1973).

---

† Control Data Corporation, 8100 34th Avenue S, Minneapolis, Minnesota 55440.

‡ Additional playback terminals will be in operation at the California Institute of Technology, Pasadena, Calif., and at Chalmers University, Gothenburg, Sweden, in 1976.

is 0.0004 in. The information density is, therefore, approximately 0.5×10^6 bpi². The tape format is shown in Fig. 5.

The block diagram of the record terminal is similar to the general diagram shown in Fig. 1. The signals for timing and formating the data are generated in the timing unit, which derives, from a 5-MHz reference signal, signals at 4 MHz, 200 and 3.84 kHz, and 50 and 1 Hz. All the counters in the divide circuits can be reset and started by an external pulse, such as a second tick from an atomic clock. The 200-kHz signal is provided for Loran C reception. The format unit encodes timing and other information, and records it on one audio track at a 3.84-kHz rate. This information consists of 16 binary-coded decimal (BCD) digits recorded every 1/60 sec, and includes the day, hours, minutes, and seconds (read from the chronolog clock), and the frame count. Driven by the 1-Hz reference from the timing generator, the chronolog clock has digits that can be set manually. The 1-bit sampled data are also encoded by the format unit by means of a digital diphase code that allows the data bits plus a clock signal to be recovered at playback. The encoded data waveform has a clock transition every 250 nsec. If the data bit is a binary 1, an extra transition is inserted between the clock transitions; if it is a 0, no extra transition is inserted. Hence, if the data were all 1's, the encoded waveform would be a 4-MHz square wave, and if they were all 0's it would be a 2-MHz square wave. This encoding procedure converts the signal band from about 0 to 2 MHz to a range from about 2 to 4 MHz, and allows the signal to be recorded directly without additional modulation. The data are organized into frames 1/60 sec long. Each frame begins with a 24-bit-period synchronization word encoded at two-thirds the normal bit rate, which cannot be confused with the data. This is followed by: a beginning-of-frame (BOF) pattern, six bits giving the frame number (0–59) within the second, 66,000 bits of data, an end-of-frame (EOF) bit pattern, and about 600 1's, which occur during the head-switching period. An 8-bit sync pattern, 11111110, is inserted every 512 μsec in the data to check synchronization on playback.†

The playback terminal, shown in Fig. 6, is controlled by a Varian‡ 620I computer. The 4-MHz crystal clock is divided to 60 Hz, and used to phase lock the head drums of the two playback tape recorders. The recorded time and the frame counts on the data tapes are compared, and the tapes are aligned by adjusting the phase of the 60-Hz signal. The video waveform received from each tape is decomposed into the recovered clock and the data stream. The recovered clock, a 4-MHz square wave, is passed through an analog filter and reclipped. This smoothed signal, called an "assured clock," is used to load the data bits into two storage buffers, each of which has 2048

---

† Specifications for this particular system are given in English units.
‡ Varian Associates, 611 Hansen Way, Palo Alto, California 94303.

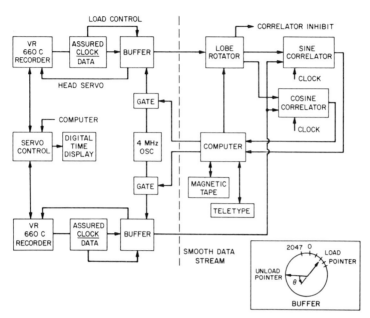

FIG. 6. NRAO Mark II data processor.

locations. Each buffer can be thought of as being arranged in a circle with a load pointer, driven by the assured clock, that moves cyclically through all the addresses. The load pointer is set to zero at the beginning of each frame; its position is checked thereafter 31 times per frame by examining the 8-bit sync pattern (11111110) which occurs every buffer cycle. Each data buffer has an unload counter, driven at a uniform rate, which follows the load counter by about half a cycle and unloads the data. The relative delay between the two data streams is adjusted by dropping pulses to the unload counter in one of the buffers. The address of each unload counter generates a 60-Hz signal that controls, in an open loop, the head drum on the tape recorder. Hence, the two data streams emerging from the buffers have a smooth time base, and the geometric delay of the interferometer has been removed.

The lobe rotator multiplies one data stream by the three-level approxima-tion to the sine and cosine of the expected fringe phase, as shown in Fig. 4b. The fringe period is generated by a computer-controlled programmable divider, which has a period increment of 50 psec.

The digital correlator measures a 95-point complex correlation function. The 96th channels of the cosine and sine correlators accumulate the number of bits correlated. These numbers vary because the lobe rotator inhibits the correlator. The data rate to the correlators is 4 Mbps regardless of the band-width of the signal recorded. The standard bandwidths are 2000, 1000, 500,

250, 62.5, 31.25, and 15.625 kHz. However, the period of the shift clock that drives the correlator, which determines the delay interval in the correlation function, is readily changed by factors of 2^N. Hence, the spectral resolution can always be made equal to 2.4 $B/95$ for uniform weighting.

The procedure for processing data on the playback system is as follows. The two data tapes are loaded onto the tape recorders and mechanically aligned until the heads scan directly over the recording tracks. A test pattern, which has a repetitive bit pattern and is recorded at the beginning of each tape, facilitates mechanical and electrical alignment. The two tapes are then positioned in time to within a few seconds of one another. The parameters of the observations (e.g., baseline or source coordinates) are read into the Varian 6201 computer from the tape that will be used to hold the measured correlation functions. The computer brings the data tapes into temporal alignment, and the data are transferred to the correlator. The computer updates the fringe phase, the fringe rate, and the geometric delay every 0.1 sec, and the correlation function and output every 0.2 sec. For continuum observations, only the central 31 delay channels of the correlator are written out on magnetic tape, while all 95 channels are written out for spectral-line observations. One standard 2400-ft computer tape holds the correlation functions for 60 min of data. The processor drops out of the data transfer mode if the frame counts on the tapes disagree. There are also some diagnostic lights to indicate the quality of the data processing. The most important one indicates when a load pointer has been jumped because of a discrepancy between the bit count and the 8-bit sync pattern. There is an oscilloscope display of the central 10 points of the real part of the correlation function versus time. Thus, fringes can be seen during playback if the correlation is greater than about 1%. Constant delay offsets can be entered by teletype during the processing, a useful procedure when a processing session is being set up and the instrumental delay is unknown.

This VLBI system is being expanded and improved: more record terminals are being added. In 1975, the capability for processing three tapes at once was provided, along with a larger correlator. For sprectral-line work, a fast Fourier-transform device will be connected to the Varian computer so that cross-power spectra will be available every 0.2 sec.

5.3.3.3. The Canadian System.   The Canadian VLBI system was developed by a research group from several Canadian institutions (National Research Council, University of Toronto, Queens University, and the Dominion Radio Astrophysical Observatory).[9] The system uses analog recording and processing techniques. It has evolved substantially since 1967 and is still being improved. The present state of the equipment is described here.

    [9] N. W. Broten, R. W. Clarke, T. H. Legg, J. L. Locke, J. A. Galt, J. L. Yen, and R. M. Chisholm, *Mon. Not. Roy. Astron. Soc.* **146**, 313 (1969).

This system uses standard television recorders, IVC† model series 800 or 900 for recording and 900 for playback, with only minor modifications. The recorders, single-head, helical-scan machines, write on 1-in.-wide magnetic tape, with a tape speed of 6.91 ips. There are two audio tracks and a head control track. Recording times are limited to 3.5 hr, and the time-base error of the recorder is ~ 1 μsec. The record terminal is shown in Fig. 7. Timing is controlled by the synchronization generator, which is driven by a reference signal at 1.008 MHz, derived from a 5-MHz signal from the atomic frequency standard. The frequencies generated are 1, 60, 600, and 15,750 Hz. The 60- and 15,750-Hz pulse trains are recorded on the video track with the

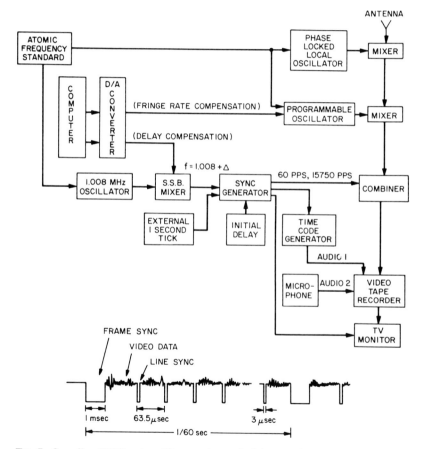

FIG. 7. Canadian VLBI system. The remote terminal does not have computer control for delay and frequency tracking.

† International Video Corporation, 990 Almanor Avenue, Sunnyvale, California 94086.

radiometer signal, which is band-limited from 0 to 4 MHz. They correspond to the frame and the line sync pulses, respectively, in conventional television terminology. However, these sync pulses are narrower than the standard for television in order to increase the data capacity and improve the timing accuracy. The line sync pulses occur every 63.5 μsec (525 lines per frame), and correspond roughly to the sync pattern in the NRAO Mark II system, which occurs every 512 μsec. The video signal and the sync pulses are frequency modulated onto a 5.5-MHz carrier and recorded. The tape recorders have both read and write heads so that a television monitor can be utilized to check the quality of the recording. Timing information is also put on the audio track by the time-code generator. This consists of one 42-bit word written every four frames, about every 67 msec. The word contains a sync code, the BCD time in hours, minutes, and seconds, and the frame count. The bit period is 1.6 msec, and the tapes can be aligned to about 1 msec by means of the sync code.

Special equipment is used at one terminal to compensate for both the interferometer geometric delay and the fringe rate. A small on-line computer calculates the delay and the fringe rate. The fringe-rate compensation is accomplished by continually adjusting the frequency of a programmable Hewlett-Packard† synthesizer, which is used as the final local oscillator. The delay tracking is done by changing the frequency of the 1.008-MHz signal driving the sync generator. This either stretches or contracts the time base in a continuous fashion to follow the changing interferometer delay. The initial delay must be set manually.

A block diagram of the playback system is shown in Fig. 8. Timing is controlled by a 1.008-MHz reference signal, which drives the two sync-pulse generators producing line and frame pulses. The speed of each tape recorder is controlled on playback by locking the frame pulses on the tape to the external frame pulses. Ahead of one of the sync-pulse generators is a phase rotator, which advances or retards the sync pulses so that one tape recording can be delayed with respect to the other. The signals coming from the tape recorders have a time-base jitter of about 1 μsec. This jitter is compensated for by a commercially available device called a time base corrector, which employs a varactor-controlled delay line to make the line sync pulses on the data coincide with the external line sync pulses. The jitter is thereby reduced to about ±20 nsec. This unit corresponds to the buffer storage in the NRAO Mark II processor. The two data streams are aligned by manually adjusting the phase shifter before the sync generator. As an aid in data alignment, an oscilloscope displays the time difference between the line pulses in the two recorders. The sync pulses are removed from each data stream before it passes to the correlator.

---

† Hewlett-Packard Co., 195 Page Mill Road, Palo Alto, California 94306.

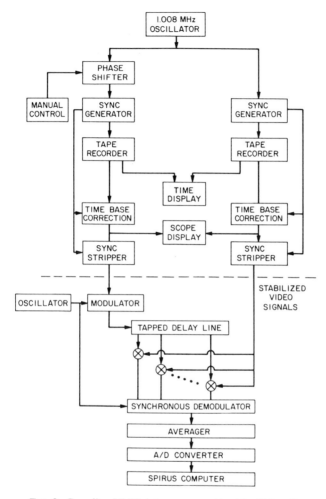

FIG. 8. Canadian VLBI data processor (double sideband).

Since the correlator is designed for continuum sources and for double-sideband recording, there are no quadrature multipliers. The correlator consists of 24 multipliers and a tapped delay line. One of the signals is modulated at about 500 Hz to avoid dc problems, and the multiplier outputs are synchronously demodulated. The averaged multiplier output signals are then passed to a smaller computer for further processing.

A spectral-line processor now under construction will produce 256-point compex spectra, by sampling each data stream at the Nyquist rate, and calculating the Fourier transform of each group of 512 data points. The Fourier transforms are generated in a digital circuit by using the Fast Fourier Transform (FFT) algorithm. With three complex multipliers, the circuit is classified as a RADIX 4 FFT device. Veenkant[10] describes the operation of these circuits. The Fourier transforms for each data group are multiplied, giving the complex cross-power spectrum; the individual spectra are accumulated over the desired integration time.

The cross-power spectrum obtained by cross multiplying the Fourier transforms of the individual signal functions is mathematically equivalent, in the continuous signal case, to that obtained by Fourier transforming the cross-correlation function of the signal functions. It is interesting to compare how the two methods can be made to give the same cross-power spectrum in the discrete signal case. The Fourier transforms of the two N-point signals $x(t_i)$ and $y(t_i)$ are

$$X(\omega) = 1/N \sum_{k=0}^{N-1} x(t_k) e^{-j\omega t_k}, \qquad Y(\omega) = 1/N \sum_{i=0}^{N-1} y(t_i) e^{-j\omega t_i}. \qquad (5.3.11)$$

If we let $t_k = k/2B$ and $t_i = i/2B$, where B is the bandwidth and $\omega = 2\pi l(2B)/N$, then the cross-power spectrum $S(\omega) = X(\omega) Y(\omega)$ is

$$S(l) = 1/N^2 \sum_{i,k=0}^{N-1} x(k) y(i) e^{-j2\pi l(k-i)/N}. \qquad (5.3.12)$$

The correlation function is

$$R(n) = \sum_{l=0}^{N-1} S(l) e^{j2\pi ln/N}. \qquad (5.3.13)$$

Substituting Eq. (5.3.12) into Eq. (5.3.13), interchanging the order of summation, and noting that

$$\sum_{l=0}^{N-1} e^{j2\pi l(k-i-n)/N} = \begin{cases} 0, & k-i-n \neq 0, \\ N, & k-i-n = 0, \end{cases}$$

we obtain

$$R(n) = 1/N \sum_{k=0}^{N-1} x(k) y(k+n), \qquad 0 < (k+n) < N-1, \qquad (5.3.14)$$

for n ranging between $-(N-1)$ and $(N-1)$. Hence, the number of data points multiplied for the nth lag is $N - |n|$, since correlation into the adjacent data blocks is not allowed. The spectrum obtained by the direct FFT is equivalent to that obtained by Fourier transforming the cross-correlation function with Bartlett or triangle weighting. The FFT technique can also be arranged, however, to be equivalent to the Fourier transform of the correlation function with any weighting.

[10] R. A. Veenkant, *IEEE Trans. Audio Electroacoust.* **20**, 180 (1972).

5.3.4. Comparison among Systems

The sensitivity of the Canadian system on continuum sources is greater by a factor of 2 than that of the NRAO Mark II system because of the clipping noise and the smaller bandwidth of the latter. The time-base stability of the NRAO system is much better than that of the Canadian system. The 20-nsec time-base jitter in the Canadian system appears as phase noise on the fringes.

The Mark I system has the disadvantage that prodigious amounts of time on a general-purpose computer are necessary for the analysis of the 10^8 bits acquired on each 3-min tape. However, with a special-purpose processor, such as the one attached to the CDC 3300 computer at Haystack, these data can be processed easily. Because this processor can be programed in software by using basic Fortran, the Mark I is the most flexible and versatile system. Its disadvantages are that the bandwidth is limited to 360 kHz, and that data cannot be taken continuously for more than 3 min unless at least two tape drives are available at each site. The information density is also quite low, compared to the densities on the Canadian and NRAO Mark II systems.

ACKNOWLEDGMENTS

The author gratefully acknowledges many illuminating discussions with B. G. Clark on the NRAO Mark II VLBI system, with J. L. Yen on the Canadian VLBI system, and with T. Hagfors.

5.4. Frequency and Time Standards†*

5.4.1. Introduction

The purpose of Sections 5.4.1–5.4.11 is to acquaint the reader with the concepts and definitions of frequency stability, and then to use these concepts in the discussion of the performance of various high-stability oscillators. Such oscillators play a central role in systems used for very long-baseline interferometer (VLBI) observations (see Chapter 5.3). A bibliography covering the subject of time and frequency standards is included at the end of this chapter.

There are two ways of looking at frequency stability—in the frequency domain and in the time domain. Spectral measurements tell us the behavior of phase (or frequency) fluctuations of an oscillator. The spectral density of phase fluctuations $S_\phi(f)$ can be visualized as the result of looking with a spectrum analyzer at the output from a perfect phase detector. The spectral density of frequency fluctuations $S_{\dot\phi}(f)$ is the result of analyzing the output from a perfect frequency discriminator. Spectral densities give us the best clues about the inside workings of an oscillator since we can often relate the spectral behavior with noise processes in various parts of the oscillator, leading us to certain models for the behavior of the oscillator. From these we often find physical processes that cause the noise.

When we talk of spectra, we must assume that the fluctuations in the signal are stationary with time; this means that we must be careful how we handle effects that produce systematic changes in frequency, such as temperature, vibration, aging, humidity, and magnetic fields. Usually, the effect of these parameters on frequency is given in the specification of an oscillator. If we know the spectrum of the perturbing parameter and the response of the oscillator to it, we can derive spectral information for the perturbed oscillator. Generally, this is not easily done, and we content ourselves with such cause-and-effect statements as "$\Delta f/f = +3$ parts in 10^{-12} for a change of $+1°C$ near 23°C."

The difficulty with spectra is that it is hard to get direct information below about 1 Hz. In principle, we could obtain the phase-fluctuation data between two oscillators by heterodyning them, recording the slow beat in digital form,

† This work was supported in part by contract NSR 09-015-098 from the National Aeronautics and Space Administration.

---

* Chapter 5.4 is by Robert F. C. Vessot.

198

and performing a Fourier transform in a computer, thereby obtaining a spectrum. Another way to obtain these low-frequency spectra is by recording time-domain data and deriving spectral information from them.

Time-domain information is obtained by measuring elapsed phase over a given interval of time. This can be done by counting the number of zero crossings of a signal in a given interval, or, for low-frequency signals, measuring the time intervals between zero crossings of the signal. We can accumulate these data, make histograms of the number of zero crossings in a given interval (or time intervals), and, assuming the distributions are Gaussian (which often they are not), obtain a value of σ, the rms or standard deviation. The relationship between σ and the time interval τ over which its value is obtained can be plotted, and we will show that this σ versus τ plot contains information about the spectrum of the phase (or frequency) fluctuations of the oscillator in question.

Having obtained the spectrum of the phase or frequency fluctuations, we find empirically that it usually consists of sections that can be well approximated by line segments of constant slope when we plot the spectrum on a log–log basis. Experience has shown that we can model the behavior of an oscillator's spectrum by a series of terms $S_y(f) = \sum_\alpha h_\alpha f^\alpha$, and obtain a good fit to the behavior of most oscillators. Some of these terms can be identified with physical noise processes, such as white phase noise due to additive thermal noise and white frequency noise due to thermal noise within the oscillating amplifier feedback loop.

For any particular oscillator, if we know its resonator Q, its power levels, and its feedback parameters, we should be able to predict its spectrum and relate the prediction to experiment.

It is important to define what we mean here by accuracy, stability, and precision of an oscillator, since these terms tend to be loosely used and lead to considerable confusion.

Accuracy is the capability of an oscillator to provide *independently* a frequency that is known in terms of the accepted definition of the second. At present, 9,192,631,770 cycles of the unperturbed hyperfine frequency separation of the cesium atom define the time interval adopted as the second.

Stability is the property of an oscillator to resist changes in its rate. A high-stability oscillator may not necessarily be an accurate one.

Precision is the ability of an oscillator to return to a previously calibrated frequency.

The discussion that follows relates to frequency stability, which we will define in two ways:

(1)   by the spectral density of phase or frequency fluctuations;

(2)   by the sample variance of fractional frequency variations measured in the time domain.

The relationship between spectral density and the time-domain variance leads to some interesting questions involving sample number, bandwidth, and averaging time. These questions can strongly influence the strategy we use to obtain experimental data.

To illustrate the above relationships, models of spectral densities will be discussed and their effect on time-domain variances described. Further details can be found in an IEEE study made in 1955–1970, and reported in 1971 by J. A. Barnes et al.[1]

5.4.2. Frequency-Domain and Time-Domain Measures of Frequency Stability and Their Relationship

Let us consider an oscillator with instantaneous output voltage

$$V(t) = [V_0 + \varepsilon(t)] \sin[2\pi\nu_0 t + \phi(t)].$$

Here, V_0 is the amplitude with small fluctuations $\varepsilon(t)$ and $|\varepsilon(t)/V_0| \ll 1$, and ν_0 is the frequency with phase fluctuations $\phi(t)$ and $|\dot{\phi}(t)/2\pi\nu_0| \ll 1$. The instantaneous phase is given by

$$\Phi(t) = 2\pi\nu_0 t + \phi(t),$$

and the corresponding instantaneous frequency is

$$\nu(t) = \frac{1}{2\pi}\frac{d\Phi}{dt} = \nu_0 + \frac{\dot{\phi}(t)}{2\pi}.$$

We adopt here the notation of Barnes et al.,[1] distinguishing between time-variable frequencies $\nu(t)$ and Fourier frequencies $\omega = 2\pi f$. This distinction must be made so that $\Phi(t)$ is a well-defined function of time.

We define spectral density $S_g(f)$ as the one-sided spectral density of a real function $g(t)$ on a per-hertz basis. The mean-square value of $g(t)$ is thus given by $\int_0^\infty S_g(f)\,df$. If $g(t)$ is the electrical current passing through a resistance R, then $R\int_0^\infty S_g(f)\,df$ is the total power dissipated by the resistor. The relationship between frequency spectral density and phase spectral density is given by $S_{\dot{\phi}}(f) = \omega^2 S_\phi(f)$. Either of these parameters is a useful definition of frequency stability; however, it is convenient to normalize the spectral density of the frequency fluctuations to the frequency ν_0 by defining

$$S_y(f) = \frac{1}{(2\pi\nu_0)^2} S_{\dot{\phi}}(f) = \frac{1}{\nu_0^2} f^2 S_\phi(f).$$

Similarly, for phase fluctuations we have

$$S_x(f) = \frac{1}{(2\pi\nu_0)^2} S_\phi(f).$$

---

[1] J. A. Barnes et al., NBS Tech. Note 394, U.S. Govt. Printing Office, Washington, D.C. (1970); also in *IEEE Trans. Instrum. Meas.* **IM-20**, 105 (1971).

These spectral densities are invariant when, with ideal equipment, the frequency of the signal is multiplied or divided.

Another measure of frequency stability is the sample variance of fractional frequency variations, and this is perhaps a more intuitive measure since it is related directly to measurements of phase over given intervals of time. We define the fractional average frequency over an interval τ at time t_k as follows:

$$\bar{y}_k = \frac{\phi(t_k + \tau) - \phi(t_k)}{2\pi \nu_0 \tau}. \tag{5.4.1}$$

When a series of measurements is taken, then $t_{k+1} = t_k + T$, $k = 0, 1, 2, \ldots$. Each average is taken over a time interval τ. The relation giving the variance of the fractional frequency differences of a set of N samples, each taken over an interval τ with repetition period T, is

$$\langle \sigma_y^2(N, T, \tau) \rangle \equiv \left\langle \frac{1}{N-1} \sum_{n=1}^{N} \left(\bar{y}_n - \frac{1}{N} \sum_{k=1}^{N} \bar{y}_k \right)^2 \right\rangle.^\dagger \tag{5.4.2}$$

The infinite time average is denoted by the brackets. *This is another definition of frequency stability.* However, we must be very careful in letting $N \to \infty$, since there are noise processes that cause $\langle \sigma_y^2(N, T, \tau) \rangle$ to diverge with increasing N. To obtain a consistent basis for defining and comparing stability, Allan[2] suggested that the time average of the variance for $N = 2$ and $T = \tau$ would be a good choice. This value is denoted henceforth as $\sigma_y^2(\tau)$, defined as

$$\sigma_y^2(\tau) \equiv \langle \sigma_y^2(2, \tau, \tau) \rangle = \langle (\bar{y}_{k+1} - \bar{y}_k)^2 \rangle / 2,$$

where the brackets again denote the infinite time average.

The relationship between $\langle \sigma_y^2(N, T, \tau) \rangle$ and $S_y(f)$ has been shown by Cutler and Searle[3] to be

$$\langle \sigma_y^2(N, T, \tau) \rangle = \frac{N}{N-1} \int_0^\infty df\, S_y(f) \left[\frac{\sin^2(\pi f \tau)}{(\pi f \tau)^2} \right] \left[1 - \frac{\sin^2(r \pi f N \tau)}{N^2 \sin^2(r \pi f \tau)} \right], \tag{5.4.3}$$

where $r = T/\tau$.

We can readily obtain some insight into this expression, at least for processes that converge with $N \to \infty$ and for contiguous measurements ($\tau = T$). The relationship between $\langle \sigma_y^2(\tau) \rangle$ and the autocorrelation function of the phase $R_\phi(\tau)$ is obtained as follows. Starting with the concept of angular frequency and the definition of the square of the standard deviation, we have

$$\langle \sigma_y^2 \rangle = \langle \bar{y}_k^2 \rangle - \langle \bar{y}_k \rangle^2, \qquad y_k = \frac{\phi(t_k + \tau) - \phi(t_k)}{2\pi \nu_0 \tau}.$$

[2] D. W. Allan, *Proc. IEEE* **54**, 221 (1966).
[3] L. S. Cutler and C. L. Searle, *Proc. IEEE* **54**, 136 (1966).

† This expression is often referred to as the Allan variance.

We can put $\langle \bar{y}_k \rangle$ equal to zero since we are dealing with deviations from the *average* frequency. We then have

$$\langle \sigma_y^2 \rangle = \frac{1}{(2\pi\nu_0)^2} \frac{1}{\tau^2} [\langle \phi(t + \tau)^2 \rangle - 2\langle \phi(t + \tau)\phi(t) \rangle + \langle \phi^2(t) \rangle].$$

Since the time average of the ensemble is not affected by a time translation (stationarity), we have

$$\langle \sigma_y^2 \rangle = \frac{1}{(2\pi\nu_0)^2} \frac{1}{\tau^2} 2[\langle \phi(t)\phi(t - 0) \rangle - \langle \phi(t)\phi(t - \tau) \rangle]$$

$$= \frac{2}{(2\pi\nu_0)^2} \frac{1}{\tau^2} [R_\phi(0) - R_\phi(\tau)].$$

From the Wiener–Khinchin Fourier-transform relationship between the spectral density of a random signal and the autocorrelation function, we have

$$S_x(\omega) = 2 \int_0^\infty R_\phi(\tau) e^{-j\omega\tau} \, d\tau, \qquad R_\phi(\tau) = \frac{1}{\pi} \int_0^\infty S_x(\omega) e^{j\omega\tau} \, d\omega.$$

By substituting R_ϕ above into the equation for $\sigma^2(\tau)$, we obtain

$$\langle \sigma_y^2 \rangle = \frac{1}{\pi} \int_0^\infty S_y(\omega) \frac{\sin^2(\omega\tau/2)}{(\omega\tau/2)^2} \, d\omega. \qquad (5.4.4)$$

The $\sin^2(\omega\tau/2)/(\omega\tau/2)^2$ term results from the transformation of the " square " time window to the frequency domain. This is the weighting function in the frequency domain (or the filter) required to make the transformation between the time and the frequency domain. Note that for finite N, the last term in Eq. (5.4.3), i.e., $\{1 - [\sin^2(r\pi fN\tau)]/[N^2 \sin^2(r\pi f\tau)]\}$, has the effect of a high-pass filter. Since N is never infinite in practice, this eliminates our worries about divergences for certain spectral types. However, $\langle \sigma_y^2(N, T, \tau) \rangle$ can depend strongly on N, and therefore pose a problem in standardizing our measurement. It is here that the great convenience of the two-sample variance becomes apparent.

5.4.3. Spectral-Density Models

The most often used spectral-density model for characterizing oscillators consists of five independent noise processes. The spectral density of fractional frequency fluctuations due to these processes is given by

$$S_y(f) = h_{-2}f^{-2} + h_{-1}f^{-1} + f_0 + h_1 f + h_2 f^2 \qquad (5.4.5)$$

in the range $0 \le f \le f_h$, where f_h is a high-frequency cutoff of an ideal sharp-cutoff, low-pass filter. Substitution of each of the five terms separately into

Eq. (5.4.4) gives us a very clear picture of the effect of the bandwidth. We see that terms in h_{-1} and h_{-2} tend to diverge at low frequencies ($f \to 0$); however, the finite number of samples N acts as a high-pass filter, excluding contributions to the integrand when $f \to 0$.

The finite bandwidth, either of the measuring apparatus or of the device itself, will have an effect on the value of σ, especially in cases where the predominating noise is in the higher Fourier frequency range. In any data giving σ, the upper frequency cutoff f_h should be specified along with N, T, and τ. If we substitute

$$S_y(f) = \begin{cases} h_\alpha f^\alpha, & 0 \leq f \leq f_h, \\ 0, & f > f_h, \end{cases} \tag{5.4.6}$$

where α is a constant, into Eq. (5.4.3), then $\langle \sigma_y^2(N, T, \tau) \rangle$ varies as $|\tau|^\mu$ for $2\pi\tau f_h \gg 1$, and for N and $r = T/\tau$ both held constant. The relationship between μ and α is shown in Fig. 1. Values of σ for integral α are given below.

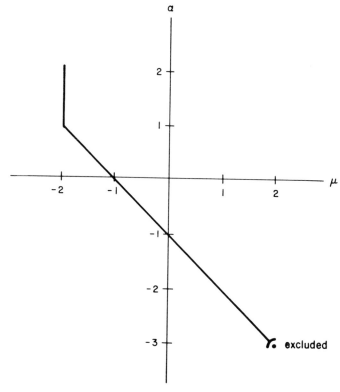

FIG. 1. Relationship between the assumed exponent α of the spectral density in Eq. (5.4.6) and the resulting exponent μ in the time-domain variance given in Eq. (5.4.3).

By obtaining μ from measurements of $\langle \sigma_y^2(N, T, \tau) \rangle$, we can thus identify α and determine h_α. The ambiguity in α for $\mu = 2$ can readily be removed by making a change in f_h. The dependence of σ_y^2 on f_h goes as $\ln(2\pi f_h \tau)$ for $\alpha = 1$, and linearly with f_h for $\alpha = 2$.

The following relationships between $\langle \sigma^2(N, T, \tau) \rangle$ and h_α are given for the five values of α most likely to be encountered. In all cases, $r = T/\tau \geq 1$ and $0 \leq f \leq f_h$.

(1) The $S_y(f) = h_{-2}/f^2$ spectral distribution is often designated as "random walk of frequency." For this case,

$$\langle \sigma_y^2(N, T, \tau) \rangle = (h_{-2}(2\pi)^2 |\tau| [r(N + 1) - 1])/12, \qquad r \geq 1,$$

$$\langle \sigma_y^2(N, \tau, \tau) \rangle = (h_{-2}(2\pi)^2 |\tau| N)/12, \qquad r = 1,$$

$$\sigma_y^2(\tau) = (h_{-2}(2\pi)^2 |\tau|)/6, \qquad r = 1, \quad N = 2.$$

The behavior of oscillators under conditions where this spectral distribution occurs can often be confused with linear drift, or aging. This is the limiting case for convergence in Eq. (5.4.2). We know of no physical processes that produce this type of spectrum.

(2) The case $S_y(f) = h_{-1}/f$ is often referred to as "flicker of frequency," in analogy with the $1/f$ or "flicker" noise-power distribution. Here,

$$\langle \sigma_y^2(N, T, \tau) \rangle = h_{-1} \frac{1}{N(N-1)} \sum_{n=1}^{N} (N - n)[-2(nr)^2 \ln(nr)$$
$$+ (nr + 1)^2 \ln(nr + 1) + (nr - 1)^2 \ln(nr - 1)], \qquad r \geq 1,$$

$$\langle \sigma_y^2(N, \tau, \tau) \rangle = h_{-1} \frac{N \ln N}{N - 1}, \qquad r = 1,$$

$$\sigma_y^2(\tau) = h_{-1} 2 \ln 2, \qquad r = 1, \quad N = 2.$$

Note that there is no τ dependence in this case. The leveling off of a σ versus τ plot is very prevalent in precision oscillators for large values of τ. The physical causes for this type of spectral distribution are as yet not well understood, although the cause can often be traced to particular components (such as transistors, resistors, and capacitors) within the oscillator feedback loop that have flicker phase noise behavior (see (4) below). In the case of self-oscillating systems such as masers, mechanical strain relaxations of frequency-perturbing elements—e.g., the cavity resonator or various tuning mechanisms (including capacitors)—can cause this type of instability. This behavior is nearly always accompanied by a long-term drift, and may result from random processes associated with such a drift.

(3)   The case where $S_y(f) = h_0$, a constant, is referred to as "white frequency noise" or "random walk of phase." For this case,

$$\langle \sigma_y^2(N, T, \tau) \rangle = (h_0/2)|\tau|^{-1}, \qquad r \geq 1,$$

$$\langle \sigma_y^2(N, \tau, \tau) \rangle = (h_0/2)|\tau|^{-1}, \qquad r = 1,$$

$$\sigma_y^2(\tau) = (h_0/2)|\tau|^{-1}, \qquad r = 1, \quad N = 2.$$

Here, σ does not depend on N or r. The physical process that leads to this behavior is white noise power within the bandwidth of the frequency-determining element of the system. For atomic-beam devices, the noise comes from shot noise due to the random arrival of particles at a detector. In optically monitored frequency-control devices, the noise is due to the random arrival of photons at the photodetector. For masers, thermal noise within the natural linewidth of the oscillating atoms will cause the same behavior. In crystal-controlled oscillators, thermal noise kT within the resonance bandwidth of the resonator is a basic limitation. The excess noise FkT due to the feedback amplifier is often used to characterize this behavior, where F is the amplifier noise figure.

The general expression relating this noise process to the oscillator quality factor Q is[4]

$$\sigma_y^2(\tau) = kT/2PQ^2\tau.$$

(4)   For $S_y(f) = h_1|f|$, or $S_x = h_1/(2\pi)^2 f$, we have the phase spectral-density analogy to flicker or $1/f$ noise; hence the term "flicker phase." In this case,

$$\langle \sigma_y^2(N, T, \tau) \rangle = \frac{h_1}{(2\pi\tau)^2} 2 \left[\frac{3}{2} + \ln(2\pi f_h \tau) \right.$$

$$\left. + \frac{1}{N(N-1)} \sum_{n=1}^{N} (N - n) \ln\left(\frac{n^2 r^2}{n^2 r^2 - 1} \right) \right], \qquad r \geq 1,$$

$$\langle \sigma_y^2(N, \tau, \tau) \rangle = \frac{h_1}{(2\pi\tau)^2} \frac{2(N+1)}{N} \left[\frac{3}{2} + \ln(2\pi f_h \tau) + \frac{\ln N}{N^2 - 1} \right], \quad r = 1,$$

$$\sigma_y^2(\tau) = \frac{h_1}{(2\pi\tau)^2} \left[\frac{9}{2} + 3 \ln(2\pi f_h \tau) - \ln 2 \right], \qquad r = 1, \quad N = 2.$$

This type of noise is observed in resistors, capacitors, and transistors, as well as in resonant cavity filters. It is possible that random strain relief can cause this behavior in passive devices. Some forms of diffusion processes across junctions of semiconductor devices may also be the cause.

---

[4] W. A. Edson, *Proc. IRE* **48**, 1454 (1960).

(5)   The last term in the series is $S_y(f) = h_2 f^2$, or $S_x(f) = h_2/(2\pi)^2$. Called "white phase noise," it is a type of noise known to occur in all electronic devices at all times because of the presence of additive thermal noise kT (per unit bandwidth). In a phasor representation, components of this noise power, added to a phasor of length (power)$^{1/2} \gg (kTB)^{1/2}$, cause equal amounts of amplitude and phase perturbations. The mean-square phase deviation is $\langle \Delta\theta^2 \rangle = kTB/P$. For a device with a noise figure F, the noise power is $FkTB$, and the mean-square phase deviation is increased by that factor. For white phase noise,

$$\langle \sigma_y^2(N, T, \tau) \rangle = \frac{h_2[N + \delta_k(r - 1)2f_h]}{N(2\pi)^2\tau^2}, \qquad r \geq 1, \quad \delta_k = \begin{cases} 0, & r \neq 1, \\ 1, & r = 1, \end{cases}$$

$$\langle \sigma_y^2(N, \tau, \tau) \rangle = \frac{h_2(N + 1)2f_h}{(2\pi)^2 N\tau^2}, \qquad r = 1,$$

$$\sigma_y^2(\tau) = \frac{h_2\, 3f_h}{(2\pi)^2\tau^2}, \qquad r = 1, \quad N = 2.$$

This type of noise is particularly troublesome in the case of oscillators operating at very low power levels. The effect of this noise can be reduced by use of a low-pass filter in a system where a low frequency or zero beat is obtained. A bandpass filter can be used when the frequency has been translated to a low, but not zero, frequency.

In all the above cases, it is important to note that μ, the exponent of τ, does *not* depend on N or on $T/\tau = r$, so long as these parameters are kept constant throughout. Tables are available[5] to convert data taken with a particular N and r to other values of N and r. These are in the form of the following ratios:

$$B_1(N, T/\tau, \mu) \equiv \frac{\langle \sigma_y^2(N, T, \tau) \rangle}{\langle \sigma_y^2(2, T, \tau) \rangle}, \qquad B_2(T/\tau, \mu) \equiv \frac{\langle \sigma_y^2(2, T, \tau) \rangle}{\langle \sigma_y^2(2, \tau, \tau) \rangle}.$$

Summary.   Frequency stability can be defined in two ways:

(1)   by the spectral densities of phase or frequency fluctuations, designated by the normalized quantities $S_x(f)$ and $S_y(f)$;

(2)   by the use of time-domain variances of an ensemble of phase or frequency measurements.

The relationship between the two definitions can be established if we specify four parameters: N, the number of samples in the distribution; τ, the time interval during which the signal phase is being measured; T, the time interval between measurements; and f_h, the noise bandwidth of the measuring system (assumed here to be a sharp cutoff).

5 J. A. Barnes, NBS Tech. Note 375, U.S. Govt. Printing Office, Washington, D.C. (1969).

We see that spectral distributions of various types each lead to a characteristic variance, and that we can estimate the spectral distribution from the behavior of variance with averaging time τ. By using the bias functions B_1 and B_2, we can relate the levels of the variances for different combinations of N and T/τ, provided we know μ, the exponent of τ, in the plots of variance versus τ.

5.4.4. Phase and Time Prediction

We can use the measure of frequency stability $\langle \sigma^2(\tau) \rangle^{1/2}$ as an equivalent measure of time stability if we multiply by τ. For time intervals τ where the σ data are reasonably good, this type of prediction scheme will be valid. We must be very cautious to extend this process only for future time intervals for which data exist on a statistically significant number of past samples of length τ. However, we can answer such questions as the following: Within one standard deviation, how far can we expect the clock to deviate in the next 1000 sec from its expected position? The expected position is, of course, the extrapolation of the mean rate we have determined from past observations. By use of the $\sigma_y^2(\tau)$ (or Allan two-sample) variance, this is easy to obtain. Cutler and Vessot[6] in Fig. 2 give these data for several types of fre-

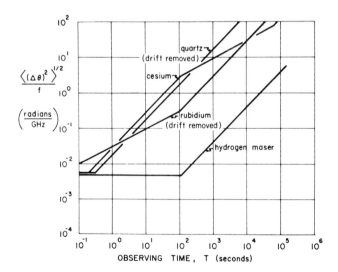

FIG. 2. Normalized phase departure versus observation time for various frequency standards [from L. S. Cutler and R. F. C. Vessot, *1968 NEREM Record*, Boston 68 (1968)].

[6] L. S. Cutler and R. F. C. Vessot, *1968 NEREM Record*, Boston 68 (1968).

quency standards, in terms of the rms phase fluctuation as a function of observing-time interval normalized to 1 GHz. Barnes and Allan[7] describe these data in terms of the rms time error, again as a function of observing time. Winkler et al.[8] and Allan and Gray[9] also offer good examples of this technique.

5.4.5. Frequency and Time Standards

Radio astronomy requires the use of high-stability oscillators both for generating stable local-oscillator signals and for keeping time. The requirements for phase stability in the local oscillator are particularly stringent in very long baseline interferometry (VLBI) experiments, where signals at microwave frequencies are translated downward to be recorded at more than one station and later analyzed by correlation techniques. The accuracy in translating frequency at each station and in timing the recorded signals depends on the performance of high-stability oscillators, often referred to as "atomic clocks."

Three types of atomic clocks are now in general use in radio astronomy. The frequency-controlling components of these are as follows:

(1) cesium-beam-tube resonator,
(2) rubidium-gas-cell resonator, and
(3) atomic-hydrogen-maser oscillator.

Both the cesium and the rubidium devices use atomic transitions to signal when the proper frequency has been applied to an assembly of atoms. The hydrogen maser is an active oscillator that emits a signal with the frequency of the hyperfine separation of the hydrogen atom in the ground state.

Rubidium and cesium are hydrogen-like atoms that have a single unpaired electron in the outermost shell. This electron is in a zero-orbital angular-momentum state (known as $^2S_{1/2}$ to spectroscopists) and has only spin-angular momentum. The nuclei of these atoms also have spin-angular momentum $I\hbar$, where I is $\frac{1}{2}$ for hydrogen, $\frac{3}{2}$ for rubidium, and $\frac{7}{2}$ for cesium, and $\hbar$ is Planck's constant divided by 2π. The angular momenta of the nucleus and the electron are coupled by the interaction of the nuclear magnetic dipole in the field of the electron magnetic dipole. They exert torque on each other, much as two bar magnets do when they are brought near each other. This torque causes a mutual precession of the magnetically coupled angular-momentum vectors, such that the total angular momentum F (in units of $\hbar$) is the vector sum of the nuclear and the electron momenta.

[7] J. A. Barnes and D. W. Allan, *Frequency* **5**, 15 (1967).
[8] G. M. R. Winkler, R. G. Hall, and D. B. Percival, *Metrologia* **6**, 126 (1970).
[9] D. W. Allan and J. E. Gray, *Metrologia* **7**, 79 (1971).

The rules of quantum mechanics will allow the atom to remain only in configurations that differ in angular momentum by one unit of $\hbar$. Since the electron spin is $\hbar/2$, only two hyperfine ground-state levels are possible in these atoms.

The presence of an external magnetic field provides a spatial quantizing axis, and the total angular-momentum vector precesses about it. The vector can be in $2F + 1$ orientations with respect to the quantizing field; each orientation, when projected on the field axis, differs from its neighbor by one unit of $\hbar$. The orientations result in different, magnetically dependent sublevels, labeled M_F according to the scheme $M_F = F, F - 1, \ldots, -F$, as shown in Figs. 3–5. Note that since F is an even number, there is always an $M_F = 0$ state, which, at low field, is quadratically magnetic-field dependent. The $M_F = 0 \to M_F = 0$ transitions, denoted by arrows, are those used for frequency

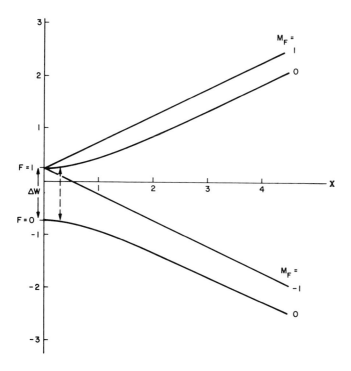

FIG. 3. Hyperfine structure of atomic hydrogen, with nuclear spin $I = \frac{1}{2}$, $\nu_0 = \Delta W/h = 1,420,405,751$ Hz, and $X = [(-\mu J/J) + (\mu I/I)]H_0/\Delta W$ calibrated in units of 0.51×10^3 Oe.

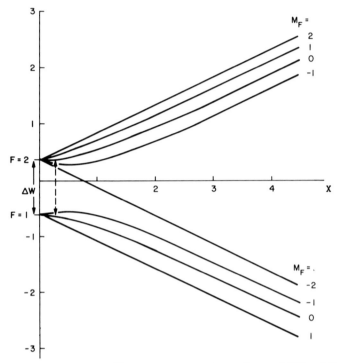

FIG. 4. Hyperfine structure of ^{87}Rb, with nuclear spin $I = \frac{3}{2}$, $\nu_0 = \Delta W/h = 6,834,682,605$ Hz, and $X = [(-\mu J/J) + (\mu I/I)]H_0/\Delta W$ calibrated in units of 2.44×10^3 Oe.

control. Table I shows the second-order frequency dependence, in units of hertz per oersted^2.

The atoms act like oscillating magnetic dipoles, and the interaction of these dipoles with magnetic fields at, or very near, the frequencies shown in the table is the basis for the atomic clocks. The linewidth of the resonance depends inversely on the amount of time the atom can be "interrogated" in an otherwise unperturbed state. The great stability of the atomic clocks results from their ability to obtain narrow resonances. In the cesium-beam tube, the atom drifts in a vacuum at about 10^4 cm/sec for a distance l cm, and consequently has a linewidth in the order of $10^4/l$ Hz, or a line Q in the order of $(\nu_{Cs}/v) \times l \approx 10^5 l$. In the rubidium gas cell, the atoms are stored in a cell containing a noninteracting buffer gas, which confines the rubidium atom through inelastic collisions that affect only slightly the state of the rubidium atom.

In the hydrogen maser, the atoms are confined in the rf field in a specially coated storage bulb that allows the hydrogen atoms to make elastic collisions with very little effect on the atomic state. The resulting signal is the radiation

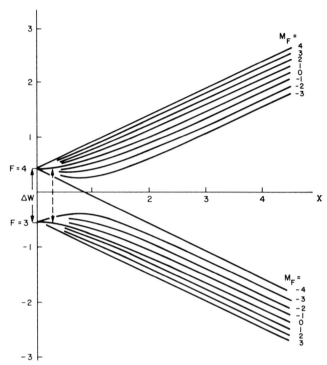

FIG. 5. Hyperfine structure of ^{133}Cs, with nuclear spin $I = \frac{7}{2}$, $\nu_0 = \Delta W/h = 9{,}192{,}631{,}770$ Hz, and $\mathbf{X} = [(-\mu J/J) + (\mu I/I)]H_0/\Delta W$ calibrated in units of 3.27×10^3 Oe.

TABLE I. Characteristics of the Hydrogen, Cesium, and Rubidium Atoms

	Nuclear spin	Electron spin	Hyperfine separation resonance frequency (Hz)	Second-order magnetic-field dependence (Hz, where H is in oersteds)
Hydrogen 1	$\frac{1}{2}$	$\frac{1}{2}$	$1{,}420{,}405{,}751.768 \pm 0.002^a$	$2750\ H^2$
Rubidium 87	$\frac{3}{2}$	$\frac{1}{2}$	$6{,}834{,}682{,}605^b$	$574\ H^2$
Cesium 133	$\frac{7}{2}$	$\frac{1}{2}$	$9{,}192{,}631{,}770^c$	$427\ H^2$

^{a} H. Hellwig, R. F. C. Vessot, M. W. Levine, P. W. Zitzewitz, D. W. Allan, and D. J. Glaze, *IEEE Trans. Instrum. Meas.* **IM-19**, 200 (1970).
^{b} P. L. Bender, E. C. Beatty, and A. R. Chi, *Phys. Rev. Lett.* **1**, 311 (1958).
^{c} This number of cycles of cesium has been adopted as the definition of the second.

from the atoms as they coherently make transitions $F = 1$, $M_F = 0 \rightarrow F = 0$, $M_F = 0$. The storage time is usually about 1 sec, which gives a line Q of about 10^9.

The frequency-control systems currently used in rubidium and cesium atomic clocks involve a quartz-crystal flywheel oscillator controlled by a frequency lock servo system. For the hydrogen maser, a phaselock system controls a high-power-level oscillator—most often, a crystal oscillator.

It is appropriate first to look briefly into the properties of crystal oscillators, since we will later see these properties at the short-term end of the frequency–stability (σ versus τ) curves of the above-mentioned atomic clocks.

5.4.6. Quartz-Crystal-Controlled Oscillators

Brandenberger et al.[10] gives a good example of the performance of a recently developed high-stability quartz crystal oscillator. His data are taken from carefully constructed 5-MHz oscillators that use components selected to provide very low flicker-phase noise and flicker-frequency noise in the oscillator output.

Data on the phase spectral density and rms frequency stability of this oscillator are shown in Figs. 6 and 7. These two figures illustrate the comple-

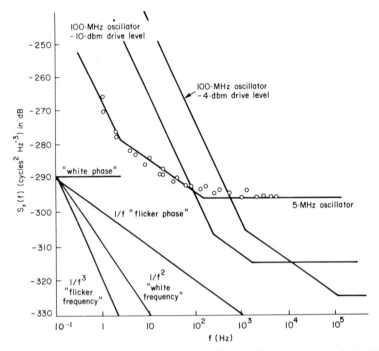

Fig. 6. Spectral densities of various crystal oscillators. The data points of the 5-MHz oscillator are from H. Brandenberger, F. Hadorn, and J. Shoaf, *Proc. Annu. Symp. Freq. Contr., 25th, Atlantic City, 1971* 226 (1971).

[10] H. Brandenberger, F. Hadorn, and J. Shoaf, *Proc. Annu. Symp. Freq. Contr., 25th, Atlantic City, 1971* 226 (1971).

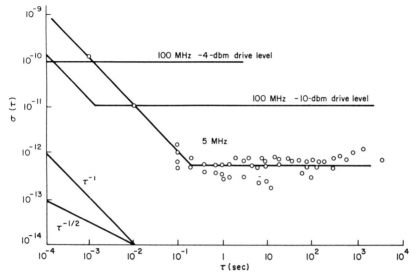

FIG. 7. Frequency stability of crystal-controlled oscillators. The data points of the 5-MHz oscillator are from H. Brandenberger, F. Hadorn, and J. Shoaf, *Proc. Annu. Symp. Freq. Contr.*, *25th*, *Atlantic City*, *1971* 226 (1971).

mentarity of frequency and time-domain data. From the phase spectral density, Brandenberger shows that a good fit can be made with three spectral components as follows:

$$S_\phi(f) = \frac{10^{-11.7} \; [\text{rad}^2 \; \text{Hz}^2]}{f^3} + \frac{10^{-12.5} \; [\text{rad}^2]}{f} + \frac{10^{-14.4} \; [\text{rad}^2 \; \text{Hz}^{-1}]}{f^0}.$$

Figure 6 shows a scale of $S_x(f)$, which normalizes this quantity to cycles^2 Hz^{-3}. The terms are identified with h_{-1}, h_1, and h_2 of Eq. (5.4.5)—i.e., flicker-frequency, flicker-phase, and white phase noise. We also note that no data are shown for averaging times exceeding 10^3 sec, probably because the aging of the oscillator, which causes a systematic frequency drift, is becoming apparent in the statistical representation of the data. For instance, an aging rate of a few parts in 10^{11} per day,[11] which is not unusual in this type of oscillator, will cause a drift of a few parts in 10^{12} in 10^4 sec. In general, statistical data can be taken in the presence of such a drift by first removing the linear drift during the total data-taking interval and then taking the frequency differences from the linear datum. Statistics made in this manner are labeled " drift removed " on the figure.

The noise processes that affect the oscillator in the range 10^2–10^4 Hz can be identified as additive white phase noise. This noise is the result of the thermal

[11] E. Haffner and R. S. Blewer, *Proc. Annu. Symp. Freq. Contr.*, *22nd*, *Atlantic City*, *1968* 136 (1968).

noise kT that accompanies the signal and of excess thermal noise $(F - 1)kT$. The term in h_0 is not visible, since it is swamped by the large flicker-frequency (h_{-1}) and flicker-phase (h_1) terms. The flicker-phase noise of the amplifier circuit and the feedback circuit in the oscillator can contribute flicker-frequency noise to the system in addition to the noise of the crystal resonator itself. Flicker-phase noise in the output is also present in the amplifier stages between the oscillator and the output. Halford *et al.*[12] discusses the reduction of this noise in transistor amplifiers.

Typical data for two other types of oscillators are shown as solid lines in Figs. 6 and 7. These data are for 100-MHz crystal oscillators operated at -10 dbm and -4 dbm[13] drive levels, the latter in a temperature-controlled oven.

The versatility of the quartz oscillator is very evident from the plots of $S_x(f)$ and $\sigma_y(\tau)$. We note that there are considerable possibilities for trading off high-frequency against low-frequency spectral performance. The choice of a particular crystal-oscillator frequency and its excitation level will depend on the performance of the atomic system by which it will be controlled.

5.4.7. The Atomic Hydrogen Maser

The hydrogen maser at present is the most suitable frequency standard for use in very long baseline interferometry, since it combines excellent long- and short-term stability, serving both as a stable reference oscillator and time reference.

The hydrogen maser is an active device in that it produces a signal directly from the radiating transitions from the upper to the lower hyperfine level in the ground state of atomic hydrogen. Figure 8 is a schematic diagram of the hydrogen maser.

The atomic hyperfine transition is used to provide the highly stable output frequency from the $F = 1$, $M_F = 0$ state to the $F = 0$, $M_F = 0$ state. At zero magnetic field and at $0°K$ (when the atom is motionless with respect to the observer), this transition occurs at a frequency $f_0 = 1,420,405,751.768$ Hz.

To obtain a beam of atomic hydrogen, we take molecular hydrogen at low pressure (10^{-2} to 5×10^{-1} Torr) and dissociate it with an rf discharge. The dissociator is a glass tube about 1 in. in diameter and 1 in. in length, surrounded by a coil that is part of an oscillator running at about 100 MHz. When operating properly, the hydrogen discharge will exhibit spectral lines of the hydrogen Balmer series and appear bright red. In the discharge plasma, the electrons collide with hydrogen molecules and excite them to energy levels

[12] D. Halford, A. E. Wainwright, and J. A. Barnes, *Proc. Annu. Symp. Freq. Contr., 22nd, Atlantic City, 1968* 340 (1968).

[13] R. A. Baugh and L. S. Cutler, *Microwave J.* **13**, 43 (1970).

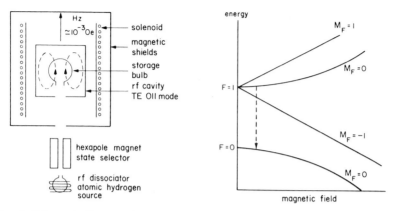

FIG. 8. Schematic diagram of the hydrogen maser and energy-level diagram of atomic hydrogen.

that are higher than the bonding energy of the molecule. The molecule can then either reradiate its energy or dissociate. Once dissociated, the atoms will stay that way unless they make three-body collisions in the volume of the tube or, as is much more probable, stick to the surface and encounter another hydrogen atom and recombine.

Hydrogen atoms in the source, at a pressure of about 5×10^{-1} Torr, are allowed to emerge from a multitube collimator to form a directed beam into the vacuum chamber and into a hexapole state-selector magnet, where the atoms in the desired $F = 1$, $M_F = 0$ and the $F = 1$, $M_F = +1$ states are physically separated from those in the two lower states ($F = 1$, $M_F = -1$ and $F = 0$, $M_F = 0$).

The separation is achieved by deflecting the atoms according to their effective dipole moments in the highly inhomogeneous field produced by the hexapole magnet. Atoms in the two upper states will seek a low energy by going to the low-field region of the hexapole, near its center. Similarly, atoms in the two lower energy states will move outward to high fields. In this way, we can direct a beam of the proper atoms into a resonant cavity, where they can deliver their energy and produce a signal.

The stability of the signal depends on the length of time the atoms can be kept in the in-phase part of a resonant rf magnetic field. This field stimulates the transition, thus releasing energy that is stored in a high Q resonator; in turn, through the stored fields, this resonator provides the stimulating fields necessary to continue the oscillating process.

The heart of the maser is the storage bulb, which serves to locate the atoms in the in-phase rf field, and also to confine the atoms by virtue of the nonrelaxing coatings on its walls. The coating, normally of FEP

Teflon,† has the remarkable property that hydrogen atoms with velocities of 2 km/sec can hit it and retain not only their energy state but also the phase‡ of their oscillating dipole interaction with the rf magnetic field. According to Heisenberg's uncertainty principle, the uncertainty of the oscillation frequency (oscillator linewidth) is inversely proportional to the length of time the atoms are stored in an unperturbed way. The storage time of the bulb is determined chiefly by kinetic-theory considerations. The atom enters the bulb through a small collimated hole and emerges at random with a probability function of time $f(t) = \gamma e^{-\gamma t}$, where γ is the bulb time constant determined by the bulb and collimator geometry. Typical values of γ are about 1–2 sec^{-1}.

Figure 8 shows the orientation of the source and magnet and the cavity with relation to the magnetic shields. Inside the innermost shield, a solenoid provides a uniform dc magnetic field at about 0.7×10^{-3} Oe, oriented to be generally parallel with the rf magnetic-field lines. According to the selection rules of quantum mechanics, this allows the two $M_F = 0$ states to be optimally coupled by the rf magnetic field and minimizes the interaction between the $F = 1$, $M_F = 1$ and $F = 0$, $M_F = 0$ states.

For the maser to oscillate, the cavity must be tuned very near the hydrogen resonance frequency. If the cavity is mistuned slightly by an amount Δf_c, the oscillation level is only very slightly reduced; however, there occurs a frequency-pulling effect, described by the expression

$$\Delta f_{\text{out}} = \Delta f_c (Q_c/Q_l).$$

Here, Q_c is the loaded Q of the cavity, and Q_l is the atomic line $Q = \pi f_0/\gamma_T$ (where γ_T is the total relaxation rate of the atoms in the bulb). In practice, the long-term stability of the maser is limited by this cavity-pulling effect chiefly because of thermally induced changes in cavity dimensions.

The cavity is usually made of a mechanically stable material that has a very low thermal coefficient of expansion, such as the glass ceramic material CER-VIT.§

The signal from the cavity is led to the phase-lock synthesizer described later in this section in Fig. 11. The spectral distribution of the phase fluctuations $S_\phi(f)$ in the signal has two major components. At frequencies very close to f_0 (within 1 mHz), we observe a $1/f^3$ distribution in $S_\phi(f)$. A typical value normalized to the hydrogen frequency f_0 is

$$S_\phi(f)/f_0 = -285 \ [\text{dB}]/f^3.$$

---

† Dupont trademark for a copolymer of polytetrafluoroethylene and hexafluoropropylene.

‡ This is mainly true; however, there is a small temperature-dependent average phase (or retardation). There is also evidence of recombination of atoms after about 10^5 bounces with the walls.

§ Trademark of Owens–Illinois.

This spectral behavior causes the flattening out of the σ versus τ plots, and is due to causal effects such as cavity-tuning or magnetic-field fluctuations; this is in contrast to fundamental processes such as thermal noise. The second component of the overall spectral density of the phase fluctuations is due to additive white phase noise and excess thermal noise of the first stages of the maser synthesizer,

$$S_\phi(f)/f_0 = (2\pi f_0)^{-2} FkT/P.$$

Under typical operating conditions, P is about 2×10^{-13} W and $F = 6$, so that

$$S_\phi(f)/f_0 = -268 \quad [\text{dB}].$$

These two noise contributions are shown in Fig. 9, along with the phase spectral density of a 100-MHz crystal oscillator.

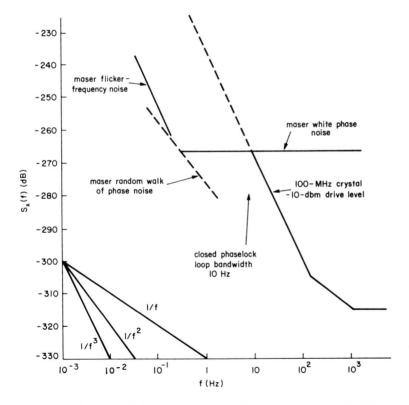

FIG. 9. Spectral density of phase noise for the hydrogen maser and a 100-MHz crystal oscillator.

The phaselock synthesizer will remove the components of additive white phase noise by narrow-band filtering, but cannot affect the $1/f^3$ components. From measurements in the time domain, we observe these components as a flattening out of the σ versus τ plot shown in Fig. 10. The level of the τ^{-1} portion of the plot is determined by the noise bandwidth.

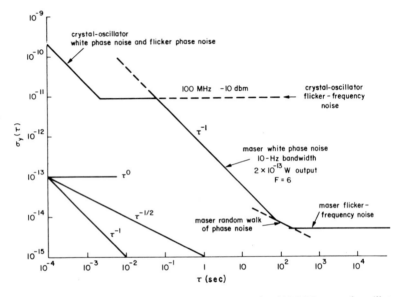

FIG. 10. Frequency stability of the hydrogen maser and a 100-MHz crystal oscillator.

To obtain output frequencies at useful levels, the maser signal is generally used to control the phase of a crystal oscillator that has good short-term stability. As an example, we will consider the 100-MHz crystal oscillator operating at -10 dbm described in Figs. 2 and 3. The characteristics of the phaselock-loop transfer function $G(f)$ will determine the overall behavior of the output signal from the locked servo loop. Figure 11 shows a block schematic of the system. It consists of a double heterodyne low-noise receiver that takes the -97-dbm output of the maser at 1,420,405,751.xxx Hz and produces a high-level signal at 405,751.xxx Hz. The phase of this signal is compared with that of a 405.751xxx-kHz signal obtained from a digital synthesizer controlled by the same voltage-controlled crystal oscillator that provides the first and second local-oscillator frequencies at 1.4 GHz and 20 MHz, respectively.

The phase information is led through an active filter with a spectral response of $g(f)$, and is used to control the phase of the crystal oscillator. The function $G(f)$ is given by $(K_1 K_0/j\omega)g(f)$, where K_1 is the output constant of

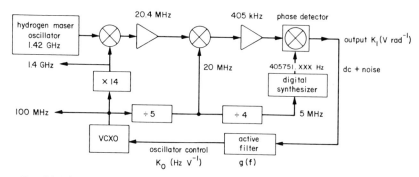

FIG. 11. Phaselock servo system. Maser output frequency is 1,420,405,751.68 Hz.

the phase detector in volts per radian, and K_0 is the control rate of the oscillator in hertz per volt. Normalized phase spectral density of the output frequency at 1.4 GHz is given by

$$S_x(f) = S_x(f)_{osc} \frac{1}{|1 + G(f)|^2} + S_x(f)_{maser} \left| \frac{G(f)}{1 + G(f)} \right|^2.$$

Here, $S_x(f)_{osc}$ is the normalized phase spectral density of the crystal oscillator and multiplier, and $S_x(f)_{maser}$ refers to the maser signal at the input to the receiver. The effects of other noise sources in the system are not included, but because of the high signal levels and low frequencies of the second local-oscillator digital synthesizer, the contribution of these effects is small compared to those of the maser and the oscillator multiplier.

In order to cope with the $1/f^3$ characteristic of the crystal oscillator, the function $g(f)$ is made to have a pole at zero frequency. The *frequency* response $K_1/j\omega$ of the phase detector supplies another pole at zero frequency, and thus $[G(f)]^2$ behaves as $1/f^4$, and the effect of the crystal oscillator $1/f^3$ noise in the servo loop is removed. The closed servo-loop bandwidth for the combination of maser and oscillator shown in Fig. 10 should be about 10 Hz.

5.4.8. The Cesium-Beam Resonator

The cesium-beam resonator uses the techniques developed for molecular and atomic beams, where atoms are allowed to drift in vacuum and behave as free, noninteracting particles that are analyzed according to their trajectories through strongly inhomogeneous magnetic fields. The cesium-beam resonator is contained in a vacuum system usually kept at a very low pressure by an internal ion pump. The resonator is shown schematically in Fig. 12. Cesium atoms are evaporated from a supply of liquid in an oven, and emerge into the vacuum system as a highly collimated, narrow, ribbon-like beam di-

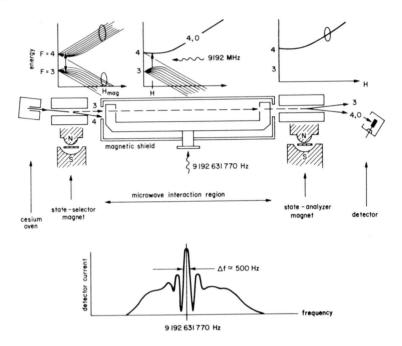

FIG. 12. Schematic diagram of the cesium-beam resonator.

rected into the state-selector magnet. This highly inhomogeneous magnetic field deflects the atoms according to their magnetic moment. As shown in Fig. 12, the atoms in the $F = 4$, $M_F = +4$ to $M_F = -3$ states (labeled by the symbol "4") are deflected toward weaker magnetic fields. Atoms in the remaining states are deflected toward stronger fields and, in this example, proceed into a magnetically shielded interaction region containing a two-section cavity. The cavity produces a microwave magnetic field parallel to a weak dc magnetic field produced by coil windings within the magnetic shield. The dc field provides a quantizing axis for the atoms, and orients them so that transitions between similar M_F levels ($\Delta M_F = 0$) are stimulated and $\Delta M_F = \pm 1$ transitions are suppressed. The dc field also splits the states sufficiently to prevent overlapping of undesired transitions at frequencies near the desired "clock" frequency and consequent deformation of the $F = 3$, $M_F = 0$ $\rightarrow F = 4$, $M_F = 0$ resonance curve.

The microwave magnetic field, if it were applied uniformly and optimally throughout the interaction region in the presence of a uniform dc magnetic field, would yield a resonance pattern $1/\tau$ Hz in width at half maximum, where τ is the transit time of the atoms in the cavity. In contrast, the two-section, or Ramsey, cavity applies two separate pulses of microwave field to the traversing atoms, and produces a pattern analogous to that of an optical two-slit diffraction process. The width at half-maximum of the central lobe of this pattern is approximately $1/2\tau$ Hz. The narrower resonance of the Ramsey resonator is also unperturbed by doppler shifts from atoms interacting with the running waves that would occur in a single, long-interaction cavity. In practice, resonator lengths range from 7.5 to 17.5 cm and correspondingly provide Ramsey resonance widths of 1300 to 360 Hz. For a given application, a compromise is reached between beam-tube linewidth and beam flux, which diminishes as the interaction length is increased.

Atoms that have been exposed to the microwave field of proper frequency and intensity will emerge from the interaction region of Fig. 12 in the $F = 4$, $M_F = 0$ state, and are sorted from the other atoms in group 3 by a state-analyzer magnet similar to the state-selector magnet. These atoms are deflected to a heated ribbon of metal, such as tungsten, that has a work function substantially greater than that of cesium. The cesium atoms readily become ionized by losing their weakly bound electrons, and the ions are passed through a mass spectrometer to separate them from ions of other species produced by the hot ribbon. The ions are accelerated to the first dynode of an electron multiplier, where the current of about 10^{-12} A is amplified by ejecting secondary electrons, which, in turn, are amplified by secondary emission in the subsequent stages of the multiplier. The output current, now at a level of about 10^{-7} A, depends strongly on the frequency of the microwave excitation and is maximum at the cesium resonance.

The chief noise mechanism in the output signal is shot noise due to the arrival of individual atoms at the detector. This noise has a "white," or constant, power spectral density that, as we will see later, causes a "white" spectral density of frequency fluctuations in the output of the frequency-lock servo system.

The cesium-beam resonator is probably the best understood and most accurate frequency standard available to date. The cesium atom has been chosen as the basis for the definition of the second. According to the present definition, 9,192,631,770 cycles of the unperturbed cesium hyperfine transition frequency elapse in 1 sec. Cesium devices are generally used where accuracy and stability over long periods of time are required, and where short-term stability and high spectral purity are not of major importance.

5.4.9. The Rubidium-Gas-Cell Resonator

The rubidium-gas-cell resonator uses the technique of optical pumping to produce an unbalanced population of atoms in the ground state of ^{87}Rb. Resonance transitions are then produced by applying a microwave resonance signal, and the degree of resonance is monitored optically.

The process of optical pumping in the ^{87}Rb resonator uses selected optical-pumping transitions connecting levels above the ground state with the lower ground-state hyperfine level. The lower level is thus kept depopulated. When microwave energy at the proper hyperfine resonance frequency is applied, the lower level is repopulated and the pumping light is absorbed. The absorption of the pumping light offers a means for detecting the resonance. This scheme is shown in Fig. 13. The pumping light is obtained from an rf plasma

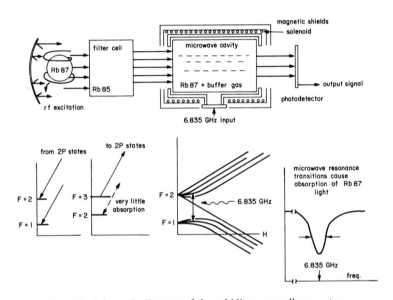

FIG. 13. Schematic diagram of the rubidium-gas-cell resonator.

discharge in a tube containing ^{87}Rb, and the light from this tube contains spectral lines involving the two ground-state levels. In order to depopulate the lower ground-state hyperfine level of the rubidium atoms in the microwave cavity, the light must first be filtered to remove the optical components involving the *upper* ground-state levels. This is done by using a highly selective optical filter composed of ^{85}Rb atoms mixed with an inert gas in a transparent cell. The ^{85}Rb isotope has optical transitions that nearly match those involving the upper hyperfine level of ^{87}Rb, while the corresponding lower hyperfine-level transitions of ^{85}Rb and ^{87}Rb do not match. The result is that light emerging from the filter cell will primarily cause energy-absorptive transitions from the lower hyperfine levels of ^{87}Rb in the microwave cavity.

The atoms in the cavity are spatially confined in a buffer gas that, by means of elastic collisions, keeps the atoms in a relatively unperturbed condition, and thus reduces the linewidth of the microwave resonance transitions. The use of buffer-gas collisions to sharpen the microwave resonance has been under continuous investigation since its invention by Dicke in 1953.[14] Considerable efforts have been made to find a combination of gases having compensating negative and positive thermal and pressure shifts to the resonance frequency in order to obtain an environmentally stable mixture.

Other causes of instability in the rubidium-gas-cell resonator result from variations in optical-pumping light intensity and spectral distribution. However, the simplicity of the resonator and its excellent signal-to-noise properties have been a strong incentive to the development of small and relatively inexpensive frequency standards with excellent short-term frequency stability.

Returning to Fig. 13, we see that when the applied microwave signal repopulates the $F = 1$, $M_F = 0$ level, light absorption takes place. As in the cesium device, the overlap of the desired $F = 2$, $M_F = 0 \rightarrow F = 1$, $M_F = 0$ transition with other $M_F = 0$ transitions is removed by the addition of a weak, homogeneous magnetic field, and the effects of external magnetic fields are minimized by shielding.

The absorption resonance is generally between 200 and 1000 Hz in width at half-maximum and involves a very large number of atoms, thus giving a signal-to-noise ratio roughly 100 times better than that of the cesium-beam resonator. The random arrival of photons gives a white noise power spectral density that, as in the case of cesium, produces a white noise spectral density in the frequency fluctuations of the frequency-lock servo system.

5.4.10. Frequency-Lock Servo Systems

Since the cesium and rubidium frequency standards use similar servo techniques, it is appropriate to discuss them together here. The basic frequency-lock servo system is shown in Fig. 14. A voltage-controlled crystal oscillator (usually a 5-MHz high-stability oscillator) drives a frequency synthesizer that produces a signal at the appropriate atomic hyperfine transition frequency. Provision is made for frequency-modulating the microwave signal, usually by phase-modulating a low-frequency stage of the synthesizer. By scanning a small part of the resonance, the modulated microwave signal produces a signal at the resonator output; the magnitude and phase of the signal are proportional to the slope of the resonance curve at the average frequency of the microwave signal. This output signal is compared in phase with the modulation signal, and a dc voltage is obtained whose magnitude and sign, corresponding to the location of the microwave signal on the resonance, can be used to control the frequency of the crystal oscillator so as to keep it at the peak of the resonance. Of course, this signal also contains noise from the resonator (shot noise from cesium atoms or photon noise from the

---

[14] R. H. Dicke, *Phys. Rev.* **89**, 472 (1953).

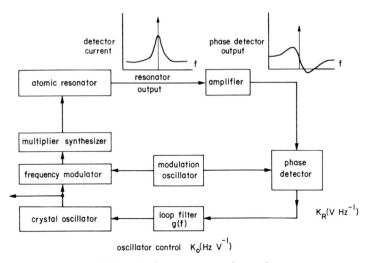

FIG. 14. Passive frequency-control servo loop.

rubidium light), as well as the noise from frequency fluctuations in the crystal-oscillator–frequency-synthesizer combination. The combination of these two noise sources (resonator and frequency generator) can be optimized by placing a filter in the servo loop so that the high-frequency noise contributions of the resonator do not corrupt the output of the oscillator, and the slow frequency variations of the oscillator are corrected by the atomic resonator.

The spectral density of the closed-loop output frequency fluctuations is given by

$$S_y(f) = S_y(f)\Big|_{\substack{\text{crystal oscillator} \\ \text{and synthesizer}}} \frac{1}{|1 + G(f)|^2} + S_y(f)\Big|_{\text{resonator}} \left|\frac{G(f)}{1 + G(f)}\right|^2,$$

where $G(f) = K_R K_0 g(f)$ is the servo-loop gain function. The term K_R is the output characteristic of the phase detector, given in terms of volts per hertz (over the linear portion near the resonance maximum). The term K_0 is the oscillator control sensitivity, expressed in hertz per volt. We can adjust the filter function $g(f)$ to optimize the overall value of $S_y(f)$. This process is shown in Fig. 15 in a time-domain plot of the crystal-oscillator stability and the equivalent stability of an oscillator wholly governed by a cesium resonator. The crossing point of the two stability functions determines the ideal closed-loop time constant of our servo loop and, consequently, the choice of the loop filter function. In most applications, the loop time constant is made adjustable to allow for operation during adverse environmental conditions on the crystal oscillator, such as vibration. Under these conditions, the crystal oscillator is controlled with a shorter time constant (or a higher frequency

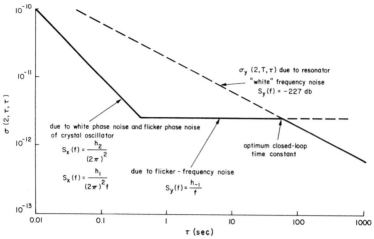

FIG. 15. Cesium-beam resonator with crystal oscillator.

cutoff in $g(f)$). The filter function $g(f)$ is generally obtained by operational integration and gain so that the resulting value of $G(f)$ is given by $G(f) = f_c/jf$.

From the closed-loop response equation, we see that this acts as a high-pass filter for the oscillator synthesizer and as a low-pass filter for the resonator.

The example in Fig. 15 describes a high-performance cesium resonator and a good-quality crystal oscillator. In the case of the rubidium-gas-cell resonator, we generally will have a white-frequency-noise fluctuation level, $S_y(f) = h_0$, 1/100 of that for cesium, and a corresponding level of the $\sigma(\tau) =$ constant $\times \tau^{-1/2}$ stability function 10 times improved. If the same crystal were used, we would make the closed-loop time constant 100 times smaller.

5.4.11. The Present State of the Art

As in all fast-moving areas of technology, it is risky to try to stop at a point in time to assess where we are. With this in mind, Fig. 16 is offered as a composite plot of cesium, hydrogen, and rubidium frequency standards. A typical, good-quality 5-MHz quartz oscillator is assumed as part of the cesium and rubidium standards. Its open-loop performance is shown labeled "drift removed."

There is a wide spread in the stability available from different types of cesium instruments. A large investment is being made by private industry to develop small, low-cost standards for airborne applications such as navigation and collision avoidance. These standards provide good stability in adverse environments and are represented at the top of the shaded area in Fig. 16. The development of high-accuracy cesium standards with good long-term stability is being pursued by the bureaus of standards of many nations to

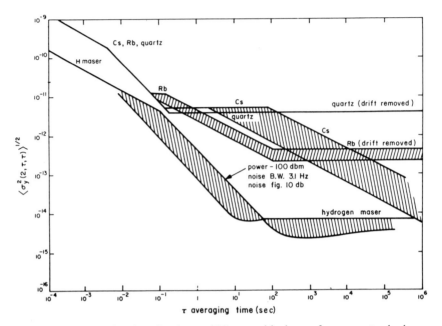

FIG. 16. Composite plot of cesium, rubidium, and hydrogen frequency standards.

provide clocks to maintain a coordinated atomic time scale. Some of these devices have very high flux and, consequently, superior short-term stability. The lower bound on the cesium plot is an estimate of present capabilities of assemblages of such clocks.

In the case of rubidium, there has been considerable effort toward miniaturization, and there are now available completely packaged standards of size comparable to high-quality quartz oscillators. These new standards are highly suitable as traveling clocks for synchronizing widely separated clocks.

Hydrogen masers offer some tradeoff of performance between long and short averaging times, as shown in the figure. The flattening out of the σ versus τ plot is due to systematic changes in frequency; this is in contrast to random processes such as thermal noise or shot noise. The systematic effect of cavity mistuning (and the consequent frequency pulling) is the chief problem, and means are being sought to stabilize the cavity and output circuitry. The pulling can also be reduced by further narrowing of the atomic resonance linewidth by extending the storage time through the use of larger storage bulbs and better coating materials.

Frequency and time metrology, at present, offers the scientist the most sensitive and accurate techniques for measuring physical quantities. These quantities can often be related to frequency through well-understood quantum-mechanical processes. Nuclear quadrupole resonances in various materials are used to measure temperature; voltage can be measured via the Josephson effect; magnetic fields are measured by the precession of atomic nuclei. These are but a few examples of frequency-metrology techniques.

The accuracy and stability of atomic clocks allows us to measure and compare distance if we tacitly assume a constant and isotropic velocity of light throughout space. At present, planetary astronomy using radar and the tracking of space probes employs time intervals as a measure of distance.

Clocks and frequency standards will become increasingly important to physics and astronomy as we continue to expand the size of our laboratory in space. Even the once "impossible" gedanken or thought experiments devised by Einstein and others to illustrate relativity using rods and clocks are now realistically feasible.

BIBLIOGRAPHY ON FREQUENCY AND TIME STANDARDS

L. Essen and J. Parry, An atomic standard of frequency and time interval, *Nature* (*London*) **176**, 280 (1955).

E. A. Gerber and R. A. Sykes, State of the art—Quartz crystal units and oscillators, *Proc. IEEE* **54**, 103 (1966).

E. Hafner, The effects of noise in oscillators, *Proc. IEEE* **54**, 179 (1966).

W. Bell and A. Bloom, Optically detected field-independent transitions in sodium vapor, *Phys. Rev.* **109**, 219 (1958).

M. Arditi and T. Carver, Pressure, light and temperature shifts in the optical detection of O–O hyperfine resonance of alkali metals, *Phys. Rev.* **124**, 800 (1961).

M. E. Packard and B. E. Swartz, The optically pumped rubidium vapor frequency standard, *IRE Trans. Instrum.* **I-11**, 212 (1962).

R. F. Lacey, A. L. Helgesson, and J. H. Holloway, Short-term stability of passive atomic frequency standards, *Proc. IEEE* **54**, 170 (1966).

R. E. Beehler, R. C. Mockler, and J. M. Richardson, Cesium beam atomic time and frequency standards, *Metrologia* **1**, 114 (1965).

D. Kleppner, H. C. Berg, S. B. Crampton, N. F. Ramsey, R. F. C. Vessot, H. E. Peters, and J. Vanier, Hydrogen maser principles and techniques, *Phys. Rev.* **138A**, 972 (1965).

R. F. C. Vessot, M. W. Levine, P. W. Zitzewitz, P. Debely, and N. F. Ramsey, Recent developments affecting the hydrogen maser as a frequency standard, *in* " Precision Measurement and Fundamental Constants" (D. N. Langenberg and B. N. Taylor, eds.), p. 27. Nat. Bur. Std., Spec. Publ. No. 343, 1971.

M. W. Levine and R. F. C. Vessot, Hydrogen-maser time and frequency standard at Agassiz Observatory, *Radio Sci.* **5**, 1287 (1970).

H. H. Hellwig, Atomic frequency standards: a survey, *Proc. IEEE* **63**, No. 2, 212 (1975).

R. F. C. Vessot, " Lectures on Frequency Stability and Checks and on the Gravitational Redshift Experiment " (B. Bertotti, ed.), p. 111. Academic Press, New York, 1974.

5.5. Very Long Baseline Interferometric Observations and Data Reduction*

5.5.1. Introduction

Very long baseline interferometry (VLBI) is similar to conventional interferometry except that the receiving elements are not connected in "real time." The received signals are recorded, undetected, on magnetic tape under the control of atomic frequency standards and reproduced at a later time and processed to give interference fringes. There are important differences, however, between VLBI and conventional interferometry. The telescopes used in VLBI can be very far apart owing to their independence, and the measurable quantities of phase, delay, and phase rate take on much larger values and play different roles in the data interpretation. Also, because the data storage is limited, the antennas are immovable, and the data acquisition is difficult, the coverage on the uv (or projected baseline) plane is quite limited, and aperture-synthesis techniques are generally not very useful. Hence, other methods must be used to ensure that each measurement contributes as much as possible to the knowledge of the brightness distribution. Finally, because the phase stability of the interferometer is imperfect, owing to the limited spectral purity of the frequency standards, special techniques are needed to cope with the added phase noise.

Five topics of importance to VLBI observations and data reduction will be discussed here. First, the measurable quantities will be defined by tracing the received signals through the VLBI system. Next, the problem of estimating the fringe amplitude of weak signals in the presence of phase noise will be treated, and expressions for the signal-to-noise ratio for so-called "broken-coherence" averaging will be derived. Third, the probability of misidentifying the interference fringes is given. Fourth, the interpretation of the data in terms of simple models of brightness distribution will be discussed. Finally, some operational problems encountered in acquiring data will be described.

The discussion is oriented toward digital one-bit-per-sample VLBI systems,[1] and single-sideband operation is assumed.

No sharp distinction is made between the processing of continuum and spectral-line data, since they are very similar and, in principle, differ by a

[1] B. G. Clark, *Proc. IEEE* **61**, 1242 (1973).

---

* Chapter 5.5 is by J. M. Moran.

Fourier transformation. The basic computation for processing the data is the calculation of the cross-correlation function between the two data streams. This function must be calculated at a number of delays because of the uncertainty in the instrumental and geometric delays for the observation. The Fourier transform of the correlation function is the cross-power spectrum, which is the important function in spectral-line measurements (see Section 5.1.2). However, even in continuum measurements, many operations on the data are more readily performed on the cross-power spectrum than on the cross-correlation function. The examination of continuum data in the frequency domain also offers new ways to extract the desired measurement parameters. For example, the geometric delay for a continuum source can be established by determining the center of the correlation function or by finding the slope of the phase of the cross-power spectrum as a function of frequency.

5.5.2. Measurement of Fringe Amplitude and Phase

Figure 1 shows a simplified diagram of a very long baseline interferometer, emphasizing the signal-processing aspects of the system. The phase shifts encountered by the received signals will be discussed first. The voltages induced owing to a point source in the receivers of the two widely separated antennas can be represented as $A_1(\omega) \exp(j\omega t)$ and $A_2(\omega) \exp[j\omega(t - \tau_g)]$, where $j = \sqrt{-1}$ and τ_g is the excess travel time, or geometric delay, given by

$$\tau_g = (D/c)[\sin \delta_B \sin \delta_S + \cos \delta_B \cos \delta_S \cos(L_S - L_B)] + \tau_B + \tau_{atm}, \quad (5.5.1)$$

where D is the baseline length, c the velocity of light, and δ_B, δ_S, L_B, and L_S the declinations and hour angles of the baseline and source, respectively. The baseline geometry is shown in Fig. 2. The term τ_B is a small one that accounts for the earth's rotation in the interval between the arrival times of a given wave crest at the two antennas,[2] and τ_{atm} is the differential atmospheric phase delay. Each signal is converted to a video band from 0 to B_0 Hz by one or more single-sideband conversions, represented here by a single mixer. The frequency-conversion schemes need not be identical. However, the difference in the number of lower sideband conversions in each of the two receivers must be an even number so that the rf components will appear at the same frequency in the two video bands. The local-oscillator signals at each station are derived from atomic frequency standards and contain the additive phase noise terms $\phi_1(t)$ and $\phi_2(t)$. The signals, after frequency conversion, are

$$
\begin{aligned}
x_1 &= A_1(\omega)G_1(\omega) \exp\{j[(\omega - \omega_1)t - \phi_1(t) - \psi_1(\omega)]\}, \\
x_2 &= A_2(\omega)G_2(\omega) \exp\{j[(\omega - \omega_2)t - \omega\tau_g - \phi_2(t) - \psi_2(\omega)]\},
\end{aligned}
\quad (5.5.2)
$$

[2] M. H. Cohen and D. B. Shaffer, *Astron. J.* **76**, 91 (1971).

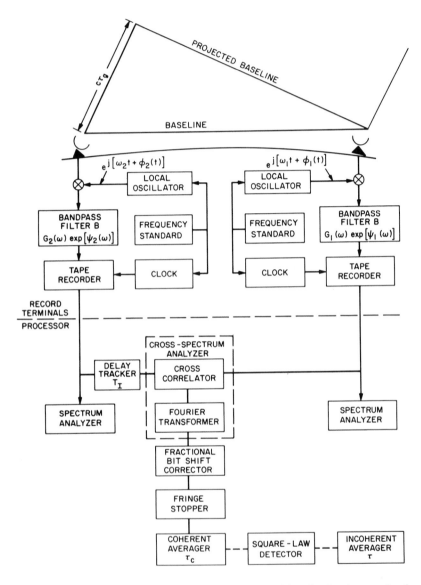

FIG. 1. Block diagram of a digital VLBI system emphasizing the signal-processing functions. For nonspectral-line work, the spectrum analyzers are not needed and the Fourier transformer can be omitted or, alternatively, another one added after the fractional-bit-shift corrector. The square-law detector and incoherent averager are not always used, because they improve the signal-to-noise ratio substantially only if τ_c is small (see Section 5.5.2).

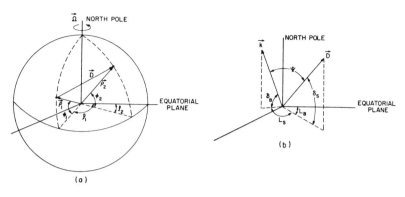

FIG. 2. (a) The baseline vector joining two telescopes located at latitudes ϕ_1 and ϕ_2 and longitudes l_1 and l_2. (b) The baseline vector $\mathbf{D}$ and the unit vector in the direction of the source $\mathbf{k}$. If the earth's rotation and the atmosphere are ignored, the geometric delay is $\tau_g = \mathbf{D} \cdot \mathbf{k}/c$, where c is the velocity of light.

where $G_1(\omega)$, $G_2(\omega)$, $\psi_1(\omega)$, and $\psi_2(\omega)$ are the amplitude and phase responses of the band-limiting filters in each receiver. Finally, the signals are recorded, along with timing information. The station clocks, driven by the frequency standards, are assumed to have constant epoch errors, τ_{c_1} and τ_{c_2}, so that the recorded signals are

$$x_1 = A_1(\omega)G_1(\omega) \exp\{j[(\omega - \omega_1)t - (\omega - \omega_1)\tau_{c_1} - \phi_1(t) - \psi_1(\omega)]\},$$

$$(5.5.3)$$

$$x_2 = A_2(\omega)G_2(\omega) \exp\{j[(\omega - \omega_2)t - \omega\tau_g - (\omega - \omega_2)\tau_{c_2} - \phi_2(t) - \psi_2(\omega)]\}.$$

In digital VLBI systems, the signals are clipped and sampled at intervals of $1/2B_0$. The playback processor advances one data stream by T_I, which is the geometric delay of a reference position in the sky, τ_{g_0}, rounded off to the nearest data sampling interval. The output of the cross-spectrum analyzer is thus

$$S_{12}(\omega, t) = x_1^\dagger(t)x_2(t + T_I) = A_1(\omega)A_2(\omega)G_1(\omega)G_2(\omega) \exp(j\Phi), \quad (5.5.4)$$

where the † denotes the complex conjugate and where

$$\Phi(\omega) = (\omega_1 - \omega_2)t - \omega\tau_g + (\omega - \omega_2)\tau_I + \Delta\phi + \Delta\psi(\omega) + \omega\,\Delta\tau_c + \gamma, \quad (5.5.5)$$

in which

$$\Delta\phi(t) = \phi_1(t) - \phi_2(t), \qquad \Delta\tau_c = \tau_{c_1} - \tau_{c_2},$$

$$\Delta\psi(\omega) = \psi_1(\omega) - \psi_2(\omega), \qquad \gamma = \omega_2\tau_{c_2} - \omega_1\tau_{c_1} = \text{const.}$$

The difference in the local-oscillator frequencies, $\omega_1 - \omega_2$, is usually chosen to be equal to $\omega\dot\tau_g$ (the dot ($\cdot$) denotes the time derivative) at the beginning of the observation. This offsetting of the local-oscillator frequencies keeps the apparent fringe rate $\dot{\Phi}(\omega)$ low, which is convenient but not always necessary. Since ωt_g is the Doppler shift between the received signals due to the earth's rotation, the effect of offsetting the local oscillators is to remove this Doppler shift and align the recorded signal bands. The effective or usable bandwidth is therefore $B_0 - \dot{\Phi}(\omega)/2\pi$. The insertion of T_I before the spectrum analyzer is necessary to keep $\partial\Phi/\partial\omega$ low, i.e., to keep the center of the cross-correlation function within the delay range of the correlator.

The fractional-bit-shift corrector multiplies the output of the spectrum analyzer by $\exp[j(\omega - \omega_2)\,\Delta\tau]$, where $\Delta\tau = \tau_{g0} - T_I$, in order to correct for the discrete delay tracking. If this correction is omitted, the phase of the output will have a saw-tooth waveform, and the time-averaged waveform and time-averaged fringe amplitude will be reduced by a factor of 0.6 at the upper edge of the band, as shown in Fig. 3. This correction is sometimes

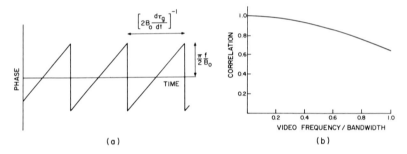

(a)   (b)

FIG. 3. (a) The residual fringe phase versus time, showing the effect of discrete delay tracking in a digital processor: f is the video frequency, B_0 the video bandwidth, and $(2B_0\,d\tau_g/dt)^{-1}$ the time in which the geometric delay changes by one sampling interval. When the fringe phase is calculated from the cross-correlation function, the half-amplitude of the saw tooth is $\pi/4$. (b) Reduction of the time-averaged correlation versus video frequency if the delay tracking is not compensated. The peak of the normalized cross-correlation function is reduced from unity to 0.88.

omitted, especially in continuum VLBI, where the fringe amplitude is reduced by a factor of only 0.88, and the time-averaged effects of the phase discontinuities are small. The phase of the output of the fractional-bit-shift corrector becomes

$$\Phi(\omega, t) = (\omega_1 - \omega_2)t - \omega(\tau_g - \tau_{g0}) - \omega_2\,\tau_{g0} + \Delta\phi + \Delta\psi(\omega) + \omega\,\Delta\tau_c, \quad (5.5.6)$$

where γ, a constant, has been absorbed into $\Delta\phi$. Note that if $\tau_{g0} = \tau_g$, $\omega_1 = \omega_2$, and $\Delta\phi = \Delta\psi = \Delta\tau_c = 0$, the phase is simply $\Phi(\omega, t) = -\omega_2\,\tau_g$. The reason ω_2, the local-oscillator frequency, and not ω, the sky frequency, appears in the phase equation is that the delay tracking removes the phase shift across the video band.

The fringe stopper reduces the phase rate $\dot{\Phi}$ to nearly zero (some other arbitrary frequency could be chosen) by multiplying by $\exp j[(\omega_2 - \omega_1)t - \omega_2 \tau_{g0}]$. The signal becomes

$$S_{12}(\omega, t) = A_1(\omega)A_2(\omega)G_1(\omega)G_2(\omega) \exp(j\Phi), \qquad (5.5.7)$$

where

$$\Phi(\omega, t) = -\omega(\tau_g - \tau_{g0}) + \Delta\phi(t) + \Delta\psi(\omega) + \omega \Delta\tau_c. \qquad (5.5.8)$$

Equations (5.5.7) and (5.5.8) give the basic response of a VLBI system to a point source as a function of frequency. If the source is resolved, $A_1(\omega)A_2(\omega)$ is replaced by the complex fringe amplitude. The residual fringe phase and residual fringe frequency are $\Phi(\omega, t)$ and $\dot{\Phi}(\omega, t)$, respectively, and the residual delay is $\partial\Phi(\omega, t)/\partial\omega$.

In general, $S_{12}(\omega, t)$ will always have some time dependence because the phase-noise term $\Delta\phi$ is not zero, and because it is difficult to "stop" the fringes exactly. Hence, it is generally useful to consider the Fourier transform of $S_{12}(\omega, t)$, called the fringe-frequency spectrum, defined by

$$S_{12}(\omega, \omega_f) = 1/T \int_0^T S_{12}(\omega, t) \exp(j\omega_f t) \, dt. \qquad (5.5.9)$$

In spectral-line measurements, the spectral analysis is usually achieved with a digital cross correlator whose output is Fourier transformed to give $S_{12}(\omega, t)$ according to the Wiener–Khinchine theorem. When the signals are recorded in the one-bit digital representation, the cross-correlation function $\rho_{12}(\tau)$ can be recovered from the cross-correlation function of the clipped signals $\rho_c(\tau)$ by the Van Vleck relation[3]:

$$\rho_{12}(\tau) = [(T_{R_1} + \bar{T}_{A_1})(T_{R_2} + \bar{T}_{A_2})]^{1/2} \sin[(\pi/2)\rho_c(\tau)], \qquad (5.5.10)$$

where T_{R_1}, T_{R_2}, $\bar{T}_{A_1}$, and $\bar{T}_{A_2}$ are the receiver noise and antenna temperatures, respectively, averaged over the band. The Fourier transform of $\rho_{12}(\tau)$, i.e., $S_{12}(\omega)$, will have units of degrees Kelvin. In virtually all cases, $\rho_c(\tau) \ll 1$, so the sine term in Eq. (5.5.10) can be replaced by its argument.

The correlation function must be measured over a range of both positive and negative delays since it is not an even function, and the resulting spectrum is, in general, complex. The resolution will therefore be equal to $2.4B_0/N$ for uniform weighting and $4.0B_0/N$ for Hanning weighting, where N is the number of delays, spaced at intervals of $(2B_0)^{-1}$, at which the correlation function is measured. Because the receiver noise voltages are uncorrelated, the expectations of the real and imaginary parts of $S_{12}(\omega, t)$ are zero in the absence of signal. Hence, there is no problem in determining the zero-power level in the cross-power spectrum.

[3] J. H. Van Vleck and D. Middleton, *Proc. IEEE* **54**, 2 (1966).

The power spectra of the signal on the individual recordings, $S_1(\omega)$ and $S_2(\omega)$, should be estimated by the usual differencing technique in order to remove the receiver noise spectra:

$$S_j(\omega) = (\overline{T}_{A_j} + \overline{T}_{R_j})(S_{on_j}(\omega) - S_{off_j}(\omega))/S_{off_j}(\omega), \qquad j = 1, 2, \quad (5.5.11)$$

where S_{on} and S_{off} are the "on" and "off" source spectra calculated from the autocorrelation functions. The spectral resolution is equal to $1.2B_0/N'$ and $2.0B_0/N'$ for uniform and Hanning weighting, respectively, where N' is the number of delay samples in the autocorrelation function. The autocorrelation functions need be measured only at positive delays since they are even functions of delay. Hence, N' should be chosen equal to $(N + 1)/2$ so that the spectral resolutions will be equal in both the cross and the individual power spectra.

The fringe-visibility spectrum, given by

$$V(\omega) = S_{12}(\omega)/[S_1(\omega)S_2(\omega)]^{1/2}, \qquad (5.5.12)$$

is independent of receiver noise temperatures and antenna collecting areas and can be measured quite accurately.

The concepts discussed above can be clarified by defining an algorithm for the estimation of the cross-power spectrum from a set of M one-bit correlation functions, $\rho_c(\tau_i, t_m)$, measured at N delays at times t_m uniformly over a time interval T. The algorithm is

$$S_{12}(\omega_k, \omega_{f_l}) = C_k \sum_m S'_{12}(\omega_k, t_m) \exp[j(\Phi^* + \omega_{f_l} t_m)], \qquad (5.5.13)$$

where

$$S'_{12}(\omega_k, t_m) = \sum_i \rho_c(\tau_i, t_m) W(\tau_i) \exp[j\omega_k(\tau_i + \Delta\tau_m)],$$

$$C_k = \frac{\pi}{2M} \left[\frac{(T_{R_1} + \overline{T}_{A_1})(T_{R_2} + \overline{T}_{A_2})}{S_{off_1}(\omega_k)S_{off_2}(\omega_k)} \right]^{1/2} \exp[j\,\Delta\psi(\omega_k)],$$

$$\Phi^* = (\omega_2 - \omega_1)t_m - \omega_2\tau_{g_0}, \qquad \Delta\tau_m = \tau_{g_0} - \tau_I,$$

$$\tau_i = i/2B_0, \qquad -(N - 1)/2 \le i \le (N - 1)/2,$$

$$\omega_k = 2\pi(k - 1)B_0/N^*, \qquad 1 \le k \le N^*,$$

$$t_m = mT/M, \quad 1 \le m \le M, \quad \text{and} \quad \omega_{f_l} = 2\pi l/TM^*, \quad -M^*/2 \le l \le M^*/2.$$

$S'_{12}(\omega_k, t_m)$ is the estimate at each time of the cross-power spectrum as a function of video frequency from 0 to B_0 Hz. The spectral weighting function is $W(\tau_i)$. Generally, N^*, the number of points in the frequency spectrum, is larger than N for purposes of interpolation, and M^*, the number of points in the fringe frequency spectrum, is larger than M for the same reason. The discrete delay tracking of the digital correlator is corrected by the term

$\exp(j\omega_k \Delta\tau_m)$. Since $\Delta\tau_m$ is a function of time, the m and i summations cannot be interchanged. $S'_{12}(\omega_k) = S'^{\dagger}_{12}(-\omega_k)$, where $\dagger$ denotes the conjugate, since $\rho(\tau_i, t_m)$ is a real function. The fringe rotation or stopping, accomplished by the summation over m, moves the spectrum at negative frequencies, $-\omega_k$, away from the fringe-frequency band of interest, so that $\langle S_{12}(\omega_k, \omega_{f_l})\rangle = 0$ for $\omega_k < 0$. Hence, it is of no interest to calculate the spectrum at negative frequencies. The fringe-frequency spectrum has a width of $2\pi M/T$ Hz centered on the expected fringe frequency $d\Phi^*/dt$. The normalization term C_k accounts for the one-bit clipping, assuming $\rho_c \ll 1$, and for the bandpass characteristics of the receivers. The relative instrumental phase $\Delta\psi(\omega_k)$ can be found from observations of continuum sources. An example of $S_{12}(\omega, \omega_f)$ obtained with an interferometer formed with the 42.7-m (140-ft) antenna of the National Radio Astronomy Observatory (NRAO)† and the 36.6-m (120-ft) antenna of the Haystack Observatory‡ is shown in Figs. 4 and 5.[4] The magnitude of S_{12} versus ω and ω_f is shown in Fig. 4, and the magnitude and phase of S_{12} versus ω at the fringe frequency of the strongest feature, in Fig. 5.

With a few modifications, data from continuum sources, whose spectra are flat across the recorded band, can be processed in the same way as data from spectral-line sources. First, the visibility can be calculated only if independent measurements of $\bar{T}_{A_1}$ and $\bar{T}_{A_2}$ are available, since $S_1(\omega)$ and $S_2(\omega)$ do not change with the strength of the source. Second, because the spectral dependence is of no interest, S_{12} should be averaged over ω_k. Hence, an algorithm for the calculation of fringe visibility for continuum sources is

$$V = \sum_{k=1}^{N^*} S_{12}(\omega_k, \omega_{f_l})/(T_{A_1}T_{A_2})^{1/2}. \tag{5.5.14}$$

If the instrumental delay is not well known, a slope will occur in the phase of S_{12} with respect to ω_k, which can be removed by multiplying S_{12} by $\exp(-j\omega_k\tau_i)$. It is useful to define the modified cross-correlation function $R_{12}(\tau_i, \omega_{f_l})$ by

$$R'_{12}(\tau_i, \omega_{f_l}) = 1/N^* \sum_{k=1}^{N^*} S_{12}(\omega_k, \omega_{f_l}) \exp(-j\omega_k\tau_i). \tag{5.5.15}$$

---

[4] From unpublished data by a group from Massachusetts Inst. of Technol. (B. F. Burke, K. Y. Lo, and G. D. Papadopoulos), Naval Res. Lab. (K. J. Johnston and S. H. Knowles), and Smithsonian Astrophys. Observatory (J. M. Moran).

---

† NRAO is operated by Associated Universities, Inc., under contract with the National Science Foundation.

‡ Haystack Observatory is operated by Massachusetts Institute of Technology with the support of the National Science Foundation.

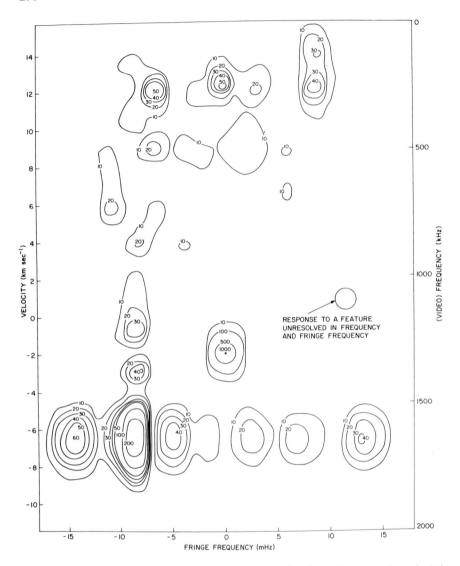

FIG. 4. Magnitude of the cross-power spectrum as a function of frequency (or velocity) and fringe frequency for a 1024-sec observation of the H_2O source in W49 N obtained, on November 12, 1971 at 23^h GMT, with the Haystack–NRAO interferometer. For this particular run, the fringe-frequency sensitivity was 18 mHz/arc sec at a position angle of 140°. Hanning weighting was used to reduce sidelobe levels. The feature at −1.9 km/sec was used as a phase reference. Note that only selected contours are plotted. [Figures 4–7: From unpublished data by a group from Massachusetts Inst. of Technol. (B. F. Burke, K. Y. Lo, and G. D. Papadopoulos), Naval Res. Lab. (K. J. Johnston and S. H. Knowles), dan Smithsonian Astrophys. Observatory (J. M. Moran).]

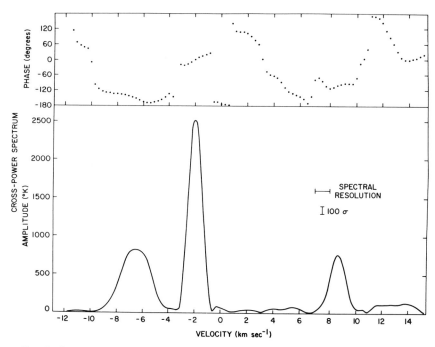

FIG. 5. Cross-power spectrum of the H_2O source in W49 N obtained, on November 12, 1971 at 23^h GMT, with the interferometer formed by the 120-ft antenna of the Haystack Observatory and the 42.7-m (140-ft) antenna of the NRAO. Since the fringe spacing was 0.003 arc sec and the integration time was 32 sec, none of the features was separated in fringe frequency. This spectrum was made from part of the data used in Fig. 4.

If the bandpass function $G_1(\omega_k)G_2(\omega_k) \exp[j\,\Delta\psi(\omega_k)]$ is unity, then

$$R'_{12}(\tau, \omega_f) = (T_{A_1}T_{A_2})^{1/2} \exp\{j[\Delta\phi - \omega_2 \tau' - \pi B_0(\tau + \tau')]\} \frac{\sin \pi B_0(\tau + \tau')}{\pi B_0(\tau + \tau')},$$
$$(5.5.16)$$

where $\tau' = \tau_g - \tau_{g_0} + \Delta\tau_c$. Hence, the phase at the peak of the correlation function ($\tau = -\tau'$) is $\Delta\phi - \omega_2 \tau'$, while the phase at $\tau = 0$ is

$$\Phi'(\tau = 0) = \tan^{-1}[\text{Im}\{R(0)\}/\text{Re}\{R(0)\}] = -\omega_m(\tau_g - \tau_{g_0} + \Delta\tau_c) + \Delta\phi,$$
$$(5.5.17)$$

where Re{ } and Im{ } denote the real and imaginary parts of the correlation function, and ω_m is the center of the rf band, i.e., $\omega_m = \omega_2 + \pi B_0$.

It is important to note that $R'_{12}(\tau, \omega_f)$, obtained by a double Fourier transformation in Eqs. (5.5.13) and (5.5.15), is not the same function as $R_{12}(\tau, \omega_f)$, obtained by simply calculating the fringe-frequency spectrum of the correlation functions $\rho_{12}(\tau, t)$. Both $R_{12}(\tau, \omega_l)$ and $R'_{12}(\tau, \omega_l)$ have the same functional form, except for the delay-smearing of the former caused by the discrete delay tracking and the effects of the bandpass functions. However, R_{12} and R'_{12} have different signal-to-noise ratios because the algorithm for R'_{12} filters out the negative-frequency sideband, where noise but no signal is present.

Simpler algorithms for continuum data can be used if precise results are not required. For example, the sideband-filtering operation is equivalent to convolution of the correlation function with

$$D(\tau) = \frac{\sin 2\pi B_0 \tau}{2\pi B_0 \tau} + j\frac{\sin^2 \pi B_0 \tau}{\pi B_0 \tau}. \tag{5.5.18}$$

Hence, a possible processing algorithm is

$$R'_{12}(\tau_i) = \sum_m [\rho_c(\tau_i, t_m) \otimes D'(\tau_i - \Delta\tau)] \exp[j(\Phi^* + \omega_{f_i} t_m)], \tag{5.5.19}$$

where $\otimes$ denotes convolution; D' can be a three-point approximation to D. Figure 6 is an example of an observation of 3C273 showing $\text{Re}\{R'_{12}(\tau)\}$, $\text{Im}\{R'_{12}(\tau)\}$, $|R'_{12}(\tau)|$, $S_{12}(\omega, \omega_f)$, and $\Phi(\omega, \omega_f)$ at the value of ω_f that maximized $|S_{12}|$. A more complete discussion of processing algorithms is given by Whitney.[5]

The processor is not always designed as in Fig. 1. In the configuration shown, the correlation functions must be output at intervals short with respect to the apparent fringe period. It is more convenient to place the fringe stopper ahead of the correlator in one of the signal streams, thereby reducing the apparent fringe rate so that the correlator need be output only a few times a second. In this mode, the fringe stopper breaks one of the data streams into two parts, multiplying one by $\cos \Phi^*$ and the other by $\sin \Phi^*$, where $\Phi^* = (\omega_2 - \omega_1)t - \omega_2 \tau_{go}$. Hence, two correlators are required, one producing $\text{Re}\{\rho_c(\tau)\}$ and the other, $\text{Im}\{\rho_c(\tau)\}$. The processing algorithms are the same except that the multiplication by $\exp(j\Phi^*)$ in Eqs. (5.5.13) and (5.5.19) is no longer needed.

5.5.3. Measurement of Fringe Amplitude in the Presence of Phase Noise

The effects of phase noise can be described by considering the fringe-frequency spectrum, defined in Eq. (5.5.9). In the spectral-line case, the cross power or fringe amplitude is a function of frequency and fringe rate, while in the continuum case, it is a function only of fringe rate, and the dependence on ω can be ignored. In the absence of oscillator phase noise $\Delta\phi$, the maximum of S_{12} with respect to ω_f at each frequency gives the residual fringe frequency $\omega(\tau_g - \tau_{go})$ and the fringe amplitude $A(\omega)$. The presence of phase noise shifts

[5] A. R. Whitney, Ph.D. Thesis, Elec. Eng. Dept., Massachusetts Inst. of Technol. (1974).

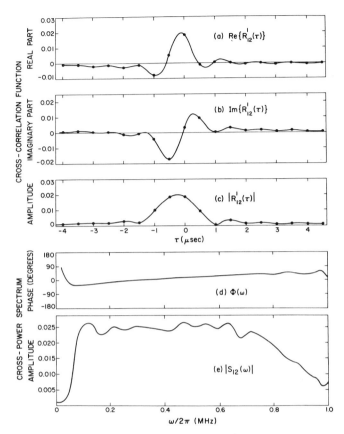

FIG. 6. The cross-correlation function [(a) real part, (b) imaginary part, (c) amplitude] and cross-power spectrum [(d) phase, (e) amplitude] observed on 3C273 at 18 cm between Haystack and NRAO with the NRAO Mark II terminals. The bandwidth was 1.0 MHz and the integration time was 200 sec. The displacement of the center of the magnitude of the cross-correlation function (c) and the slope of the phase of the cross-power spectrum (d) are equivalent indications of a delay error.

the fringe frequency of the peak of S_{12} and reduces the peak amplitude. Figure 7a presents an example of a fringe-rate spectrum calculated from a 256-sec sample of $S_{12}(\omega, t)$ on W49 N at 22 GHz.[4] Also shown, in Fig. 7b, are the rms values of the phase noise for different averaging times, which were calculated by dividing the sample of $S_{12}(\omega, t)$ into sections of length T, and fitting a straight line to the phase in each section. The rms phase errors were averaged and converted to a fractional frequency fluctuation ($\Delta f/f$) by the formula ($\Delta f/f$) = $\Delta\Phi/(fT)$, where $\Delta\Phi$ is the average rms phase error and f is the sky frequency.

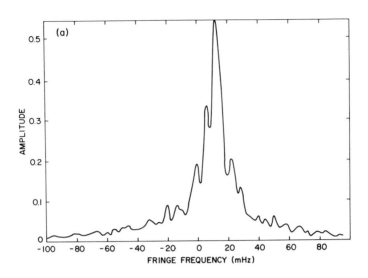

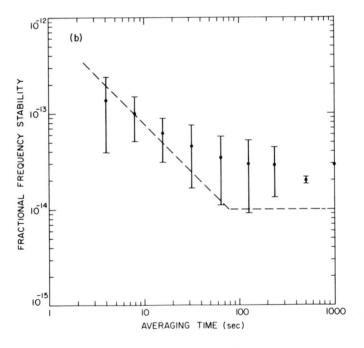

FIG. 7. (a) A fringe-frequency spectrum calculated by Fourier transforming $\exp[j\Phi(\omega, t)]$ for a 256-sec segment of data taken, on November 12, 1971 at 23^h GMT, on the -1.9-km/sec feature in the H_2O spectrum of W49 N with the Haystack–NRAO interferometer at 22,235 MHz. Hydrogen-maser frequency standards were used at both stations. (b) The fractional frequency stability of the interferometer calculated from $\Phi(\omega, t)$ for different integration times. Dashed line represents the expected laboratory stability of two hydrogen masers.

The coherence time of the interferometer, T_c, can be defined as the integration time for which the peak of $S_{12}(\omega, \omega_f)$ drops to half the true amplitude in the absence of phase noise, $A(\omega)$. Since the effect of $\Delta\phi(t)$ is to spread the fringe power over a range of fringe frequencies, $A(\omega)$ can be estimated from the fringe-frequency spectrum by using Parseval's theorem.[6] Hence,

$$A^2(\omega) = \int_{-\omega_0}^{\omega_0} |S_{12}(\omega, \omega_f)|^2 \, d\omega_f, \qquad (5.5.20)$$

where a small noise-bias term, which will be discussed later, has been ignored. The fringe-frequency window ($-\omega_0$ to ω_0) must be wide enough to pass all the fringe power, i.e., $\omega_0 > 2\pi/T_c$, and the measurements of $S_{12}(\omega, t)$ used in Eq. (5.5.9) must be made over a time interval short with respect to T_c in order for Eq. (5.5.20) to be valid. Equation (5.5.20) is a form of broken-coherence or semicoherent signal averaging.

An equivalent procedure for broken-coherence averaging is to average the squares of a series of coherent measurements of fringe amplitude each made over time T_c. The signal-to-noise ratio for the estimate of fringe amplitude can be calculated as follows. Let the fringe amplitude be a vector $\mathbf{A}$ with zero phase, i.e., $A_x = A$ and $A_y = 0$, where $A = (T_{A_1} T_{A_2})^{1/2}$ for a point source. The noise due to the signal and receivers is represented as a vector $\mathbf{n}$ whose components n_x and n_y are uncorrelated Gaussian random variables with zero means and variances, given by[7]

$$\sigma_x^2 = (T_s^2 + T_{A_1} T_{A_2})/2B\tau_c, \qquad (5.5.21)$$

$$\sigma_y^2 = (T_s^2 - T_{A_1} T_{A_2})/2B\tau_c, \qquad (5.5.22)$$

where $T_s^2 = (T_{A_1} + T_{R_1})(T_{A_2} + T_{R_2})$. The terms σ_x and σ_y should be multiplied by $\pi/2$ if one-bit digital recording is used,[8] as well as by a factor whose magnitude is about unity and whose exact value depends on the shape of the filters.[9] In the spectral-line case, B is the resolution and T_{A_1} and T_{A_2} are functions of frequency. In the continuum case, B is equal to the bandwidth B_0, and T_{A_1} and T_{A_2} are the average antenna temperatures. The observed quantity is $\mathbf{Z} = \mathbf{A} + \mathbf{n}$, whose components have the probability distribution

$$p(Z_x, Z_y) = \frac{1}{2\pi\sigma_x \sigma_y} \exp\left[-\frac{(Z_x - A)^2}{2\sigma_x^2} - \frac{Z_y^2}{2\sigma_y^2}\right]. \qquad (5.5.23)$$

When $T_{A_1} T_{A_2} \gg T_{R_1} T_{R_2}/2B\tau_c$, the variances of the measured amplitude $|\mathbf{Z}|$ and the phase $\varepsilon = \tan^{-1}(Z_y/Z_x)$ are

$$\sigma_Z^2 \sim \sigma_x^2 = (T_s^2 + T_{A_1} T_{A_2})/2B\tau_c, \qquad (5.5.24)$$

[6] B. G. Clark, K. I. Kellermann, C. C. Bare, M. H. Cohen, and D. L. Jauncey, *Astrophys. J.* 153, 705 (1968).

[7] A. E. E. Rogers, *Radio Sci.* 5, 1239 (1970).

[8] W. R. Burns and S. S. Yao, *Radio Sci.* 4, 431 (1969).

[9] M. E. Tiuri, *IEEE Trans. Antennas Propag.* AP-12, 930 (1964).

$$\sigma_\varepsilon^2 \sim \sigma_y^2/A^2 = (T_s^2 - T_{A_1}T_{A_2})/2B\tau_c T_{A_1}T_{A_2}. \qquad (5.5.25)$$

Note that as $(T_{A_1}T_{A_2})$ increases, such that $(T_{A_1}T_{A_2}) \gg (T_{R_1}T_{R_2})$, $\sigma_Z^2/(T_{A_1}T_{A_2})$ tends to a constant value $(2B\tau_c)^{-1}$, while the variance of the phase σ_ε^2 continues to decrease, i.e.,

$$\sigma_\varepsilon^2 \to [(T_{R_1}/T_{A_1}) + (T_{R_2}/T_{A_2})]/2B\tau_c. \qquad (5.5.26)$$

The problem of measuring low-intensity signals can be discussed by assuming that $T_A \ll T_R$, so that $\sigma_x^2 \sim \sigma_y^2 \sim T_R/2B\tau_c \equiv \sigma^2$, where $T_A = (T_{A_1}T_{A_2})^{1/2}$ and $T_R = (T_{R_1}T_{R_2})^{1/2}$. In this case, the probability distributions of Z and ε become[10]

$$p(Z) = \frac{Z}{\sigma^2} \exp\left(-\frac{Z^2 + A^2}{2\sigma^2}\right) I_0\left(\frac{ZA}{\sigma^2}\right), \qquad Z > 0, \qquad (5.5.27)$$

$$p(\varepsilon) = \frac{1}{2\pi} \exp\left(-\frac{A^2}{2\sigma^2}\right)\left\{1 + \left(\frac{\pi}{2}\right)^{1/2} \frac{A\cos\varepsilon}{\sigma} \exp\left(\frac{A^2\cos^2\varepsilon}{2\sigma^2}\right)\right.$$

$$\times \left.\left[1 + \mathrm{Erf}\left(\frac{A\cos\varepsilon}{\sqrt{2}\sigma}\right)\right]\right\}, \qquad (5.5.28)$$

where $I_0(\ )$ is the modified Bessel function of order zero given by

$$I_0(x) = (1/2\pi)\int_0^{2\pi} \exp(x\cos\phi)\,d\phi, \qquad (5.5.29)$$

and $\mathrm{Erf}(\ )$ is the error function given by

$$\mathrm{Erf}(x) = (2/\sqrt{\pi})\int_0^x \exp(-\phi^2)\,d\phi. \qquad (5.5.30)$$

The expectations of Z, Z^2, and Z^4 are

$$\langle Z \rangle = \left(\frac{\pi}{2}\right)^{1/2} \sigma \exp\left(\frac{-A^2}{4\sigma^2}\right)\left[\left(1 + \frac{A^2}{2\sigma^2}\right)I_0\left(\frac{A^2}{4\sigma^2}\right) + \frac{A^2}{2\sigma^2}I_1\left(\frac{A^2}{4\sigma^2}\right)\right], \quad (5.5.31)$$

$$\langle Z^2 \rangle = A^2 + 2\sigma^2, \qquad (5.5.32)$$

and

$$\langle Z^4 \rangle = A^4 + 8\sigma^2 A^2 + 8\sigma^4, \qquad (5.5.33)$$

where $I_1(\ )$ is the modified Bessel function of order 1 and $\langle\ \rangle$ denotes expectation. The probability distributions $p(Z)$ and $p(\varepsilon)$ are plotted in Figs. 8a and 8b for several values of A/σ; $p(Z)$ is sometimes called the Rice

[10] A. Papoulis, "Probability, Random Variables and Stochastic Processes," p. 196. McGraw-Hill, New York, 1965.

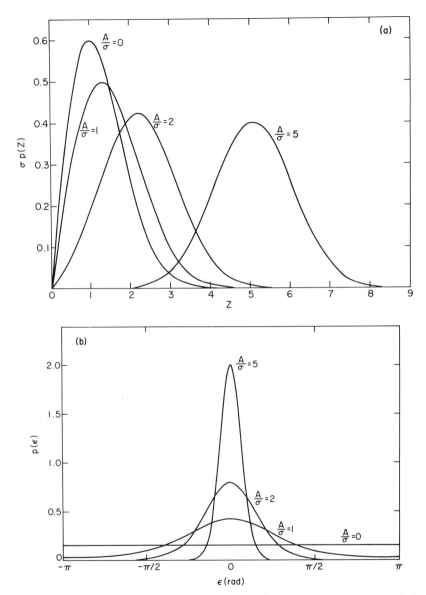

FIG. 8. (a) Probability distributions of fringe amplitude $p(Z)$ for various values of A/σ. Curves are defined in Eq. (5.5.27). (b) Probability distributions of phase $(p\varepsilon)$ for various values of A/σ. Curves are defined in Eq. (5.5.28).

distribution. In the "strong"signal case, $A \gg \sigma$, the distributions of Z and ε approach Gaussian distribution:

$$p(Z) \cong \frac{1}{(2\pi)^{1/2}} \frac{1}{\sigma} \left(\frac{Z}{A}\right)^{1/2} \exp\left[-\frac{(Z-A)^2}{2\sigma^2}\right], \qquad (5.5.34)$$

$$p(\varepsilon) \cong \frac{1}{(2\pi)^{1/2}} \frac{A}{\sigma} \exp\left(-\frac{A^2\varepsilon^2}{2\sigma^2}\right). \qquad (5.5.35)$$

Hence,

$$\langle Z \rangle \cong A[1 + (\sigma^2/2A^2)], \qquad (5.5.36)$$

$$\sigma_Z = (\langle Z^2 \rangle - \langle Z \rangle^2)^{1/2} \cong \sigma, \qquad (5.5.37)$$

$$\langle \varepsilon \rangle = 0, \qquad (5.5.38)$$

and

$$\sigma_\varepsilon \cong \sigma/A, \qquad (5.5.39)$$

In the "weak" signal case, $A \ll \sigma$,

$$p(Z) \cong \frac{Z}{\sigma^2} \exp\left(-\frac{Z^2}{2\sigma^2}\right) \left[1 - \frac{1}{2}\frac{A^2}{\sigma^2} + \frac{1}{4}\left(\frac{ZA}{\sigma^2}\right)^2\right], \qquad (5.5.40)$$

$$p(\varepsilon) \cong \frac{1}{2\pi} + \frac{1}{(2\pi)^{1/2}} \frac{A}{\sigma} \cos \varepsilon. \qquad (5.5.41)$$

Hence,

$$\langle Z \rangle \cong \left(\frac{\pi}{2}\right)^{1/2} \sigma\left(1 + \frac{A^2}{4\sigma^2}\right), \qquad (5.5.42)$$

$$\sigma_Z \cong \sigma\left(2 - \frac{\pi}{2}\right)^{1/2} \left(1 + \frac{A^2}{4\sigma^2}\right), \qquad (5.5.43)$$

$$\langle \varepsilon \rangle = 0, \qquad (5.5.44)$$

and

$$\sigma_\varepsilon \cong \frac{\pi}{\sqrt{3}} \left[1 - \left(\frac{9}{2\pi^3}\right)^{1/2} \frac{A}{\sigma}\right]. \qquad (5.5.45)$$

The noise-to-signal ratio σ_Z/A and σ_ε are plotted versus A/σ in Fig. 9. The signal suppression evident in the expression for $\langle Z \rangle$ in the weak-signal case [Eq. (5.5.42)] means that in the spectral-line case, the presence of a spectral feature whose width exceeds the frequency resolution can be detected in a spectrum more easily by inspection of the fringe phase than by inspection of the fringe amplitude.

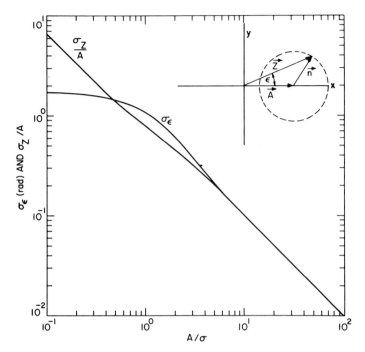

FIG. 9. σ_ε, the standard deviation of the fringe phase in radians, and σ_Z/A, the standard deviation of the measured fringe amplitude divided by the signal amplitude, versus A/σ, where $\sigma = T_R(2B\tau)^{-1}$, T_R is the geometric mean receiver temperatures, B the bandwidth, τ_c the integration time for the case $T_A \ll T_R$, and T_A the geometric mean antenna temperature. Curves are from Eqs. (5.5.37), (5.5.39), (5.5.43), and (5.5.45). σ_ε is always less than $\pi/\sqrt{3}$ since ε ranges between $-\pi$ and π. Also shown is a diagram of the signal and noise vectors.

Consider now a series of N measurements of Z, each made over a time period τ_c short with respect to the coherence time of the interferometer. The likelihood function[11] of a set of measurements Z is

$$L = \prod_{i=1}^{N} p(Z_i), \qquad (5.5.46)$$

where $\prod$ denotes the product and $p(Z)$ is given by Eq. (5.5.27). If σ is known from measurements where no signal is present, then the maximum of L with respect to A is obtained when

$$A - \frac{1}{N}\sum_{i=0}^{N} Z_i \frac{I_1(Z_i A/\sigma^2)}{I_0(Z_i A/\sigma^2)} = 0, \qquad (5.5.47)$$

[11] H. L. Van Trees, "Detection, Estimation and Modulation Theory." Wiley, New York, 1968.

where I_1 is a modified Bessel function of order 1, which leads to an approximate estimate of A^2, denoted A_e^2, of

$$A_e^2 = \left[1/N \sum_{i=1}^{N} Z_i^2 \right] - 2\sigma^2 \qquad (5.5.48)$$

in the weak-signal case. The $2\sigma^2$ term represents the bias arising from the fact that the fringe amplitudes are always positive. In the strong-signal case, where $p(Z)$ is approximately Gaussian, the maximum-likelihood estimate of A is the average of the Z_i's. However, since the use of the optimum estimator is not essential in the strong-signal case, it is helpful here to consider the estimate given by Eq. (5.5.48) for both strong- and weak-signal cases. Equation (5.5.48) gives an unbiased estimate of A^2 for both cases. From Eqs. (5.5.32), (5.5.33), and (5.5.48), we have $\langle A_e^2 \rangle = A^2$ and $\langle A_e^4 \rangle = A^4 + 4\sigma^2(A^2 + \sigma^2)/N$, so that the signal-to-noise ratio (SNR) is

$$\text{SNR} = \frac{\langle A_e^2 \rangle}{(\langle A_e^4 \rangle - \langle A_e^2 \rangle^2)^{1/2}} = \frac{\sqrt{N} A^2}{2\sigma^2} \left(1 + \frac{A^2}{\sigma^2} \right)^{-1/2}. \qquad (5.5.49)$$

The coherent integration time for the sample measurements τ_c defines an equivalent fringe-filter width $B_f = \tau_c^{-1}$. The total integration time is then $\tau = N/B_f$, and hence, $\sigma = T_R(B_f/2B)^{1/2}$. Recalling that $A = T_A = (T_{A_1} T_{A_2})^{1/2}$ allows Eq. (5.5.49) to be written as

$$\text{SNR} = \frac{T_A^2}{T_R^2} \left[\left(\frac{B^2 \tau}{B_f} \right) \Big/ \left(1 + \frac{2 T_A^2 B}{T_R^2 B_f} \right) \right]^{1/2}, \qquad (5.5.50)$$

which reduces to

$$\text{SNR} \cong \frac{1}{\sqrt{2}} \frac{T_A}{T_R} (B\tau)^{1/2}, \qquad T_A \gg T_R/(2B\tau)^{1/2}, \qquad (5.5.51)$$

and

$$\text{SNR} \cong \left(\frac{T_A}{T_R} \right)^2 \left[B\tau \left(\frac{B}{B_f} \right) \right]^{1/2}, \qquad T_A \ll T_R/(2B\tau)^{1/2}, \qquad (5.5.52)$$

where $T_A \ll T_R$. Thus, in the strong-signal case, the signal-to-noise ratio is the same as the one that would have been achieved by coherent averaging except for a factor of 2, which can be regained by use of a more optimum estimator of A. In general, B_f should be made as narrow as possible while still passing all the fringe power. If it is necessary to increase B_f to equal B, then the system is equivalent to the Hanbury Brown–Twiss interferometer.[12] This was recognized by Clark.[13]

[12] R. Hanbury Brown and R. Q. Twiss, *Phil. Mag.* **45**, 663 (1954).
[13] B. G. Clark, *IEEE Trans. Antennas Propag.* **AP-16**, 143 (1968).

If we assume a 4:1 signal-to-noise ratio for detection, then we see from Eq. (5.5.52) that the minimum detectable fringe amplitude is

$$T_A(\min) = 2T_R N^{-1/4}(B\tau_c)^{-1/2}. \tag{5.5.53}$$

Hence, incoherent averaging is not very effective in lowering the minimum detectable fringe amplitude, because of the $N^{-1/4}$ dependence.

5.5.4. Misidentification of Signal

It is often necessary to search a wide range of fringe rate or delay, or both, in order to find interference fringes. This is usually done by scanning the array of measured values of $S_{12}(\omega, \omega_f)$ to find the largest one. The exact value of fringe rate can be found by interpolation. In the absence of signal, the measurements will be Rayleigh distributed:

$$p(Z | A = 0) = (Z/\sigma^2) \exp(-Z^2/2\sigma^2), \qquad Z \geq 0. \tag{5.5.54}$$

If q is the largest of n independent samples of Z with this distribution, it will have the distribution

$$p(q | A = 0) = (nq/\sigma^2)[1 - \exp(-q^2/2\sigma^2)]^{n-1} \exp(-q^2/2\sigma^2). \tag{5.5.55}$$

Figure 10 shows $p(q | A = 0)$, while $\langle q \rangle$ and σ_q are listed in Table I for various values of n. As expected, the mean of q increases with n, while the rms deviation decreases.

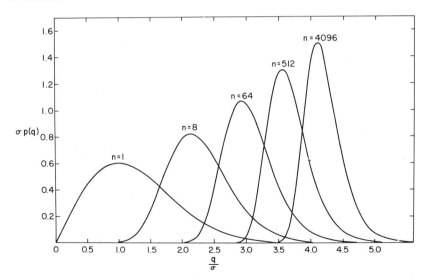

FIG. 10. Probability distribution of q, the largest of n independent Rayleigh-distributed random variables, defined by Eq. (5.5.55). Table I gives values of $\langle q \rangle$ and σ_q.

TABLE I.   Mean and Standard Deviation of the Largest of n
Rayleigh-Distributed Random Variables for $\sigma = 1$

n	$\langle q \rangle$	σ_q	n	$\langle q \rangle$	σ_q
1	1.253	0.655	256	3.482	0.353
4	1.964	0.558	1024	3.862	0.321
16	2.559	0.463	4096	4.207	0.297
64	3.054	0.398	8192	4.369	0.289

If there is signal present, then the sample at the fringe rate of the signal will have the distribution $p(Z)$ given by Eq. (5.5.27). The probability that one or more of the noise samples will exceed the amplitude of the signal sample is PE, given by

$$PE = 1 - \int_0^\infty p(Z)[1 - \exp(-Z^2/2\sigma^2)]^{n-1} \, dz. \qquad (5.5.56)$$

This probability of error is plotted in Fig. 11 as a function of A/σ and n. A rather large value of A/σ is needed to ensure that a noise peak is not mis-identified as the signal. For example, if $n = 50$, A/σ should be greater than 5.5 to make $PE \leq 10^{-3}$.

5.5.5. Measurement of Source Brightness Distribution

In this section, the problem of measuring the brightness distribution of a discrete radio source is discussed. The related problem of determining the exact position of the radio source will not be dealt with here, since it requires a careful analysis of the atmosphere, the baseline parameters, and many fine details, such as the retarded-baseline term τ_B in Eq. (5.5.1).

The four measurable quantities available from an interferometer as a function of frequency, hour angle, and polarization are the fringe amplitude (i.e., the correlated flux density), the fringe phase (corrupted by the noise from the local oscillator and the atmosphere), the rate of change of fringe phase with time (i.e., the fringe frequency), and the delay (i.e., the rate of change of phase with frequency). By rewriting Eq. (5.5.7), the interferometer response to a point source becomes

$$S_{12}(\omega, t) = A(\omega) \exp[j\omega(\tau_g - \tau_{g_0})] = A(\omega) \exp[j\,\Delta\Phi(\omega, t)], \quad (5.5.57)$$

where $A(\omega)$ is the fringe amplitude in units of antenna temperature, brightness temperature, or, most properly, flux density, and $\Delta\Phi$ is the residual fringe phase. The instrumental gain and phase terms have been ignored. The reference position used in the data processing to calculate τ_{g_0} is presumably near the centroid of the source. The Taylor expansion of τ_g is

$$\tau_g \approx \tau_{g_0} + [\partial\tau_{g_0}/\partial(\alpha \cos \delta_S)](\alpha \cos \delta_S - \alpha_S \cos \delta_S) + (\partial\tau_g/\partial\delta)(\delta - \delta_S), \quad (5.5.58)$$

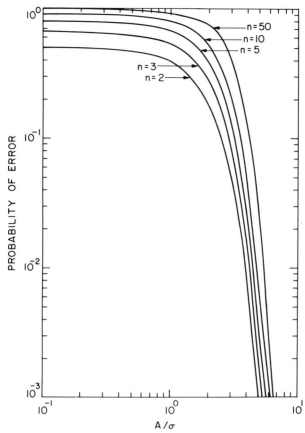

FIG. 11. Probability that the amplitude of one or more samples in the n-point fringe-frequency spectrum will exceed that of the sample at the fringe frequency of the signal versus A/σ, where A is the signal amplitude and σ is the standard deviation of the components of the noise vector. Curves are defined by Eq. (5.5.56).

where α_S and δ_S are the right ascension and declination of the reference position. It is useful to define the offset angles

$$\theta_x = \alpha \cos \delta_S - \alpha_S \cos \delta_S, \qquad \theta_y = \delta - \delta_S. \qquad (5.5.59)$$

Hence, from Eqs. (5.5.1) and (5.5.58), the phase $\Delta\Phi$, the delay $\Delta\tau = \tau_g - \tau_{g0}$, and the fringe frequency $\Delta\dot{\Phi}$ due to a position offset are

$$\Delta\Phi = K_\Phi[(\cos \delta_B \sin L)\theta_x + (\sin \delta_B \cos \delta_S - \cos \delta_B \sin \delta_S \cos L)\theta_y], \qquad (5.5.60)$$

$$\Delta\tau = K_\tau[(\cos \delta_B \sin L)\theta_x + (\sin \delta_B \cos \delta_S - \cos \delta_B \sin \delta_S \cos L)\theta_y], \qquad (5.5.61)$$

$$\Delta\dot{\Phi} = K_{\dot{\Phi}}[(\cos L)\theta_x + (\sin \delta_S \sin L)\theta_y], \qquad (5.5.62)$$

where $K_\Phi = \omega D/c$, $K_\tau = D/c$, $K_{\dot\Phi} = \omega D \cos \delta_B \Omega/c$, and $L = L_S - L_B$. The rotation frequency of the earth is designated by Ω. The uncertainties in the measured values of these quantities due to the receiver noise can be calculated from Eq. (5.5.39). The rms phase error in the case of a large signal-to-noise ratio is

$$\sigma_\Phi = \sigma_\varepsilon \cong T_R/T_A(2B\tau_c)^{1/2}. \qquad (5.5.63)$$

Since $\Delta\tau = \partial \, \Delta\Phi/\partial\omega$, the rms delay error, which can be thought of as the error in determining the slope of phase versus frequency, is

$$\sigma_\tau = \sqrt{12}\sigma_\varepsilon/2\pi B = (\sqrt{6}/2\pi)T_R\, T_A^{-1}B^{-3/2}\tau_c^{-1/2}. \qquad (5.5.64)$$

Similarly, the rms fringe-frequency error, the error in determining the slope of phase versus time, is

$$\sigma_{\dot\Phi} = \sqrt{12}\sigma_\varepsilon/\tau_c = \sqrt{6}T_R\, T_A^{-1}B^{-1/2}\tau_c^{-3/2}. \qquad (5.5.65)$$

The sensitivities of phase, delay, and fringe rate to a position offset $\Delta\theta$ are neglecting geometric factors,

$$\Delta\theta \text{ (phase)} = \sigma_\Phi/K_\Phi \sim (\sigma_\Phi/2\pi)(\lambda/D), \qquad (5.5.66)$$

$$\Delta\theta \text{ (delay)} = \sigma_\tau/K_\tau \sim \sqrt{12}(\sigma_\Phi/2\pi)(f/B)(\lambda/D), \qquad (5.5.67)$$

$$\Delta\theta \text{ (fringe frequency)} = \sigma_{\dot\Phi}/K_{\dot\Phi} = \sqrt{12}(\sigma_\Phi/2\pi)(\tau_0/2\pi\tau_c)(\lambda/D_H), \qquad (5.5.68)$$

where $\lambda = 2\pi c/\omega$, $D_H = D \cos \delta_B$, $f = \omega/2\pi$, and $\tau_0 = 2\pi/\Omega$. Hence, delay and fringe frequency are very much less sensitive to position offsets (by factors of B/f and τ_c/τ_0, respectively) than is phase, but they can be used together to determine coarse position offsets. Equations (5.5.61) and (5.5.62) show that the directions of maximum sensitivity of $\Delta\tau$ and $\Delta\Phi$ are nearly orthogonal over large parts of the sky, with the result that a source position can be determined from a single observation.[14] It has been suggested that a relative map of a field of radio sources could be made by analyzing the cross correlation as a function of delay and fringe frequency.

Phase data are difficult to use directly because of the phase ambiguities (since phase is modulo 2π), and because of the phase noise due to the local oscillators and the atmosphere. However, phase data can be used effectively for relative position measurements. For example, if two antennas are available at each site and observe two radio sources separated by a small angle, then each of the two long baseline interferometers observes one of the sources. Since the local oscillators at each site are derived from the same frequency standard, and since the atmospheric delay will be about the same for both sources, the relative phase, i.e., the phase measured by interferometer #1 on

[14] M. H. Cohen, A. T. Moffet, D. Shaffer, B. G. Clark, K. I. Kellermann, D. L. Jauncey, and S. Gulkis, *Astrophys. J. Lett.* **158**, L83 (1969).

source 1 minus that measured by interferometer #2 on source 2, is approximately free of instrumental effects. When measured as a function of hour angle, this relative phase can be used to determine the angular separation between the sources to very high accuracy, i.e., that accuracy imposed by Eq. (5.5.66).

Another example of the use of fringe-phase data is provided by the case of spectral-line emission from molecules such as OH and H_2O in galactic HII regions. The problem is to measure the relative positions among the various velocity features in a source such as W49 N, for which a sample spectrum is shown in Fig. 5. The coherent integration time can be extended far beyond the coherence time of the interferometer by using the phase of a strong feature as a reference to remove the instrumental phase. In the two-dimensional fringe-frequency and frequency spectrum resulting from 1000 sec of integration shown in Fig. 4 for W49 N, the feature at -1.9 km/sec was used as a phase reference. The source is assumed to be composed of a collection of unresolved, spatially separated point sources. To begin with, assume that only one source is present at each frequency. The fringe frequency of each spectral feature is found by locating the peak in the fringe-frequency spectrum. If N measurements are taken over a wide range in hour angle, then the relative position for each feature, given by θ_x and θ_y, can be determined unambiguously by least-mean-squares fitting of the data to Eq. (5.5.62). The error expected in the absence of correlation between parameters can be found from Eqs. (5.5.62) and (5.5.65) to be

$$\sigma_{\theta_x} = \sigma_{\dot{\phi}}/\pi(D/\lambda)\Omega \cos \delta_B (2N)^{1/2}, \qquad (5.5.69)$$

$$\sigma_{\theta_y} = \sigma_{\dot{\phi}}/\pi(D/\lambda)\Omega \cos \delta_B \sin \delta_S (2N)^{1/2}. \qquad (5.5.70)$$

When there is more than one feature at a given frequency, their hour-angle dependence can be analyzed separately. Source-distribution maps from fringe-frequency data have been made by Moran et al.,[15,16] Johnston et al.,[17] and others. In practice, the fringe-frequency spectrum defined in Eq. (5.5.9) must be wide enough to include all the source features. Also, the signal-to-noise ratio must be high enough so that a noise peak is not mistaken for the signal. This probability of error, given in Section 5.5.4, increases with the width of the fringe-frequency window. Because of the finite range of the fringe-frequency spectrum, which limits the range of the possible measured values of the fringe frequency, it is possible to obtain quite respectable position errors for nonexistent features.

[15] J. M. Moran, B. F. Burke, A. H. Barrett, A. E. E. Rogers, J. A. Ball, J. C. Carter, and D. D. Cudaback, *Astrophys. J. Lett.* **152**, L97 (1968).
[16] J. M. Moran et al., *Astrophys. J.* **185**, 535 (1973).
[17] K. J Johnston et al., *Astrophys. J. Lett.* **166**, L21 (1971).

After the coarse fringe-rate map has been constructed, a more accurate one can be made from the phase data in some circumstances. The difficulty in using the phase data arises from the fact that with noise-corrupted data, two or more positions for a feature may be equally probable if the data are not sampled frequently enough. Moran[18] has shown that if the average interval in the data points along the uv-plane locus is less than $(2\sigma_\theta)^{-1}$, where σ_θ is the position error from fringe rate measurements, then the position can be determined unambiguously. The error in the relative position will then be

$$\sigma'_{\theta_x} \approx \sigma_\Phi/(2N)^{1/2}\pi(D/\lambda)\cos\delta_B,\tag{5.5.71}$$

$$\sigma'_{\theta_y} \approx \sigma_\Phi/(2N)^{1/2}\pi(D/\lambda)(2\sin^2\delta_B\cos^2\delta_S + \cos^2\delta_B\sin^2\delta_S)^{1/2}.\tag{5.5.72}$$

The structure of extended sources that are not collections of point sources can also be studied but, in general, cannot be determined uniquely by VLBI techniques. The response to a point source at frequency ω and offset position θ_x and θ_y is, from Eqs. (5.5.57) and (5.5.58),

$$S_{12}(u, v) = A(\theta_x, \theta_y)\exp[j2\pi(u\theta_x + v\theta_y)],\tag{5.5.73}$$

where

$$u = (\omega/2\pi)[\partial\tau_g/\partial(\alpha\cos\delta)],\tag{5.5.74}$$

$$v = (\omega/2\pi)(\partial\tau_g/\partial\delta),\tag{5.5.75}$$

which become

$$u = w\cos\delta_B\sin L,\tag{5.5.76}$$

$$v = w(\sin\delta_B\cos\delta_S - \cos\delta_B\sin\delta_S\cos L),\tag{5.5.77}$$

where $w = D\omega/2\pi c = D/\lambda$. The response to an extended but discrete source that is smaller than the individual antenna beam is

$$S_{12}(u, v) = \iint A(\theta_x, \theta_y)\exp[j2\pi(u\theta_x + v\theta_y)]\,d\theta_x\,d\theta_y.\tag{5.5.78}$$

This is the Fourier-transform relation, which is the basis of aperture-synthesis interferometry. The terms u and v are the projected baseline lengths in the θ_x and θ_y directions, respectively, in units of wavelength. In the Fourier-transform integral, it is useful to think of u and v as spatial frequencies with units of cycles per radian. If r and ψ are the polar coordinates associated with θ_x and θ_y, and if $A(\omega, r, \psi) = A(\omega, r, -\psi)$, then S_{12} will be a real function. Furthermore, if the source is circularly symmetric, $S_{12}(s)$ and $A(r)$ are Hankel transforms[19] of one another:

$$S_{12}(s) = 2\pi\int A(r)J_0(2\pi sr)r\,dr, \qquad A(r) = 2\pi\int S_{12}(s)J_0(2\pi sr)s\,ds,\tag{5.5.79}$$

where $s = (u^2 + v^2)^{1/2}$, and J_0 is the zeroth-order Bessel function.

[18] J. M. Moran, *Proc. IEEE* **61**, 1236 (1973).

[19] R. N. Bracewell, "The Fourier Transform and Its Applications," p. 244. McGraw-Hill, New York, 1965

Three conditions must be met for Eq. (5.5.78) to be valid. First, the source must be incoherent, so that S_{12} can be written as the summation of independent radiators. If the source has spatial-coherence properties, as may well be the case with molecular maser sources and pulsars, the synthesized image will be an apparent image, which will generally be smaller than the true image, i.e., the true power distribution in the source. Second, the source must be in the far field of the interferometer; if it is not, additional phase corrections must be made for the fact that the rays reaching each antenna are not parallel. That is, the distance to the source, R, must satisfy the condition $R \gg D^2/\lambda$, which, for example, with $D = 10^9$ cm and $\lambda = 1$ cm, gives $R \gg 0.3$ parsec. Finally, the source must be small enough that the first-order expansion of the delay in Eq. (5.5.58) is accurate. More details of the aperture-synthesis problem can be found in Born and Wolf[20] and Fomalont.[21]

Two problems limit the usefulness of the aperture-synthesis approach to data analysis. First, the uv-plane is poorly sampled. Ideally, one would like measurements uniformly spaced on the uv-plane, with the interval given by the spatial sampling theorem, i.e., $\Delta s = 1/\Delta\theta$, where $\Delta\theta$ is the angular extent of the source.[22] The effect of limited sampling of the uv-plane can be seen by examining the interferometer's equivalent antenna or response pattern, which is defined as

$$P(\theta_x, \theta_y) = 1/N \sum_i \cos 2\pi(u_i \theta_x + v_i \theta_y). \tag{5.5.80}$$

The response pattern of a two-element interferometer will have objectionably large sidelobes. In a multiantenna experiment, the uv-plane coverage is much improved, since if K antennas are involved, $K(K - 1)/2$ baselines are sampled in one observation (see Section 5.2.3). Figure 12 shows the projected baseline loci for an interferometer observing a source at $40°$ declination with the following five antennas: the Owens Valley Radio Observatory (OVRO) 39.6-m (130-ft) in Big Pine, California; the NRAO 42.7-m (140-ft) in Green Bank, West Virginia; the Haystack Observatory 36.6-m (120-ft) in Tyngsboro, Massachusetts; the Maryland Point Observatory 25.9-m (85-ft) at Maryland Point, Maryland; and the Harvard College Observatory 25.9-m (85-ft) in Fort Davis, Texas. Here, aperture synthesis becomes a possibility. Figure 13 shows the response pattern given by Eq. (5.5.80) in the vertical and horizontal planes for a set of data collected on the baselines in Fig. 12. The second problem is that phase information is not generally available. To obtain usable phase information would require phase stability over a whole day,

[20] M. Born and E. Wolf, " Principles of Optics," 3rd ed., Chapter 10. Pergamon, Oxford, 1965.
[21] E. B. Fomalont, *Proc. IEEE* **61**, 1211 (1973).
[22] R. N. Bracewell, *Proc. IRE* **46**, 97 (1958).

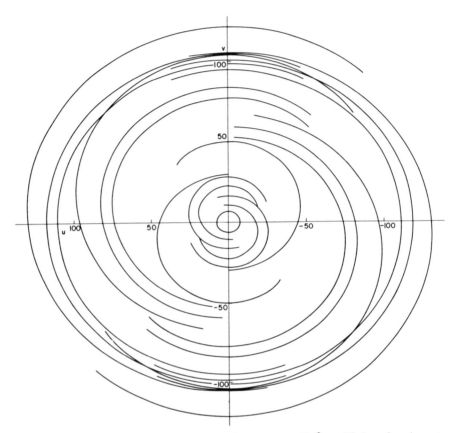

FIG. 12. The projected baseline coverage, for a source with $\delta_S = 40°$, by a five-element interferometer at a frequency of 30 GHz. The conjugate loci are not shown. Coverage is shown for all times when the source is above 20° elevation angle. Stations: Haystack Observatory, NRAO, OVRO, Harvard College Observatory field station, and the Maryland Point Observatory.

which is very difficult. Some phase information, called "closure phase," is available when more than two antennas are used, because of the requirement that the algebraic sum of the fringe phases among baselines, which vectorially add to zero, is zero for a point source. The utilization of these closure-phase data is described by Rogers.[23]

The general problem posed by the lack of phase information is that a whole class of source distributions has the same fringe amplitude but different phase functions. To see this, note that $|S_{12}(u, v)|^2$ from Eq. (5.5.78) is

$$|S_{12}(\mathbf{s})|^2 = \int A(\mathbf{\theta}' - \mathbf{\theta}) \exp[j2\pi\mathbf{s} \cdot (\mathbf{\theta}' - \mathbf{\theta})] \, d\mathbf{\theta}', \qquad (5.5.81)$$

[23] A. E. E. Rogers et al., Astrophys. J. 193, 293 (1974).

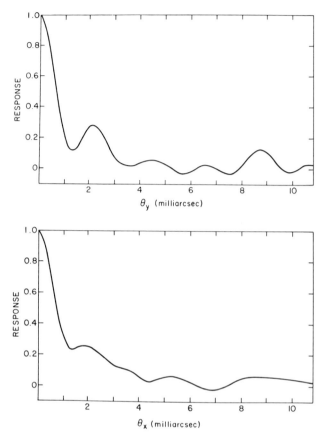

FIG. 13. Response pattern, from Eq. (5.5.80), at position angles 0° and 90° for the interferometer system with baselines shown in Fig. 12; $\lambda = 1$ cm.

where $\boldsymbol{\theta} = \theta_x \hat{\mathbf{x}} + \theta_y \hat{\mathbf{y}}$ and $\mathbf{s} = u\hat{\mathbf{x}} + v\hat{\mathbf{y}}$. Hence, all distributions with the same spatial autocorrelation function are indistinguishable without phase information. For example, if the θ_x and θ_y axes are reversed, the fringe amplitude remains unchanged. However, Bates[24] has shown that often there are only a small number of realizable brightness distributions that can produce the same fringe-amplitude distribution.

The absence of phase data and the poor sampling of the uv-plane have led most workers to fit model brightness distributions to the measured fringe-amplitude data. If only a few data points are available, then only the second moment or effective size of the source can be determined. In Fig. 14, the fringe amplitude of a uniformly bright disk, a Gaussian disk, and a ring are

[24] R. H. T. Bates, *Mon. Not. Roy. Astron. Soc.* **142**, 413 (1969).

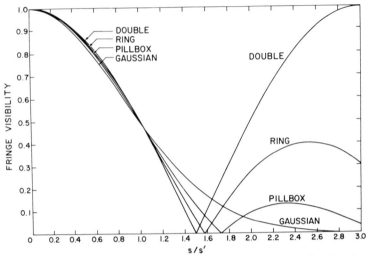

FIG. 14. Fringe visibility versus s/s', where s is the projected baseline length, and s' is the projected baseline length at which the visibility is 0.5 for each source model. (For the double-source model, the direction of the projected baseline is parallel to the vector joining the components.) The visibilities are substantially different only for $s/s' > 1$. The diameters of the models are pillbox (uniformly bright disk), $0.705/s'$; ring, $0.483/s'$; Gaussian-tapered disk (full width at half-power), $0.443/s'$; double source, separation $= 0.333/s'$.

plotted together. They are almost indistinguishable out to the first null. An example of fitting a complex model to fringe amplitude data is shown in Fig. 15.[25] The fitting of complex models is a nonlinear problem, and finding the best one is difficult. Because of the large amount of computer time required, it is not possible to find all the models that reasonably fit the data.

5.5.6. Operational Considerations

The selection of two or more antennas to be used in a VLBI experiment depends on the resolution and projected baseline coverage, as well as on other considerations. The approximate parameters of the baseline joining two antennas can be calculated from their geographic coordinates. By using the standard earth ellipsoid adopted by the International Astronomical Union in 1964,[26] the length of the vector ρ from the center of the earth to the radio telescope (see Fig. 1) can be calculated by

$$\rho = A + B \cos 2\phi + C \cos 4\phi + D \cos 6\phi + H, \qquad (5.5.82)$$

[25] T. H. Legg, N. W. Broten, D. N. Fort, J. L. Yen, F. V. Bale, P. C. Barber, and M. J. S. Quigley, *Nature (London)* **244**, 18 (1973).

[26] "The American Ephemeris and Nautical Almanac for the Year 1973," p. 543. U.S. Govt. Printing Office, Washington, D.C., 1971.

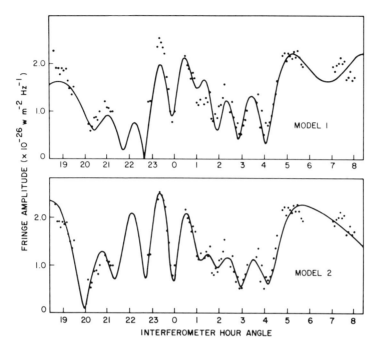

FIG. 15. Fringe amplitude versus interferometer hour angle for 3C84 at 10,680 MHz in November 1972, from the data of T. H. Legg *et al.* [*Nature (London)* **244**, 18 (1973)]. The interferometer elements were the 45.7-m (150-ft) antenna of the Algonquin Radio Observatory, Lake Traverse, Canada, and the 25.6-m (84-ft) antenna of the Radio and Space Research Station, Chilbolton, England. Two models were fitted to the data. In model 1, the source was assumed to consist of a large resolved component and four unresolved components (diameters < 0.0002 arc sec), and 12 parameters were determined: the flux and the position of each of the four components. In model 2, the source was assumed to consist of four partially resolved components of elliptical shape, and 24 parameters were determined: the flux, the major axis, the minor axis, the position angle of the major axis, and the position of each of the four components.

where $A = 6,367,489.4$ m, $B = 10,692.6$ m, $C = 22.4$ m, $D = 0.05$ m, and H is the height above mean sea level. The geodetic latitude ϕ is related to the geocentric latitude ϕ' by

$$\phi' - \phi = \alpha \sin 2\phi + \beta \sin 4\phi + \gamma \sin 6\phi, \qquad (5.5.83)$$

where $\alpha = 692.743$ arc sec, $\beta = 1.163$ arc sec, and $\gamma = 0.0026$ arc sec. The baseline vector for two telescopes with radius vectors $\boldsymbol{\rho}_1$ and $\boldsymbol{\rho}_2$ is

$$\mathbf{D} = \boldsymbol{\rho}_1 - \boldsymbol{\rho}_2, \qquad (5.5.84)$$

which has cartesian coordinates

$$D_x = \rho_1 \cos \phi_1' \cos l_1 - \rho_2 \cos \phi_2' \cos l_2, \qquad (5.5.85)$$

$$D_y = \rho_1 \cos \phi_1' \sin l_1 - \rho_2 \cos \phi_2' \sin l_2, \qquad (5.5.86)$$

$$D_z = \rho_1 \sin \phi_1' - \rho_2 \sin \phi_2', \qquad (5.5.87)$$

where l_1 and l_2 are the antenna longitudes from Greenwich. The polar co-ordinates are defined by

$$\delta_B = \tan^{-1} D_z/(D_x^2 + D_y^2)^{1/2}, \qquad (5.5.88)$$

$$L_B = \tan^{-1}(D_y/D_x), \qquad (5.5.89)$$

$$D = (D_x^2 + D_y^2 + D_z^2)^{1/2}. \qquad (5.5.90)$$

Some observers prefer to use L_1, the interferometer hour angle, instead of L_B, defined as the hour angle when the x component of the projected baseline is maximum (i.e., East–West resolution is maximum); that is,

$$L_1 = \tan^{-1}(D_x/D_y) = L_B - (\pi/2). \qquad (5.5.91)$$

Note that if the latitudes of the antennas are the same, $\delta_B = 0$; and if the longitudes are the same, $\delta_B = (\pi - \phi_1 - \phi_2)/2$.

Of particular concern when VLBI measurements are being made is whether the data recorded will actually produce interference fringes. With this type of interferometry, it is possible to "misplace" the fringes and never be able to find them. For example, if the clock delay is inadvertently set wrong at one station (by 1 sec) and this fact remains unknown, the fringes will probably never be discovered. With long-baseline interferometry, no "quick-look" capability is available to detect errors and problems. The experiment must be done correctly without reference to a sample of the final results.

A checklist of five questions is presented as a guide to help ensure that an experiment will succeed:

(1)   Is the antenna pointed at the desired radio source?
(2)   Is the signal connected all the way to the tape recorder?

These two are elementary requirements. In the clipped digital-recording system, the total-power information is lost, and there is no way of determining from the individual data tapes whether or not, in the case of continuum radio sources, the telescope was actually pointed at the source. The spectrum of the recorded signal is simply the normalized spectrum of the low-pass filter in the video converter. The gain of most clippers is so high that their output will give both 1's and 0's randomly, even when no input signal is connected.

(3)   Are all the local oscillators phase locked to a properly operating frequency standard, and is the local-oscillator frequency set correctly?

If the frequency is set wrong by 10 Hz, for example, the interference fringes will be moved in apparent fringe frequency by 10 Hz and probably never found. The local oscillators are generally offset to compensate for the natural fringe rate, especially if the rate is large with respect to the recording band-width. The fraction of the data equal to the ratio of the fringe rate to the recording bandwidth is lost. Indeed, if the fringe rate exceeds the bandwidth, the recorded signal bands would not overlap at all, and no amount of post-correlation fringe rotation can produce fringes. If a frequency synthesizer serving a local oscillator is inadvertently left locked to its internal crystal standard, instead of to the atomic frequency standard, the experiment can be seriously degraded. The local-oscillator chains at the different sites do not have to be identical, and are usually not. However, in a frequency-conversion stage in which the lower sideband is selected, the frequency axis is reversed. It is important that the direction of the frequency axis at video be the same at all terminals; i.e., the number of lower sideband conversions must be either an odd or an even number at all stations.

(4)   Is the timing set correctly?

Sometimes ac power can be lost for a few milliseconds without being noticed, although such conditions are usually detected with alarm circuits. The power supply of the atomic clock is generally floated across batteries for immunity from ac power dropouts.

(5)   Is the feed polarization set correctly?

The polarizations are generally required to be the same. If the antennas are linearly polarized and set orthogonal, the interferometer will respond only to the circularly polarized component of the source. If an antenna is equatorially mounted, its plane of polarization is fixed on the sky as it tracks a source. If it has an alt-azimuth mount, the plane of polarization continually rotates on the sky. Hence, if antennas with alt-azimuth mounts are used, their feeds must be rotated continually if their planes of polarization are to be kept parallel. The position angle α of an antenna whose polarization vector is vertical with respect to the local horizon is given by

$$\tan \alpha = \sin L \cos \phi'/(\cos \delta_S \sin \phi' - \sin \delta_S \cos \phi' \cos L), \quad (5.5.92)$$

where L is the local hour angle of the source, δ_S the source declination, and ϕ' the latitude of the observing site. The rate of change of polarization position angle can be large near the local zenith. The problem of rotating the plane of polarization can be avoided by using circularly polarized feeds. These feeds are normally made by combining the signals from two orthogonal linear feeds in quadrature so that the rotation of the linear plane of polarization on the sky will produce a shift in the fringe phase. This will introduce an instrumental fringe rate of about 0.01 mHz. When circularly polarized feeds are used, the number of reflections in the antenna optics must be counted, since each one reverses the sense of polarization. Hence, a cassegrain feed and a prime-focus feed should be oppositely polarized.

5.6. Estimation of Astrometric and Geodetic Parameters*

5.6.1. Introduction

The technique of very long baseline interferometry (VLBI) has enormous potential for important applications in astrometry and geodesy. Considerable progress has already been made in the exploitation of this technique for these applications, but the ultimate accuracy inherent in the method has not even been approached: VLBI is in a stage of rapid improvement and probably will not achieve a mature status for some years to come. In this chapter, we confine discussion to the information content of the VLBI observables and to the deduction of the relevant astrometric and geodetic quantities from this content. References to results already obtained are included.

5.6.2. VLBI Observables

Interferometric observations of a point source of continuum radio radiation can yield fringe phase and fringe amplitude as functions of time and frequency. Because the observations, for practical reasons, are usually broken up into relatively short time intervals and relatively narrow frequency intervals, it is convenient to consider two additional ("derived") quantities: the time derivative of the fringe phase, the so-called fringe rate, and the angular frequency derivative of the fringe phase, the so-called differenced group delays or, simply, group delay. In similar language, the fringe phase and the fringe rate, after division by the angular frequency, are the differenced phase delays and the differenced phase delay rates, respectively. Despite the obvious lack of parallelism, we adopt the currently used terms: fringe amplitude, fringe phase, fringe rate, and group delay.

The manner in which the fringe phase ϕ, the fringe rate $\dot{\phi}$, and the group delay τ are determined from the true observables—the recorded signals—is discussed elsewhere in this volume, as are the expressions for the uncertainties in these determinations (see Chapter 5.5). A discussion of the problems associated with the extraction of fringe amplitude from the recorded signals can also be found in Chapter 5.5 and elsewhere.[1,2] Since our principal

---

[1] A. R. Whitney, Ph.D. Thesis, Dept. Elec. Eng., Massachusetts Inst. of Technol. (1974).
[2] J. J. Wittels *et al.*, *Astrophys. J.* **196**, 13 (1975).

---

* Chapter 5.6 is by Irwin I. Shapiro

concern here is with geodetic and astrometric measurements, we shall concentrate mainly on ϕ, $\dot\phi$, and τ. We may express these "second-generation" observables theoretically. To do so, we choose a coordinate system with origin at the solar-system barycenter, and we assume the validity of general relativity. The reason for this seemingly bizarre choice for a system to describe quantities inferred from earth-based observations is simple: In order to combine in a comprehensive analysis diverse astronomical observations, such as of interplanetary time delays and of pulsar signals, and to determine quantities which are each sensitive to more than one of the observables, it is far easier to do all the computations in one coordinate system than to do each in a system convenient for one particular observable and then to convert each to a common system. The opportunities for error in the latter procedure are plentiful because of the importance of relativistic effects in modern observations.

In terms of the so-called isotropic Schwarzschild coordinates, one can show that for two elements i and j of an interferometer array:

$$\phi_{ij}(t) \simeq \omega\tau_{ij}(t) - 2\omega\tau^p_{ij} + 2\pi n, \tag{5.6.1}$$

$$\frac{d\phi_{ij}(t)}{dT_i} \simeq \omega \frac{d\,\Delta t}{dt} - \frac{\omega}{c^2}[\dot{\mathbf{R}} \cdot (\dot{\mathbf{r}}_j - \dot{\mathbf{r}}_i) + \ddot{\mathbf{R}} \cdot (\mathbf{r}_j - \mathbf{r}_i)] + \omega(t^c_{ij} + t^a_{ij} - t^p_{ij}), \tag{5.6.2}$$

$$\tau_{ij}(t) \simeq \Delta t - \frac{1}{c^2}[\dot{\mathbf{R}} \cdot (\mathbf{r}_j - \mathbf{r}_i)] - \frac{1}{c^2}[\ddot{\mathbf{R}} \cdot \mathbf{r}_j + \dot{\mathbf{R}} \cdot \dot{\mathbf{r}}_j]\,\Delta t + \tau^c_{ij} + \tau^a_{ij} + \tau^p_{ij}, \tag{5.6.3}$$

where

$$T_i(t) = t + \text{UTC} - \text{A1} - 32.15 - \frac{1}{c^2}\dot{\mathbf{R}} \cdot \mathbf{r}_i$$

$$+ \text{LPT} + \sum_{k=0}^{K} \alpha_{ik}(t - t_0)^k \quad [\text{sec}], \tag{5.6.4}$$

$$\Delta t(t) \simeq \frac{1}{c}(\mathbf{r}_i - \mathbf{r}_j) \cdot \hat{\mathbf{e}}_s - \left[\frac{1}{c}(\dot{\mathbf{R}} + \dot{\mathbf{r}}_j) \cdot \hat{\mathbf{e}}_s\right]\left[\frac{1}{c}(\mathbf{r}_i - \mathbf{r}_j) \cdot \hat{\mathbf{e}}_s\right]$$

$$\times \left[1 - \frac{1}{c}(\dot{\mathbf{R}} + \dot{\mathbf{r}}_j) \cdot \hat{\mathbf{e}}_s\right], \tag{5.6.5}$$

$$\tau^c_{ij} \simeq \sum_{k=0}^{K} \{\alpha_{ik}(t - t_0)^k - \alpha_{jk}(t - t_0 - \Delta t)^k\}, \tag{5.6.6}$$

and where t is coordinate time, T_i the clock reading at site i, ω the angular (radio) frequency to which the fringe phase is referred, n an integer indicating the ambiguity in fringe phase, c the speed of light, $\mathbf{r}_i(t)$ the geocentric position of the interferometer element i at coordinate time t, $\mathbf{R}(t)$ the barycentric position of the center of mass of the earth at coordinate time t, $\hat{e}_s$ the unit vector in the direction of the observed (point) source of the radio radiation (we ignore here possible parallax effects), τ_{ij}^c the difference in the deviations of the clocks at sites i and j from "true" atomic time at the respective sites, τ_{ij}^a the delay introduced by the neutral atmosphere, and τ_{ij}^p the delay caused by the difference in the integrated charged particle content along the paths from the source to the two sites. This last delay has an effect on ϕ_{ij} and $\omega\tau_{ij}$ of nearly equal magnitude but opposite sign in our weak-magnetic-field, dilute-plasma case. The main contribution to τ_{ij}^p comes from the earth's ionosphere (Chapter 2.1) since nearer the source, presumably, the charged-particle contents along the two paths are very nearly equal. In the expression for the coordinate-time dependence of the time, T_i, kept at site i, we note that UTC denotes coordinated universal time which the site attempts to maintain, A1 denotes atomic time as kept by the United States Naval Observatory, LPT represents the long-period ($\gg 1$ day) terms in the expression for atomic time in terms of coordinate time, 32.15 sec represents the epoch offset of A1 and coordinate time chosen for convenience in other astronomical applications, and the Kth order polynomial is a simple model representing the deviation of the actual clock performance from the desired, referenced to t_0 which may be the midpoint of a short ($\simeq 1$ day) series of interferometric observations. The quantity Δt represents the coordinate-time interval between the arrival of the signal at the two sites.

A superposed dot in the equations signifies differentiation with respect to coordinate time. The i-j asymmetry in the equations is introduced because (i) the delay is measured with respect to time as kept at site i, and (ii) the fringe rate is defined in terms of the derivative of the fringe phase with respect to time as kept at site i. For a clear, elementary discussion of these observables, see Chapters 5.1 and 5.5.

From the point of view of this chapter, the terms τ_{ij}^p and τ_{ij}^a are sources of noise. The effect of the former can be virtually eliminated through exploitation of the frequency dependence of the index of refraction of a plasma (see Chapter 2.1). If the VLBI measurements are made simultaneously in two frequency bands centered at f_1 and f_2 ($f_2 > f_1$), then the standard deviation $\sigma(\tau_f)$ in the determination of τ freed from charged particle effects is related to the standard deviations $\sigma(\tau_k)$ of the group delays measured at the frequency bands f_k ($k = 1, 2$) by the expression

$$\sigma(\tau_f) \simeq (f_2^2 - f_1^2)^{-1}\{f_1^4\sigma^2(\tau_1) + f_2^4\sigma^2(\tau_2)\}^{1/2}. \tag{5.6.7}$$

For $f_2 \simeq 4f_1$ and $\sigma(\tau_2) \simeq \sigma(\tau_1)$, we find $\sigma(\tau_f) \simeq 1.07\sigma(\tau_k)$. Thus, the degradation in the measurement accuracy caused by the charged particles will be about 7% when this procedure is used. A further improvement, by an order of magnitude or more, would be possible if the fringe phase could be determined unambiguously, for example by sufficient accuracy in the measurements of τ at the two frequency bands. For such a circumstance, we find

$$\sigma(\tau_f) \simeq [2\pi(f_2{}^2 - f_1{}^2)]^{-1}\{f_1{}^2\sigma^2(\phi_1) + f_2{}^2\sigma^2(\phi_2)\}^{1/2}, \qquad (5.6.8)$$

where $\sigma(\phi_k)$, $k = 1$, 2, is the standard deviation of the measurement of fringe phase in the kth frequency band. When converted to comparable units, $\sigma(\phi_k)/\sigma(\tau_k)$ is approximately equal to B_k/f_k, where B_k is the rms about the mean of the synthesized bands used to determine the group delay (see Section 5.5.5). Since B_k/f_k is usually small, use of fringe phase, if possible, is advantageous in the determination of τ_f.

The calibration of the electrical path length of the neutral atmosphere (Section 2.5.4) presents the main limitation on the accuracy achievable with VLBI in astrometric and geodetic applications. A new method, not yet tried in a VLBI experiment, shows the most promise for this path length calibration. This method[3] involves monitoring the brightness temperature of the atmosphere at and near the 22 GHz water-vapor spectral line along the line-of-sight from each interferometer antenna to the source. The measured brightness temperatures at each site are used to infer the electrical path length by means of an empirical, linear algorithm. The algorithm, in turn, is based on a statistical analysis of large sets of actual radiosonde data from which both the implied electrical path lengths and the brightness temperatures at the various frequencies were calculated. The linear relationships deduced led to errors in the determinations of zenith electrical path lengths of about 1 cm rms in winter and 1.4 cm rms in summer, both for the Northeastern United States.

5.6.3. Information Content of Observables

The VLBI observables are affected not only by the propagation medium and the relative behavior of the clocks at the two sites, as indicated explicitly above, but also by the source's structure and motion, the solid-earth tides, the crustal motions, polar motion, speed of rotation, nutation, and precession of the earth, and other small physical effects. These effects due to the source and to the earth (see Section 5.6.4) primarily affect Eqs. (5.6.1)–(5.6.3) through the time dependences of $\hat{\mathbf{e}}_s$ and $\mathbf{r}_i$, respectively. Given suitable parametrizations for all of these influences, useful estimates of the parameters can be obtained from analysis of the VLBI observations, provided the data are of sufficient accuracy, span a sufficiently long interval of time, and are of such a character as to eliminate degeneracy—the complete masking of the effects of one parameter by those of a combination of other parameters. To reproduce the explicit parametrizations employed for each of these

---

[3] L. W. Schaper, D. H. Staelin, and J. W. Waters, *Proc. IEEE* **58**, 272 (1970).

effects on the VLBI observables would be to proceed well beyond the scope of this chapter. Instead, we will consider an oversimplified situation which is nonetheless amenable to a presentation of the most important ideas. Thus, we consider the earth to be a rigid body rotating with a constant and known angular velocity vector, and each source to be at infinite distance, of infinitesimal extent, and with an unvarying output of radio radiation that travels in vacuum between source and site. If, further, the clocks at the two sites differ only in their epoch setting and in their rate ($\alpha_{ik} \neq \alpha_{jk}$, $k = 0, 1$), we can conclude that for group-delay measurements, for example, the time dependence will be a diurnal sinusoid superimposed on a linear term that, in general, neither intercepts the origin nor has zero slope. The constant, or intercept of the linear term, represents the additive effects of the clock-epoch offset and the product of the polar, or axial, components of the baseline vector and the source-position unit vector [see Eq. (5.5.1)]. The slope of the linear term represents, of course, the clock-rate offset. The amplitude of the sinusoid measures the product of the equatorial component of the baseline and of the source-position unit vector, as also illustrated in Eq. (5.5.1); the phase of the sinusoid is a measure of the "tilt" of the baseline with respect to the plane normal to $\hat{e}_s$. In this model, then, any number of group-delay observations of a single source can provide no more than four quantities: the intercept and slope of the straight line, and the phase and amplitude of the superposed sinusoid. These four quantities constitute the information content of the group-delay observable. The situation for the fringe-phase observable is similar. For the fringe-rate observable, three quantities provide the total information content; all sensitivity to the time-independent terms in ϕ vanish upon differentiation.

Considering the information content of these observables, how may we determine the three baseline components, the two source position components, and the two clock difference characteristics? The unknowns seem to combine to seven and the knowns to only four for even the group-delay observable—clearly an untenable situation. Actually, the unknowns total one less since we are free to choose our origin of right ascension. The origin of declination is provided by the plane normal to the (given) angular velocity vector. To play a "winning game," we try the strategy of observing additional sources: Each new source for which we obtain group-delay observations can provide up to three new quantities, since the slope of the linear term is the only one independent of the source. But each new source adds only two new unknowns. Observations of a minimum of three sources are thus required to solve for all relevant parameters.[4] Two caveats must be mentioned:

(i)   With fringe phase instead of group delay used as the observable, the

[4] I. I. Shapiro and C. A. Knight, in "Earthquake Displacement Fields and the Rotation of the Earth" (A. Beck, ed.), p. 284. Reidel Publ., Dordrecht, 1970.

game can not be won completely if an unknown constant phase is introduced with each source. In such an event, the number of unknowns added for each new source just equals the number of knowns. Further, unless the fringe phases observed for a source can be followed through a substantial fraction of the diurnal cycle without introduction of 2π ambiguities, little useful information can be deduced about baseline components or source positions.

(ii)   With fringe rate alone used as the observable, it is not possible, in principle, to determine either the clock offset, as mentioned, the polar component of the baseline, or the declination of all sources. The declination of one source—a near equatorial one is the best single choice—may be fixed in accord with, say, optical observations, or an overall constraint may be applied to insure that, for example, the weighted mean of the source declinations equals a preset constant determined from optical data.

Although omitted here for brevity, it can be shown that for the necessary number of group-delay observations of three sources, the relations in terms of the unknowns can, in fact, be inverted to yield solutions for all of the relevant quantities; there are no degeneracies if the observations of each source span a reasonable fraction of the diurnal cycle.

Analyses similar to those given above are easily carried out for combinations of observables and for interferometer arrays with three or more elements. In brief, these show that, to be able to solve for all relevant parameters, at least two sources must be observed no matter how many elements in the array.

5.6.4. Astrometric and Geodetic Parameters

Using the principles described in the previous sections, and extensions thereof, we can utilize VLBI observations to deduce the positions and structures of extragalactic sources, to test the predictions of general relativity, to determine satellite orbits, to estimate the relative positions of radio transmitters placed on the moon by the Apollo astronauts, and to determine various other quantities of geophysical and astronomical interest. Here we describe some of these methods and give references to the major results so far obtained.

5.6.4.1. Source Position.   Fringe rates and group delays have been used successfully to determine positions of sources of compact continuum radiation since 1969.[1,5-7] The accuracy achieved has so far progressed from uncertainties of several arc seconds down to uncertainties of about one-tenth of an arc second. In the most recent such determinations,[8] the authors

[5] M. H. Cohen and D. B. Shaffer, *Astron. J.* **76**, 91 (1971).

[6] H. F. Hinteregger *et al.*, *Science* **173**, 396 (1972).

[7] M. H. Cohen, *Astrophys. Lett.* **12**, 81 (1972).

[8] A. E. E. Rogers *et al.*, *Astrophys J.* **186**, 806 (1973).

made use of a weighted-least-squares estimator to find the source positions from sets of redundant data obtained at a radio frequency near 7850 MHz ($\lambda \simeq 3.8$ cm) with a 3900-km-baseline interferometer—the so-called "Goldstack" interferometer formed by the 64-m-diameter antenna of the Jet Propulsion Laboratory in Goldstone, California and the 36.6-m-diameter antenna of the Haystack Observatory in Westford, Massachusetts.

Fringe phases were first utilized to obtain estimates of the relative positions of closely spaced, discrete water-vapor masers ($\lambda \simeq 1.3$ cm) in our galaxy. These emission regions all were visible in the antenna beam of each element of the interferometer employed. Even though the separate, 3-min VLBI observations were spaced $\frac{1}{2}$ h apart, it was possible to determine accurate relative positions of the distinct regions from the fringe phases. We may outline the method briefly as follows. The relative fringe phase, $\Delta\phi(t)$, between a pair of the discrete sources, measured at time t, may be expressed as

$$\Delta\phi(t) = \Phi(\Delta\alpha, \Delta\delta; t) + \varepsilon + 2\pi n \qquad (5.6.9)$$

in the strong signal case, where ε ($\ll 2\pi$) is the Gaussianly distributed error in the (ambiguous) phase determination, n is an integer, and Φ represents the theoretical value of the relative phase given as a function of the differences, $\Delta\alpha$ and $\Delta\delta$, in the right ascension and declination coordinates, respectively, of the pair of sources (other dependences have been omitted for brevity). The conditional probability density for $\Delta\alpha$ and $\Delta\delta$, given the set of independent relative fringe-phase estimates $\Delta\phi(t_{ij})$ and the assumption that all values of n are equally probable, can be written as

$$p(\Delta\alpha, \Delta\delta | \Delta\phi) \propto p_0(\Delta\alpha, \Delta\delta) \times \prod_{i=1}^{I} \prod_{j=1}^{J_i} \sum_{n=-\infty}^{\infty} \frac{1}{\sigma_{ij}}$$
$$\times \exp\{-[\Delta\phi(t_{ij}) - 2\pi n - \Phi(\Delta\alpha, \Delta\delta; t_{ij})]^2/2\sigma_{ij}^2\}, \quad (5.6.10)$$

where σ_{ij} is the standard deviation of the relative fringe phase from the jth of J_i observations on the ith of I independent baselines, and where $p_0(\Delta\alpha, \Delta\delta)$ represents the joint *a priori* probability density of $\Delta\alpha$ and $\Delta\delta$ determined from the unambiguous fringe rate data as described in Chapter 5.5. The relative position is then taken to be that set ($\Delta\alpha$, $\Delta\delta$) for which the right side of Eq. (5.6.10) is a maximum. In practice, this estimate is easily obtained by a systematic evaluation of p over successively finer grids in the $\Delta\alpha$ $\Delta\delta$-plane. Because of its relatively slow variation with $\Delta\alpha$ and $\Delta\delta$, p_0 serves mainly to delimit the area of search for the maximum of p. Further details are given by Reisz *et al.*[9] who determined the separation of two components 0.3 arc sec apart in the radio source W3 (OH), with an error ellipse whose

[9] A. C. Reisz *et al.*, *Astrophys. J.* **186**, 537 (1973).

major and minor axes were 0.0003 and 0.0001 arc sec, respectively. A three-element interferometer with maximum antenna spacing of 845 km was used in this 1971 experiment: Haystack, the National Radio Astronomy Observatory's 42.7-m-diameter antenna in Green Bank, West Virginia, and the Naval Research Laboratory's 25.9-m-diameter antenna at Maryland Point, Maryland.

For a pair of sources whose angular separation in the sky is not too great, a two-element interferometer can be used to monitor the fringe phase for each source, without the introduction of any 2π ambiguities, by switching rapidly back and forth from observations of one to observations of the other. Aside from the obvious conditions on mutual visibility, the limitation on angular separation is set primarily by the slew rates of the antennas forming the interferometer. By subtraction of the fringe phase functions for the two sources the resultant differenced fringe phase is essentially freed from the effects of any clock wandering and, for neighboring sources, from almost all of the effects of the propagation medium. Analysis of this differenced phase observable can therefore yield a very high accuracy in the estimate of the relative source positions, approaching that achieved for the discrete water-vapor maser regions. The difference observable, $\Delta\phi$, may be written as

$$\Delta\phi(t) \simeq (\omega B/c)\{\Delta\alpha \cos D \cos \delta \sin(A_0 + \Omega t - \alpha)$$
$$- \Delta\delta \cos D \sin \delta \cos(A_0 + \Omega t - \alpha) + \Delta\delta \sin D \cos \delta\},$$
$$(5.6.11)$$

where $B \equiv |\mathbf{r}_2 - \mathbf{r}_1|$, α and δ are the coordinates of the reference source, A_0 and D, respectively, the right ascension, or hour angle, at epoch and the declination of the baseline, and Ω the angular velocity of the earth. We have omitted terms of higher order in $\Delta\alpha$ and $\Delta\delta$, as well as others indicated in Eq. (5.6.1).

For some pairs of sources we find, for example, that $\Delta\delta = O(\Delta\alpha^2)$; in such cases Eq. (5.6.11) for the differenced fringe phase, accurate to $O(\Delta\delta)$, must be augmented by the term $-\frac{1}{2}\Delta\alpha^2 \cos D \cos \delta \cos(A_0 + \Omega t - \alpha)$ inside the braces. A similar modification is required for $\Delta\alpha = O(\Delta\delta^2)$. As a simple illustration, consider the effects on the estimates of $\Delta\alpha$ and $\Delta\delta$ of an error in baseline direction for the case in which $\Delta\delta = O(\Delta\alpha^2)$:

$$\delta \Delta\alpha \simeq \Delta\alpha \, \delta D \tan D, \qquad\qquad (5.6.12)$$

$$\delta \Delta\delta \simeq \Delta\alpha \, \delta A \cot \delta, \qquad\qquad (5.6.13)$$

where δA and δD are the baseline direction uncertainties. Thus, for baselines nearly East–West (small D) and high declination sources, the estimates of $\Delta\alpha$ and $\Delta\delta$ are both insensitive to baseline direction errors and to equivalent errors in the length of the day and polar motion, discussed in Section 5.6.4.5.

The pair of compact continuum radio sources 3C345 and NRAO 512 are separated in the sky by only about 10^{-2} rad, with the separation in declination being only 5×10^{-4} rad; therefore $\Delta\delta \simeq O(\Delta\alpha^2)$. Four separate sets of observations that utilized this switching technique with the Goldstack interferometer ($\lambda \simeq 3.8$ cm) were carried out on this source pair between 1971 and 1974, and yielded differences in position with an rms scatter about the mean of under 1.5 and 2.0 milliarc sec in right ascension and declination, respectively. These results obtained by the VLBI Group of the Goddard Space Flight Center, the Haystack Observatory, and the Massachusetts Institute of Technology, can be used to set an upper limit of about 0.0005 arc sec/yr on the relative proper motion of this pair and, approximately, on the "absolute" proper motion of either source, since it is highly improbable that their individual proper motions could be coordinated. Continuation of such measurements, with improvements in accuracy, could lead in the relatively near future to bounds on, or measurements of, proper motion of use in the determination of the distance scale for extragalactic objects.

If two or more antennas are available at each site of an interferometer array, then the switching described above can be avoided, and any two or more sources that are mutually visible from every site can be monitored continuously and the difference fringe phases freed from clock effects if the same clock, or frequency standard, is used at a given site to govern the local-oscillator signals for each of the antennas at that site. The only combination of this type so far utilized successfully for precision astrometry has been a four-antenna combination: two antennas at each end of the 845-km baseline formed by the Haystack Observatory and the National Radio Astronomy Observatory in Green Bank, West Virginia. These observations, carried out in 1972 at a radio frequency of 8105 MHz ($\lambda \simeq 3.7$ cm) by the above-mentioned VLBI Group, yielded postfit residuals for the difference fringe phases of about 20 psec rms ($\sigma_\phi \simeq 50°$); because of systematic effects not yet completely understood, the uncertainty of the relative source positions is at the level of a few hundredths of an arcsecond. Using such data in conjunction with an a priori covariance matrix, based on other observations, allows the baselines to be estimated as well as the source positions.

This "four-antenna" technique can also be applied to the detection of changes in relative positions of source pairs due to the changing deflection of radio waves by the sun's gravitational field as the line of sight to one or the other of the sources of a given pair passes near the sun. Such a test was performed in September–October 1972 with the Haystack–NRAO configuration discussed above[10]; observations of the pair of extragalactic continuum sources 3C273B and 3C279 yielded nearly the predicted change in relative

[10] C. C. Counselman, III et al., Phys. Rev. Lett. **33**, 1621 (1974).

position for the period surrounding the occultation of the latter by the sun on October 8. In particular, this change was found to be 0.99 ± 0.03 times that deduced from the theory of general relativity; a significant limitation was provided by the fluctuations in the solar corona which occasionally wreaked havoc on the fringe phases for observations made with 3C279 within a few degrees of the sun.

5.6.4.2. Source Structure. Almost all of the compact extragalactic radio sources observed with the VLBI technique have exhibited fine structure; for many, this structure changes very rapidly with time.[11-13] One of the main astrometric challenges posed by these sources is the mapping of their fine structure and internal kinematics so as to provide a solid base for the theoretical understanding of their behavior. For radio continuum observations, this mapping, in principle, encompasses the determination of only the brightness distribution of a source and its changes with time. The brightness distribution for a very distant source whose radiation is spatially incoherent is related to the interferometer fringe amplitude and phase through the well-known Fourier transform relation (see Sections 5.1.2 and 5.5.5)

$$V(u, v) = \iint B(x, y)e^{i(ux + vy)} \, dx \, dy, \tag{5.6.14}$$

where V is the so-called visibility function whose amplitude and phase are the fringe amplitude and fringe phase, respectively. The parameters u and v represent the resolution, say in fringes per arcsecond, of the interferometer in the East–West and North–South directions, respectively (Section 5.1.1). The parameters x and y, defined along corresponding directions in the sky, represent the Cartesian coordinates of the source on the plane of the sky with respect to some defined origin near the center of the source (Section 5.5.5). Since the source is always confined to a small region of the sky, the above relation can be inverted to infer uniquely the brightness distribution, provided $V(u, v)$ is sampled in accord with the requirements of the two-dimensional analog of the Nyquist theorem. For any given two-element interferometer, only an ellipse, or arc thereof, in the uv-plane can be sampled. Further, only the amplitude of $V(u, v)$ can be determined usefully at these sample points; uncorrelated fluctuations in the propagation paths through the atmosphere over the two sites makes the fringe phase virtually worthless for the determination of source structure. For an interferometer array with three or more elements, however, the sum of the fringe phases around a closed "loop" of baselines in the array is virtually freed from propagation

[11] C. A. Knight et al., Science 172, 52 (1971).
[12] A. R. Whitney et al., Science 173, 225 (1971).
[13] M. H. Cohen et al., Astrophys. J. 170, 207 (1971).

effects and clock errors. This "closure" phase ϕ_c, which contains information only on source structure, can be written for three baselines as[14]

$$\phi_c \simeq \phi_{12}(t_1) - \phi_{13}(t_1) + \phi_{23}(t_2), \qquad (5.6.15)$$

where $\phi_{ij}(t_i)$ signifies the fringe phase measured on the ij baseline referred to the time of arrival t_i of the signal at site i.

No VLBI array yet available is able, through the earth's rotation, to sample the required parts of the uv-plane to produce a unique brightness distribution for any source. With the limited available fringe-amplitude and closure-phase data, one usually resorts to parametrized models of the brightness distribution using the data to estimate the parameters via a suitable algorithm such as the maximum-likelihood estimator. The models may consist of the characteristics of two or more point sources of radiation, Gaussianly distributed sources of radiation, truncated one- or two-dimensional series representations of the brightness distribution, etc.—the limit is set only by the imaginations of the model makers and the capabilities of the computer! One may also parametrize the visibility function itself and, after estimation of the parameters through comparison with the available data, invert Eq. (5.6.14) to obtain the brightness distribution.[14] The difficulty with these types of approaches is that the results can be no better than the models, and may, in fact, bear little resemblance to the actual brightness distributions. In the future, larger, better arranged arrays will come into use, and the modeling problems will be much alleviated.

5.6.4.3. Satellite Orbits.   The VLBI technique can also be used to determine accurately the orbits of earth satellites if they emit radio radiation. For satellites at synchronous altitude, even effective radiated powers as low as a milliwatt, spread over a 10-MHz bandwidth, yield signals comparable in strength to those received from extragalactic radio sources.

The reduction of the recorded VLBI signals to produce the observables (see Section 5.6.2) is more complicated for satellites than for natural sources. The motion of the satellite during the recording must be accounted for explicitly and with reasonable accuracy to avoid "washing out" of the fringes: In the cross correlation of the signals (see Chapter 5.5), the a priori model must include a preliminary orbit of the satellite.

The final processing of the observables also requires modification. One must account in the equations of Section 5.6.2 for the changing parallax of the source introduced by its orbit. Although it appears that, for this application, a geocentric frame would be more appropriate, the reference frame we have chosen produces no particular difficulties since the vectors can

[14] A. E. E. Rogers *et al.*, *Astrophys. J.* **193**, 293 (1974).

be expanded in such a manner that the appropriate common parts cancel. The parameters to be estimated must be augmented to allow for the six independent orbital elements; the basic estimator may remain unchanged. If the a priori orbit was not of sufficient accuracy, the cross correlation and final processing can be repeated with the original a priori orbit replaced by the orbit obtained from the first iteration.

This application of VLBI was first tried in 1969. From only 7 hr of observations of the TACSAT I communications satellite which was in a synchronous, nearly equatorial orbit, an extremely accurate orbit was produced: six-place accuracy in both semimajor axis and eccentricity, as judged by the consistency of the results obtained from the group-delay and fringe-rate data, separately.[15] The observations were made from three sites (Haystack, Green Bank, and the Owens Valley Radio Observatory, Big Pine, California (longest baseline 3500 km)) over a 10 MHz bandwidth centered at about 7.3 GHz.

The VLBI technique for satellite-orbit determination can be compared with conventional radio tracking methods and with laser ranging.[15] Here, we shall only intercompare the utility for geodetic applications of VLBI observations of satellites and extragalactic radio sources. The celestial sources have one essential advantage. They have negligibly small proper motions, and hence provide an excellent approximation to an inertial reference frame. Having the radiation sources on the satellite allows the baselines to be located with respect to the center of mass of the earth. Ties to this center degrade with the decrease in parallax accompanying an increase in the altitude of the satellite relative to the length of the baseline; observations of extragalactic sources are completely insensitive to any parallel displacement of the baseline. Similarly, observations of satellites provide sensitivity to the earth's gravitational potential. This sensitivity, however, is a double-edged sword since, for the determination of baselines, deficiencies in the theoretical model of the potential are a hindrance.

The most promising approach may be to make use simultaneously of both natural and artificial sources. The satellites can then be located accurately with respect to a stellar frame; relative errors might be kept as low as a milliarcsecond if the satellite passes close to, and slowly by, one of the natural sources that constitute the inertial frame. One can even envision a hierarchy of ground terminals for geodetic use: The most sensitive installations can be used to observe both satellites and natural sources, while small, transportable terminals observe only the stronger signals from satellites. With the orbits determined precisely, relative to the inertial frames, by the large installations, the VLBI observations by the portable ones can be used

[15] R. A. Preston et al., Science **178**, 407 (1972).

to determine geodetic ties directly. For this latter purpose, the satellites might be equipped with corner reflectors so as to be suitable targets for laser observations as well. An important role in the geodetic applications might also be played here by radio emissions from spacecraft in interplanetary flight and in orbit about, or emplaced on, other planets.

5.6.4.4. Lunar Applications. The Apollo astronauts left a number of transmitters on the lunar surface. Five are still operating and may continue to do so for many years. They provide the opportunity, through differential interferometry,[16] to set up a selenodetic reference system, to measure accurately the moon's libration, and to relate the moon's orbital position to the inertial reference frame formed by the extragalactic radio sources.

The signals from these so-called ALSEP transmitters are characterized by a carrier and narrow-band modulation, with the former containing a small fraction of the power. The presence of a carrier allows a significant simplification in the VLBI apparatus: Instead of recording the signals themselves, one can compare at each site the received signals from any pair of ALSEPs and record only the cycle count of the suitably multiplied difference frequency. Voice communication or teletype between sites can then be used to establish immediately whether all systems are operating properly.

Since the ALSEP transmissions are at S-band frequencies ($\lambda \simeq 13$ cm), the ionosphere affects noticeably even the difference fringe phases. The transmissions from the different ALSEPs span an interval of 4 MHz, and these separations can be exploited to reduce the sensitivity of the results to the ionosphere when, for example, all five ALSEPs are observed simultaneously.

Analysis of the potential of such differential VLBI observations of the ALSEP transmitters for the establishment of a set of selenodetic reference points indicates that the uncertainties in the determinations of the vectors between the ALSEPs should be reducible to about one meter. At the present level of development of the technique, only dekameter accuracy has been obtained.[17] The uncertainties in the determinations of the moon's position with respect to the extragalactic radio sources might be reducible to the milliarcsecond level, but no results have yet been obtained. The parameters describing the moon's libration have also not yet been deduced from the VLBI observations, but should, when the technique is developed, yield values perhaps comparable in accuracy to those obtainable from laser observations of the retroreflectors emplaced on the lunar surface by the astronauts.

5.6.4.5. Geophysical Applications. The various motions of the earth's crust can be separated and measured with great accuracy via VLBI. We dis-

[16] C. C. Counselman, III, H. F. Hinteregger, and I. I. Shapiro, *Science* **178**, 607 (1972).

[17] C. C. Counselman, III, H. F. Hinteregger, R. W. King, and I. I. Shapiro, *Science* **181**, 772 (1973).

cuss briefly these motions, the methods of measurement, and the levels of accuracy so far attained.

The precession and nutation of the earth, due primarily to the solar and lunar torques exerted on the earth's equatorial bulge, are manifested by a movement of the earth's axis of rotation in an inertial frame. The coordinates of the sources (but not the lengths of the arcs between them!) will therefore appear to change if the equator of date is always used to define the origin of declination. From the pattern of change, the parameters characterizing the precession and nutation can be estimated. In actual practice, this parametrization is added to the theoretical model of the observable, and values for all relevant parameters are determined simultaneously. The period of the precession is about 26,000 yr, whereas that of the principal nutation is about 18.7 yr; thus, a number of years of observation are required for an adequate separation of the contributions of each to the observed pattern. Similar statements apply to the other periodicities present in the usual series expansion of the nutation. Except for the relative insensitivity of the VLBI measurements to the instantaneous orientation of the ecliptic, the VLBI determination of the precession would surpass the optical in a few years, despite the far longer span of the optical observations. VLBI can also be used to determine the orientation of the ecliptic (the earth's orbital plane) with respect to the frame formed by the extragalactic sources. The method requires point sources of radiation on or near other planets such as could be provided by the radio transponders on spacecraft landers or orbiters. Sets of differential VLBI observations of such spacecraft and extragalactic sources that at times lie nearly along the same line of sight can be used to infer the orientations of the orbital planes of the earth and the other planets. Observations of pulsars can, in addition, be used to determine the orientation of the ecliptic with respect to the earth's equator.[18]

There are a number of geophysical effects, most as yet poorly understood, that cause variations in the length of the day and in the orientation of the earth's crust with respect to its axis of rotation. These two types of changes are usually referred to as variations in universal time (UT1) and polar motion. Measurements of UT1 and polar motion can be made with VLBI because of their effects on the baseline vector: The direction (but not the length!) of this vector will be changed. This effect of UT1 and polar motion on the baseline contrasts with the effect of precession and nutation which cause an apparent change in the direction of the sources. Since the direction of a vector can be described by only two parameters, it is not possible to determine variations in UT1 and both components of polar motion using only a single two-element interferometer.

Variations in UT1 and in one component of polar motion have been

[18] I. I. Shapiro, *Trans. Amer. Geophys. Un.* **51**, 266 (1970).

determined with VLBI from a series of observations with the two-element Goldstack interferometer.[19] Since this interferometer has a very small North–South component, the changes in UT1 affected mainly the right ascension, or hour angle, of the baseline, and the changes in polar motion affected mainly the declination. The accuracies achieved seem to have surpassed those attained via conventional optical techniques.

The tidal bulges raised on the earth by the moon and sun also affect the baseline, both directly and indirectly, through changes in the earth's orientation caused, for example, by the effect of torques exerted on the bulge. The largest effect is due to the semidiurnal tidal component which can introduce a maximum offset in the baseline of an amplitude of several tens of centimeters. This component has apparently already been detected from analysis of VLBI observations[19] but the measurements are not yet accurate enough to deduce a useful value for the relevant Love numbers. The effects of lateral inhomogeneities in the earth, and of ocean loading on the solid-earth tides, should both be accessible for study with suitable placement of the interferometer elements and with increased accuracy in the measurements and in the calibration of the propagation medium.

The vector baselines have been determined by the VLBI technique for sites separated by distances of from 845 up to 8000 km,[1,6,19] and also for sites separated by distances of 16 km[20] and less. Three-site "closure" experiments were also performed over transcontinental distances.[1] The accuracies achieved on the short baselines, on the order of 5 cm in length, are nearly sufficient for monitoring accurately crustal motions across faults. The baseline accuracies achieved over the transcontinental and intercontinental distances were 20 cm or worse, depending on the signal-to-noise ratio available for the different interferometers used. The lower limit was actually set by systematic errors that can be, but have not yet been, sharply reduced by the use of appropriate calibration techniques (see Section 5.6.2).

To make important contributions to the measurement of crustal distortions, or plate tectonics, one must reduce the errors in baseline determination to the centimeter level through calibration and increases in signal-to-noise ratios. The calibration of the propagation medium is only a problem in regard to the neutral atmosphere; all other but the clock effects on the VLBI measurements can be considered as "signal" for the purposes of this chapter. Increases in signal-to-noise ratios seem most readily obtainable by the use of wideband instrumentation recorders,[21] which could provide a 50 MHz or greater bandwidth compared with the 360 kHz or less used in the

[19] I. I. Shapiro *et al.*, *Science* **186**, 920 (1974).

[20] J. B. Thomas, J. L. Fanselow, P. F. MacDoran, D. J. Spitzmesser, and L. Skjerve, Jet Propulsion Lab. Tech. Rep. **32-1526**, p. 36 (1974).

[21] H. F. Hinteregger, Ph.D. Thesis, Dept. of Phys., Massachusetts Inst. of Technol. (1972).

experiments described above. When such improved systems come into use, a new era could begin in our understanding of the fine structure of the earth's crustal motions and in elucidating, for example, the possible relations of strain accumulation in the crust and polar motion to earthquakes and earthquake prediction.

6. COMPUTER PROGRAMS FOR RADIO ASTRONOMY

6.1. Radial-Velocity Corrections for Earth Motion*

6.1.1. Introduction

In 1842, Christian Doppler published a paper quantitatively describing the relationship between line-of-sight velocities and wave frequencies. At the time, the wave theory of light introduced by Huygens was becoming accepted over Newton's corpuscular theory because of its ability to explain interference phenomena. Doppler was investigating implications of the wave nature of light.

Using the analogy of wave trains in the ocean moving toward (or against) a moving ship, Doppler derived formulas relating the wave frequency perceived by an observer on a moving ship to that seen by a stationary observer. He correctly noted that the shift in frequency depends not only upon the relative velocity between the wave source and the ship, but also upon whether the ship moves with respect to the wave source, or the source with respect to the ship. For if the ship moves with a velocity v_s toward the wave source, and if the waves move through the ocean with a velocity v, then the effective velocity of the wave train with respect to an observer aboard ship would be $v_s + v$, and the observed frequency f of the wave train is

$$f = (v_s + v)/\lambda_0 = f_0(1 + v_s/v), \qquad (6.1.1)$$

where f_0 and λ_0 are the actual frequency and wavelength, respectively, of waves emitted by the source. Thus, the observer would perceive a higher wave frequency owing to the ship's motion. If the wave source moves toward the stationary ship, the physical separation between wave crests would decrease to $\lambda_0(1 - v_s/v)$, and the observer would also perceive a higher wave frequency

$$f = v/\lambda = f_0/(1 - v_s/v), \qquad (6.1.2)$$

but different than that given by Eq. (6.1.1). Such frequency changes due to relative velocities between source and observer are known categorically as *Doppler effects*. It is important to note that the effect here is quantitatively

---

* Chapter 6.1 is by M. A. Gordon

277

different depending upon whether the source moves relative to the observer, or the observer moves relative to the source.

Doppler believed the equations describing water waves also described electromagnetic waves propagating through the ether, and, in his paper, did not hesitate to apply these equations to astronomy. He suggested that all stars had the same intrinsic color as the sun, and that the observed variations in the color of stars reflected not differences in their temperature and chemical composition, but differences in their radial velocities with respect to the earth. This suggestion created bitter opposition to his ideas, which persisted long after Doppler's death, until careful spectroscopic observations showed that there were indeed differences in the temperature and composition of stars, and that differences in apparent color were not due wholly to velocity effects. (Fizeau had privately pointed out that, owing to the continuous nature of stellar radiation, Doppler effects probably could not cause significant changes in apparent colors of stars.)

6.1.2. Special Relativity

Doppler's equations assume fundamental concepts regarding the vector interaction of the velocities of light, source, observer, and medium (ether). In his 1849 experiment to measure the velocity of light, Fizeau investigated, but failed to find, any effect of the earth's motion upon the velocity of light. Yet, Doppler's ideas required variations in the effective propagation velocity. Such early experiments cast doubt as to whether light waves propagated in ether the same way as ocean waves in water and acoustic waves in gas.

In 1887, Michelson and Morley performed their sophisticated experiment which indicated that the velocity of light was a constant, unaffected by vector addition or subtraction of the earth's velocity. These results led Lorentz (and, independently, FitzGerald) in 1895 to propose that one leg of the Michelson–Morley apparatus changed length as a function of velocity such that, if the velocity of light indeed changed because of the earth's motion, the arm of the measuring device would change length correspondingly, thus preventing the investigators from detecting the changes in the velocity of light. This suggestion led Lorentz to consider further the nature of velocity and electromagnetic phenomena, which in 1904 resulted in a paper which paralleled the more complete exposition by Einstein. In 1905, Einstein published his now famous paper in which he postulated that the velocity of light was independent of the motion of its source. In his discussion, he derived the "Doppler effect" of special relativity to be

$$f = f_0 \frac{1 - (v/c) \cos \phi}{[1 - (v^2/c^2)]^{1/2}}, \tag{6.1.3}$$

where ϕ is the angle between the velocity and the line of sight between source and observer, v is the velocity of the source in the observer's reference frame, and c is the speed of light. Unlike Doppler's original formulation, this equation always predicts a frequency shift even if the light source (or observer) moves at right angles (cos $\phi = 0$) to the line of sight joining the two objects.

If the source and observer move toward each other, cos $\phi = 1$, and Eq. (6.1.3) becomes

$$f = f_0 \left(\frac{1 - (v/c)}{1 + (v/c)} \right)^{1/2}, \tag{6.1.4}$$

which is quite different from Doppler's original formulation given in Eqs. (6.1.1) and (6.1.2). This difference results from the fact that light waves and water waves do not propagate in the same way.

6.1.3. Conventional Tabulation of Redshifts

Astronomical spectroscopy began considerably before Einstein's paper appeared, and thus wavelength (frequency) shifts of spectral lines are always catalogued as velocity effects calculated by Doppler's original formulations, with the assumption that sources move relative to the earth. Optical astronomers calculate radial velocities from their most basic measurements: the dispersion of wavelengths on a spectroscopic plate. Hence,

$$v_{opt} = c[(\lambda - \lambda_0)/\lambda_0], \tag{6.1.5}$$

where v_{opt} is positive for a receding source if λ is the *observed* wavelength. The astronomical redshift parameter z is defined to be

$$z = (\lambda - \lambda_0)/\lambda_0. \tag{6.1.6}$$

The system generally used by radio astronomers differs from that used by optical astronomers. Presumably, the difference stems from the fact that, unlike the case for the optical spectrometer, the radio spectrometer sorts information by frequency. Hence, the convention arose that

$$v_{radio} = c[(f_0 - f)/f_0]. \tag{6.1.7}$$

To date, neither optical nor radio astronomers conventionally use special relativity [Eq. (6.1.4)] to calculate radial velocities from observations.

Because of the differences between radial velocities quoted by radio and optical astronomers, errors[1] are apt to result when one uses optically determined redshifts of high velocity objects like quasars to position narrow-band radio spectrometers. And, occasionally, v_{opt} or z are calculated with a com-

---

[1] C. Heiles and G. K. Miley, *Astrophys. J.* **160**, L83 (1970).

bination of rest wavelengths measured in a vacuum (λ_0) and observed wavelengths measured in the air (λ) of the astronomical spectrography.

Finally, with low-dispersion optical spectrographs, the "line" spectra are in fact often blends of several lines. Thus, radial velocities calculated from blends erroneously presumed to be single transitions often contain substantial errors. This problem is a well-known one, and a variety of techniques have been used to minimize resulting errors.

6.1.4. The Aberration of Light

Early astronomers noted that the stars seemed to change position with season. This effect arises because the motion of the earth changes the aspect angle of an observer with respect to the direction of arrival of rays from stars. If θ' is the apparent angle between the direction of the earth's velocity v and the star, and θ is the actual angle, simple trigonometry shows that

$$\tan \theta' = \sin \theta / [\cos \theta + (v/c)], \qquad (6.1.8)$$

where c is the velocity of light.

For the most adverse case, position errors as large as 41 arc sec can occur.

6.1.5. Velocity Reference Frames

In considering the effects of relative velocities of source and observer, we must choose some standard reference velocity. The observer moves on the surface of the earth, which moves around the moon–earth barycenter, which moves around the solar system barycenter, which moves with respect to local stars, which move around the galactic center, etc. For standardization, the observer usually references his velocity measurements to one of two conventional reference velocities.

6.1.5.1. Heliocentric System.   Perhaps the most obvious velocity-reference system is that of the sun. In the late nineteenth century, when stellar spectroscopy began, the orbital elements of the earth's motion about the sun were known to accuracies considerably greater than could be measured from stellar absorption spectra. Consequently, radial velocities measured from the earth could be easily referenced to the sun without loss of accuracy. As mentioned in Section 6.1.3, one of the largest sources of error in these measurements is due to the possibility of unresolved blends of spectral lines. The relative intensity of blended components is probably the same amongst stars of the same spectral type, and here the relative differences in radial velocities can be accurately measured. For example, the radial velocities of F and B stars can be measured relative to the sun with considerable accuracy. Stars of other types may have blends of different relative intensities to those of the sun, and serious measurement errors can arise because of the differences in line shapes.

The intrinsic absolute accuracy of optical radial velocities measured from stellar absorption spectra is therefore limited to perhaps something less than 1 km/sec, even for measurements taken by the best spectrographs by the most experienced observers. (Measurement errors of 0.2 km/sec are often quoted, but such numbers refer to precision rather than accuracy.) Such accuracy does not warrant distinction between the sun and the solar system barycenter. Thus, the term *heliocentric velocity* assumes the sun to be the focus of the earth's orbital motion. (Motion with respect to the solar-system barycenter is discussed in Section 5.6.2.) The reader may wish to refer to work by Campbell and Moore[2] and by Petrie[3] for more detailed information about stellar velocity measurements, and to work by Herrick[4] for a convenient table to reduce radial velocities to the sun.

6.1.5.2. Local Standard of Rest. Analysis of velocities determined by stellar spectroscopy shows that the sun has a systematic motion with respect to neighboring stars. The situation is not simple, because this systematic motion is also a function of the spectral type of stars used for the radial velocity measurements. If we define a *kinematic* reference frame moving with the average velocity of these neighboring stars, then this *local standard of rest* can be said to be a function of stellar type. Clearly, some convention is necessary.

Standard solar motion is defined to be the average velocity of spectral types *A* through *G* as found in general catalogs of radial velocity, without regard to luminosity class. Here, the motion of the sun is 19.5 km/sec toward 18^h right ascension (RA) and $30°$ declination (δ) for the epoch 1900.0, which corresponds to galactic coordinates (l, b) of ($56°$, $23°$). *Basic solar motion*[5] is the most probable velocity of stars in the solar neighborhood, thereby being heavily weighted by the radial velocities of the most common spectral types *A*, *gK*, and *dM* in the vicinity of the sun. In this system, the sun moves at 15.4 km/sec toward the direction (l, b) of ($51°$, $23°$).

The conventional reference frame used for galactic studies is essentially that of standard solar motion. The convention of *Local Standard of Rest* (LSR) assumes the sun to move at the rounded velocity of 20.0 km/sec toward 18^h RA and $30°$ δ (1900.0). Since the stars used in the determination include the earlier spectral types of the distribution, their ages are comparatively younger, and, presumably, their velocities are closer to that of the interstellar gas.

For some purposes, it is necessary to know the motion of the Local Stan-

[2] W. W. Campbell and J. H. Moore, *Publ. Lick Observ.* **16** (1928).

[3] R. M. Petrie, *in* " Basic Astronomical Data " (K. A. Strand, ed.), p. 64. Univ. of Chicago Press, Chicago, Illinois, 1963.

[4] S. Herrick, Jr., *Lick Obs. Bull.* 470, **17**, 35 (1935).

[5] A. N. Vyssotsky and E. N. Janssen, *Astron. J.* **56**, 58 (1951).

dard of Rest about the galactic center. Observations[6] of globular clusters show the sun to move at 167 ± 30 km/sec toward an (l, b) of $(90°, 0°)$. (With these uncertainties, the motions of the sun and the Local Standard of Rest are indistinguishable.) Similar observations[7] of galaxies in the Local Group show the sun to move at 292 ± 32 km/sec toward $(106°, -6°)$. The reason for the difference between the two velocities is unclear, but based on such observations, the Local Standard of Rest is usually assumed to move at 250 km/sec toward $(90°, 0°)$ about the *Galactic Standard of Rest*.

6.1.6. Calculation of Radial Velocities

The calculation of the radial velocity between a position on the earth's surface and a distant astronomical object involves a number of velocity terms. Just how many terms need to be included is determined by the accuracy required. Table I lists a number of these terms and their approximate maxi-

TABLE I.   Velocities Involved in the Radial-Velocity Computation

Component	Approximate maximum velocity (km/sec)
Source with respect to Local Standard of Rest (LSR)	—
LSR with respect to solar system barycenter	20
Solar system barycenter with respect to earth–moon barycenter	30
Earth–moon barycenter with respect to earth center	0.1
Earth center with respect to telescope	0.5
Planetary perturbations upon earth's orbit	0.013

mum size. In practice, of course, the contributions of these velocity components will be less according to direction cosines.

Ball[8] has written a program to calculate the velocity of a telescope with respect to the LSR. This program includes terms described in Table I except those due to planetary perturbations of the earth's orbit, and the program omits the motion of the sun around its barycenter. Appendix A lists this

[6] T. D. Kinman, *Mon. Not. Roy. Astron. Soc.* **119**, 559 (1959).
[7] M. L. Humason and H. D. Wahlquist, *Astron. J.* **60**, 254 (1955).
[8] J. A. Ball, M.I.T. Lincoln Lab. Tech. Note TN 1969-42, Lexington, Massachusetts (16 July 1969).

subroutine, which can be used to compute the radial-velocity correction with sufficient accuracy for most spectroscopy of astronomical objects. The absolute accuracy is approximately 0.02 km/sec. For times less than a year or so, the relative accuracy is 0.005 km/sec. For more detailed discussion of the astronomical considerations involved in this computation problem, the reader is referred to standard books.[9-12] A tabulation of radial velocity components has been prepared by McRae and Westerhout.[13]

The computer program called DOP, listed in Appendix A, may be used to calculate the velocity component of the observer with respect to the Local Standard of Rest as projected onto a line specified by the right ascension and declination (RAHRS,RAMIN,RASEC;DDEG,DMIN,DSEC) for the epoch of date with time specified as follows:

NYR   last two digits of the year (for 19XX A.D.);
NDAY   day number (time UT);
NHUT, NMUT, NSUT   time (hours, minutes, seconds UT).

The location of the observer is specified by the latitude (ALAT), the geodetic longitude in degrees (OLONG), and the elevation in meters above sea level (ELEV). The program gives as output the local mean sidereal time in days (XLST), the velocity component in kilometers per second of the sun's motion with respect to the Local Standard of Rest as projected onto the line of sight to the source (VSUN), and the total velocity component in kilometers per second (V1). Positive velocity corresponds to increasing distance between source and observer.

The program DOP in Appendix A is written as a subroutine in a standard version[14] of the Fortran language with the following exception: certain variables are calculated in double precision as declared by nonstandard statements as noted by comments in the program.

ACKNOWLEDGMENTS

I thank M. S. Roberts and B. G. Clark for their vital comments.

[9] R. J. Trumpler and H. F. Weaver, "Statistical Astronomy," p. 251. Univ. of California Press, Berkeley, California, 1953.
[10] D. Mihalas and P. M. Routly, "Galactic Astronomy." Freeman, San Francisco, California, 1968.
[11] W. M. Smart, "Text-Book on Spherical Astronomy." Cambridge Univ. Press, London and New York, 1962.
[12] "Explanatory Supplement to the Astronomical Ephemeris." H.M. Stationery Office, London, 1968.
[13] D. A. McRae and G. Westerhout, Table for the Reduction of Velocities to the Local Standard of Rest. The Observatory, Lund, Sweden (1956).
[14] Amer. Std. Ass., Sect. Committee X3, *Commun. Ass. Comput. Machinery* 7, 590 (1964).

6.2. The Fast Fourier Transform*

6.2.1. Introduction

The Fourier transform is a particularly useful computational technique in radio astronomy. Since the publication in 1965 of an ultrafast calculational algorithm, the so-called fast Fourier transform,† this computation has become particularly efficient. This chapter presents the basic idea behind the fast Fourier transform (FFT), and shows how this algorithm is programmed on a digital computer. Because of the requirement for computational speed, a number of programs are given. These include short, moderately efficient subroutines for the transform of one-dimensional, complex data (FOURG and FOUR1). With the addition of a subroutine (FXRL1) to either of the above routines, real, one-dimensional data may be transformed in half the time with half the memory storage. Additional subroutines (CFFT2, RFFT2, and HFFT2) permit the transform of two-dimensional data. Finally, a program is given for transforming real, symmetric data for which only the cosine (or sine) transform is desired (FORS1). Listings of these programs, written in American National Standard Fortran,‡ are given in the Appendixes.

The Fourier integral

$$G(\omega) = 1/(2\pi)^{1/2} \int_{-\infty}^{\infty} dt \, e^{i\omega t} g(t) \tag{6.2.1}$$

may be well approximated by the trapezoidal sum

$$G(k \, \Delta\omega) = [\Delta t/(2\pi/N)^{1/2}][1/N^{1/2}] \sum_{j=0}^{N-1} g(j \, \Delta t) e^{2\pi i jk/N}, \qquad k = 0, \ldots, N-1, \tag{6.2.2}$$

---

† Basic references for the fast Fourier transform are collected in special issues of the *IEEE Transactions on Audio and Electroacoustics* for June 1967 and June 1969.

‡ In order to make the program listings in this chapter more generally useful, the versions of these subroutines in Appendixes A through H are written in a standard version of the Fortran language, the so-called American National Standard Fortran specified by the American Standards Association, Sectional Committee X3, *Commun. Ass. Comput. Machinery* 7, 590 (1964).

---

* Chapter 6.2 is by Norman Brenner

where it is assumed that j, k, and N are integers, and

(1) $g(t)$ is sampled at interval Δt and $G(\omega)$ at interval $\Delta \omega$,
(2) $g(t)$ is periodic outside the range $t = 0$ to $N \Delta t$ and $G(\omega)$ is periodic outside the range $\omega = 0$ to $N \Delta \omega$, and
(3) $\Delta t \, \Delta \omega = 2\pi/N$.

For further details, see Cooley et al.[1]

For simplicity of notation, we shall write $g(j)$ for $g(j \Delta t)$, $G(k)$ for $G(k \Delta \omega)$, $\mathbf{1}^{jk/N}$ for $e^{2\pi i jk/N}$, and drop $\Delta t/(2\pi/N)^{1/2}$. Thus, we call

$$G(k) = 1/N^{1/2} \sum_{k=0}^{N-1} g(j)\mathbf{1}^{jk/N}, \qquad k = 0, \ldots, N-1 \qquad (6.2.3)$$

the discrete Fourier transform (DFT) of the sampled function $g(j)$. The inverse DFT is

$$g(j) = 1/N^{1/2} \sum_{k=0}^{N=1} G(k)\mathbf{1}^{-jk/N}, \qquad j = 0, \ldots, N-1, \qquad (6.2.4)$$

as is easily shown by substituting Eq. (6.2.4) into Eq. (6.2.3) and interchanging the order of summation.

The time required to evaluate Eq. (6.2.3) may naively be estimated as that of N^2 complex multiplications and N^2 complex additions. For, if we assume that the complex exponentials $\mathbf{1}^{jk/N}$ are precalculated, each of the N values of $G(k)$ is computed by an N-term series. This method of computation of Eq. (6.2.3) takes no advantage of the fact that the samples $g(j)$ are equally spaced. The technique of the fast Fourier transform (FFT), by relying heavily on the equal spacing of the samples and on the periodicity of the integrand, reduces the running time by orders of magnitude.

In order to explain the FFT technique most clearly, we digress to introduce the concept of a two-dimensional Fourier transform:

$$G(\omega_1, \omega_2) = 1/2\pi \int_{-\infty}^{\infty} dt_1 \int_{-\infty}^{\infty} dt_2 \, g(t_1, t_2)e^{i\omega_1 t_1 + i\omega_2 t_2}. \qquad (6.2.5)$$

Sampling Eq. (6.2.5) in a similar way to Eq. (6.2.1), in steps of Δt_1 and $\Delta \omega_1 = 2\pi/N_1 \Delta t_1$, etc., we have

$$G(k_1, k_2) = 1/(N_1 N_2)^{1/2} \sum_{j_1=0}^{N_1-1} \sum_{j_2=0}^{N_2-1} g(j_1, j_2)\mathbf{1}^{(j_1 k_1/N_1) + (j_2 k_2/N_2)}, \qquad (6.2.6)$$

with $k_1 = 0, \ldots, N_1 - 1$ and $k_2 = 0, \ldots, N_2 - 1$.
A naive estimate of the running time of Eq. (6.2.6) is on the order of

[1] J. W. Cooley, P. A. W. Lewis, and P. D. Welch, *IEEE Trans. Audio Electroacoust.* **AU15**, 79 (1967).

$(N_1 N_2)^2$ operations (complex multiplications and additions). In fact, by performing the inner summation first in its entirety,

$$T(k_1, j_2) = 1/N_1^{1/2} \sum_{j_1=0}^{N_1-1} g(j_1, j_2) \mathbf{1}^{j_1 k_1/N_1}, \qquad (6.2.7)$$

and then performing the outer summation,

$$G(k_1, k_2) = 1/N_2^{1/2} \sum_{j_2=0}^{N_2-1} T(k_1, j_2) \mathbf{1}^{j_2 k_2/N_2}, \qquad (6.2.8)$$

we drastically reduce the number of operations. Equation (6.2.7) can be evaluated in $N_1^2 N_2$ operations and Eq. (6.2.8) in $N_1 N_2^2$, a total of $N_1 N_2 (N_1 + N_2)$ $\ll (N_1 N_2)^2$ for Eq. (6.2.6).

The essence of the FFT technique is that we can treat the one-dimensional DFT [Eq. (6.2.3)] as though it were a pseudo-two-dimensional one, and then reduce the running time by performing the inner and outer summations separately. Specifically, in Eq. (6.2.3) we must have a number of points N which is composite; i.e., N has two or more factors. (We will see later that the more factors N has, the shorter the running time.)

Let us assume that

$$N = f_1 f_2. \qquad (6.2.9)$$

Now, without physically rearranging them, let us give the elements of the vector $g(j)$ new, two-dimensional names $g(j_1, j_2)$, where

$$j = j_1 f_2 + j_2, \qquad j_1 = 0, \ldots, f_1 - 1, \quad j_2 = 0, \ldots, f_2 - 1. \qquad (6.2.10)$$

That is, the rectangular array $g(j_1, j_2)$ of sides f_1 and f_2 is stored rowwise in the vector $g(j)$. Similarly, give the elements of the vector $G(k)$ the new names $G(k_1, k_2)$, where

$$k = k_2 f_1 + k_1, \qquad k_1 = 0, \ldots, f_1 - 1, \quad k_2 = 0, \ldots, f_2 - 1. \qquad (6.2.11)$$

Note that $G(k_1, k_2)$ is stored columnwise.

Substituting Eqs. (6.2.9), (6.2.10), and (6.2.11) into Eq. (6.2.3), we have

$$G(k_1, k_2) = 1/(f_1 f_2)^{1/2} \sum_{j_1=0}^{f_1-1} \sum_{j_2=0}^{f_2-1} g(j_1, j_2) \mathbf{1}^{(j_1 k_1/f_1) + (j_2 k_1/N) + (j_2 k_2/f_2)}. \qquad (6.2.12)$$

Equation (6.2.12) is evaluated in three steps:

$$T(k_1, j_2) = 1/f_1^{1/2} \sum_{j_1=0}^{f_1-1} g(j_1, j_2) \mathbf{1}^{j_1 k_1/f_1}, \qquad (6.2.13)$$

$$U(k_1, j_2) = T(k_1, j_2) \mathbf{1}^{j_2 k_1/N}, \qquad (6.2.14)$$

$$G(k_1, k_2) = 1/f_2^{1/2} \sum_{j_2=0}^{f_2-1} U(k_1, j_2) \mathbf{1}^{j_2 k_2/f_2}. \qquad (6.2.15)$$

As expected, Eqs. (6.2.13) and (6.2.15) are just DFTs, of length f_1 and f_2, respectively. Together they take $f_1 f_2 (f_1 + f_2)$ operations, while the applications of the phaseshift factors in Eq. (6.2.14) take another $f_1 f_2$ complex multiplications. However, since $f_1 f_2 (f_1 + f_2 + 1) \ll N^2 = (f_1 f_2)^2$, this is far less running time than that needed for Eq. (6.2.3).

This is the FFT "trick." Making no approximations to Eq. (6.2.3), it exploits the periodicity of the complex exponential function to gain speed. For, if it were not true that $\mathbf{1}^{f_1 f_2 j_1 k_2 / N} = 1$, then it would be impossible to split Eq. (6.2.3) into Eqs. (6.2.13)–(6.2.15). For comparison, the kernel of the Hankel transform is a Bessel function, which is not periodic; hence, there exists no analogous fast Hankel transform.

Further speedups can be easily achieved by applying the FFT trick to Eqs. (6.2.13) and (6.2.15), each an ordinary DFT. It is easy to show that if

$$N = f_1 f_2 \cdots f_m, \tag{6.2.16}$$

then the final operation count by the FFT method is only

$$N(f_1 + f_2 + \cdots + f_m + m) \ll N^2. \tag{6.2.17}$$

For example, if $N = f^m$, this is $N(f + 1) \log_f N$ as compared to N^2.

Whenever the number of sampled function points is 100 or more, it would generally be faster to use FFT than any method running in a time interval proportional to N^2.

In order to explore the options available in designing an FFT program, we now give an explicit factorization scheme for the array indices. This leads to an explicit iterative method rather than the implicit, recursive method alluded to below Eq. (6.2.15). Let N be factored as in Eq. (6.2.16). In place of the indices j and k ($0 \leq j, k \leq N - 1$), introduce new indices j_α, k_β, where $0 \leq j_\alpha, k_\alpha \leq f_\alpha - 1$, and

$$j = \sum_{\alpha=1}^{m-1} j_\alpha f_{\alpha+1} \cdots f_m + j_m = \sum_{\alpha=1}^{m} j_\alpha N / p_\alpha,$$

$$k = \sum_{\beta=2}^{m} k_\beta f_{\beta-1} \cdots f_1 + k_1 = \sum_{\beta=1}^{m} k_\beta p_{\beta-1}, \tag{6.2.18}$$

where $p_\alpha = \prod_{\gamma=1}^{\alpha} f_\gamma$. We introduce the "multiradix index" notation

$$j = \{j_1 j_2 \cdots j_m\}, \qquad k = \{k_m k_{m-1} \cdots k_1\} \tag{6.2.19}$$

as shorthand for Eq. (6.2.18). In general, an arbitrary combination of sub-indices between braces, such as $j = \{j_5 j_7 j_2 j_1\}$, has the meaning of a single, compound index

$$j = j_5 f_7 f_2 f_1 + j_7 f_2 f_1 + j_2 f_1 + j_1. \tag{6.2.20}$$

This is analogous to the normal decimal (single radix) notation

$$1234 = 1 \times 10 \times 10 \times 10 + 2 \times 10 \times 10 + 3 \times 10 + 4.$$

It is easy to see that the range of a combination of subindices such as j in Eq. (6.2.20) is from 0 to $f_5 f_7 f_2 f_1 - 1$. One may also group subindices together into larger subindices with larger ranges; for example, it is simple to show that, in Eq. (6.2.19),

$$j = \{\{j_1 j_2 j_3\} j_4 j_5 \{j_6 \cdots j_m\}\},$$

where the index $\{j_1 j_2 j_3\} = j_1 f_2 f_3 + j_2 f_3 + j_3$, as usual.

Now, substitute Eq. (6.2.18) into Eq. (6.2.3):

$$G(\{k_m \cdots k_1\}) = 1/N^{1/2} \sum_{j_1=0}^{f_1-1} \cdots \sum_{j_m=0}^{f_m-1} g(\{j_1 \cdots j_m\}) \mathbf{1}^{\sum_{1 \leq \beta \leq \alpha \leq m} j_\alpha k_\beta p_\beta - 1/p_\alpha}. \quad (6.2.21)$$

The summation in the exponent may be rewritten in two ways, by summing first on α or on β:

$$\sum_{1 \leq \beta \leq \alpha \leq m} j_\alpha k_\beta p_{\beta-1}/p_\alpha = \sum_{\alpha=1}^{m} (j_\alpha/p_\alpha)\left(\sum_{\beta=1}^{\alpha} k_\beta p_{\beta-1}\right) = \sum_{\alpha=1}^{m} j_\alpha \{k_\alpha \cdots k_1\}/p_\alpha \quad (6.2.22a)$$

$$= \sum_{\beta=1}^{m} (k_\beta p_{\beta-1})\left(\sum_{\alpha=\beta}^{m} j_\alpha/p_\alpha\right) = \sum_{\beta=1}^{m} k_\beta \{j_\beta \cdots j_m\} p_{\beta-1}/N. \quad (6.2.22b)$$

These two rules are called "decimation in time" and "decimation in frequency." A characteristic of an FFT program depends on which of these two rules is used.

Now, split Eq. (6.2.21) into a series of $m + 1$ partial transformations. After the αth stage, we have $G^{(\alpha)}(k_1, \ldots, k_\alpha, j_{\alpha+1}, \ldots, j_m)$, where, using Eq. (6.2.22a):

$$G^{(0)}(j_1, \ldots, j_m) = (1/N^{1/2})g(j_1, \ldots, j_m),$$

$$G^{(\alpha)}(k_1, \ldots, k_\alpha, j_{\alpha+1}, \ldots, j_m)$$
$$= \sum_{j_\alpha=0}^{f_\alpha-1} G^{(\alpha-1)}(k_1, \ldots, j_\alpha, \ldots, j_m) \mathbf{1}^{j_\alpha \{k_\alpha \cdots k_1\}/p_\alpha}, \quad (6.2.23)$$

$$G(k_1, \ldots, k_m) = G^{(m)}(k_1, \ldots, k_m).$$

Alternately, the complex exponential may be $\mathbf{1}^{k_\alpha \{j_\alpha \cdots j_m\} p_{\alpha-1}/N}$ by (6.2.22b).

Note that

$$\{k_\alpha \cdots k_1\} = k_\alpha p_{\alpha-1} + \{k_{\alpha-1} \cdots k_1\}, \quad (6.2.24)$$

so that

$$\mathbf{1}^{j_\alpha \{k_\alpha \cdots k_1\}/p_\alpha} = \mathbf{1}^{(j_\alpha k_\alpha/f_\alpha) + (j_\alpha \{k_{\alpha-1} \cdots k_1\}/p_\alpha)};$$

i.e., the complex exponential consists of the kernel of a DFT of length f_α that converts subindex j_α into k_α, times a phase shift factor.

Ostensibly, $G^{(\alpha)}$ is an m-dimensional array, and is thus awkward to manipulate in a computer language like Fortran (m nested DO-loops must be written, etc.). In fact, the m subindices may be conveniently grouped into the subindices actually used in Eq. (6.2.23): k_α (or j_α), $K_\alpha = \{k_{\alpha-1} \cdots k_1\}$, which ranges from 0 to $p_\alpha - 1$, and $J_\alpha = \{j_{\alpha+1} \cdots j_m\}$, ranging from 0 to $N/p_\alpha - 1$. Now, Eq. (6.2.23) can be rewritten as a DFT plus phase shift of a merely three-dimensional array $G^{(\alpha-1)}$ along one of its subscripts:

$$G^{(\alpha)}(K_\alpha, k_\alpha, J_\alpha) = \sum_{j_\alpha=0}^{f_\alpha-1} G^{(\alpha-1)}(K_\alpha, j_\alpha, J_\alpha) \mathbf{1}^{(j_\alpha k_\alpha/f_\alpha) + (j_\alpha K_\alpha/p_\alpha)}. \quad (6.2.25)$$

Alternately, the last complex exponential may be $\mathbf{1}^{k_\alpha J_\alpha p_\alpha - 1/N}$.

Equation (6.2.25) is much easier to implement in Fortran than Eq. (6.2.23).

Finally, we must specify a rule by which the elements of the m-dimensional array $G^{(\alpha)}$ are mapped into an equivalent vector, as all computer memories, are in fact linearly addressed. Equation (6.2.19) shows the mapping rules that must be used for the arrays $G^{(0)}(j_1, \ldots, j_m)$ and $G^{(m)}(k_1, \ldots, k_m)$. Important characteristics of the FFT program depend on the choice of rule for the general case $G^{(\alpha)}(k_1, \ldots, k_\alpha, j_{\alpha+1}, \ldots, j_m)$.

6.2.1.1. The Natural Transform.   The mapping rule chosen is

$$L_\alpha = \{k_\alpha \cdots k_1 j_{\alpha+1} \cdots j_m\}. \quad (6.2.26)$$

That is, element $G^{(\alpha)}(k_1, \ldots, k_\alpha, j_{\alpha+1}, \ldots, j_m)$ is found in vector position $G^{(\alpha)}(L_\alpha)$. It is important to notice that the index L_α is composed in a simple way of the important indices j_α, J_α, k_α, and K_α in Eq. (6.2.25):

$$L_\alpha = \{k_\alpha K_\alpha J_\alpha\}, \qquad L_{\alpha-1} = \{K_\alpha j_\alpha J_\alpha\}, \quad (6.2.27)$$

so that Eq. (6.2.25) may be written

$$G^{(\alpha)}(L_\alpha) = \sum_{j_\alpha=0}^{f_\alpha-1} G^{(\alpha-1)}(L_{\alpha-1}) \mathbf{1}^{(j_\alpha k_\alpha/f_\alpha) + (j_\alpha K_\alpha/p_\alpha)}. \quad (6.2.28)$$

An FFT program based directly on Eqs. (6.2.27) and (6.2.28) is given in Appendix B. This program, FOURG, wastes both space and time to a degree: it requires a working storage array the same size as the data array, and the DFT of length f_α is performed by the same general (and inefficient) code whether $f_\alpha = 2, 3$, or larger. In many practical uses of FFT, however, the FFT calculation takes much less time than the other operations to be performed on the data (input/output, etc.). However, when the FFT calculation is the limiting constraint and space in memory is limited, it would be better to use the following mapping rule.

6.2.1.2. The In-Place Transform.   The mapping rule is

$$L_\alpha = \{ j_m \cdots j_{\alpha+1} k_\alpha \cdots k_1 \}. \tag{6.2.29}$$

Equation (6.2.28) is again the partial transformation rule, but Eq. (6.2.30) replaces Eq. (6.2.27):

$$J_\alpha' = \{ j_m \cdots j_{\alpha+1} \}, \qquad L_\alpha = \{ J_\alpha' k_\alpha K_\alpha \}, \qquad L_{\alpha-1} = \{ J_\alpha' j_\alpha K_\alpha \}. \tag{6.2.30}$$

J_α' is the so-called digit reversal of J_α in the mixed radix notation $(f_1, \ldots, f_m)$. Program FOUR1 in Appendix C is based on Eqs. (6.2.28) and (6.2.30) and needs no working storage, for it is seen that one can compute Eq. (6.2.28) by holding J_α' and K_α fixed and computing on the vector subset of $G^{(\alpha-1)}$ with varying j_α. The results are then placed in the output array $G^{(\alpha)}$ at locations with varying k_α. From Eq. (6.2.30), these locations are the same as those with varying j_α, so the same storage area may be used for both arrays $G^{(\alpha-1)}$ and $G^{(\alpha)}$. Unfortunately, the rule for L_0 in Eq. (6.2.29) is the digit reversal of the rule [Eq. (6.2.19)] for j. Hence, program FOUR1 must initially permute the input array $g(j)$ before assigning it to $G^{(0)}$; this permutation is done in place by the method described in Section (6.2.3). It would be simple to write an in-place FFT program using Eq. (6.2.22b) that transforms normally ordered data into permuted data, and a complementary program using Eq. (6.2.22a) to inverse transform permuted data into normally ordered data. The time for doing bit reversal explicitly (about 25% of the total time in program FOUR1) would then be saved.

Program FOUR1 is restricted to transforming a power-of-2 number of points for four reasons:

(1) the permutation of $g(j)$ by digit reversal in place is easiest when all the f_α are the same;

(2) the running time of an FFT, as given in Eq. (6.2.17), is minimized if all f_α are equal to e; in practice, when all $f_\alpha = 2, 3,$ or 4;

(3) when $f_\alpha = 2$, the exponential $1^{j_\alpha k_\alpha / f_\alpha}$ can assume only the values $+1$ or -1, with an elimination of several complex multiplications; similar savings occur if $f_\alpha = 4$ or 8;

(4) most experimental situations may be manipulated to produce a string of data of any length; a power-of-2 length is then no hardship.

6.2.1.3. The Fixed-Indexing Transform.   The last major variation of the FFT is given by the mapping rule

$$L_\alpha = \{ k_\alpha \cdots k_1 j_n \cdots j_{\alpha+1} \} \qquad \text{or} \qquad L_\alpha = \{ k_\alpha K_\alpha J_\alpha' \}, \qquad L_{\alpha-1} = \{ K_\alpha J_\alpha' j_\alpha \}.$$

Inserting this rule in Eq. (6.2.28), we see that the inputs $G^{(\alpha-1)}$ (i.e., for varying j_α) are separated by 1, and the outputs $G^{(\alpha)}$ (i.e., for varying k_α) are separated by N/f_α. Thus, when all factors f_α are the same, the indexing is the same for every partial transformation. This is a major advantage if the FFT is to be implemented in hardware; similarly, Singleton[2] organizes his FFT of tape-stored data around this rule. The disadvantages of this rule are that the data must be permuted initially, and that at least $N[1 - (1/f_\alpha)]$ elements of working storage are needed.

6.2.1.4. The Transform of Data with Many Zeros. Occasionally, a small number N_1 of input data are to be transformed to a large number N_2 of output values. The standard technique is to append $N_2 - N_1$ zeros to the input and use the normal FFT. However, many of the operations are superfluous since they operate on zero data. When both N_1 and N_2 are powers of two, Markel[3] has given a simple trick which reduces the running time by a factor of approximately $\log_2 N_2/[(\log_2 N_1) + 1]$. Program FOUR1 may be modified to achieve this speedup by inserting the new statement

ISTEP = MIN0(ISTEP,IMAX/N1)

immediately following the definition of ISTEP. The total number of data in FOUR1, called N, is, of course, N_2; N_1, called N1 above, should be included in the calling sequence.

Similarly, if only the first N_3 ($\ll N_2$) output values are desired, a time factor of $\log_2 N_2/[(\log_2 N_3) + 1]$ may be saved by changing the statement

DO 80 I1 = 1,ITWOL,IP0

to

I1MAX = MIN0(ITWOL,IP0*N3)
DO 80 I1 = 1,I1MAX,IP0

Similar changes apply to the other programs.

6.2.1.5. The Transform of Real Data. If the input $g(j)$ is real, then the transform $G(k)$ is complex; but only $(N/2) + 1$ values of $G(k)$ are independent for, from Eq. (6.2.3), the complex conjugate

$$G^*(k) = G(-k), \tag{6.2.31}$$

where $-k$ is taken modulo N. Therefore, it is sufficient to retain in storage only $G(0)$ through $G(N/2)$. These complex values take up slightly more storage than the N real values input from $g(j)$.

If N, the number of input data, is even, a very simple trick may be used to transform N real data in half the space and half the time of N complex data. First, from the real array $g(j)$, construct a complex array $c(j)$ of half the length by

$$c(j) = g(2j) + ig(2j + 1),   j = 0, \ldots, (N/2) - 1. \tag{6.2.32}$$

[2] R. C. Singleton, *IEEE Trans. Audio Electroacoust.* **AU15**, 91 (1967).
[3] J. Markel, *IEEE Trans. Audio Electroacoust.* **AU19**, 305 (1971).

Now, transform $c(j)$ into $C(k)$:

$$C(k) = 1/(N/2)^{1/2} \sum_{j=0}^{(N/2)-1} c(j)\mathbf{1}^{jk/(N/2)}$$

$$= (2/N)^{1/2} \sum_{j=0}^{(N/2)-1} [g(2j) + ig(2j+1)]\mathbf{1}^{2jk/N}, \qquad k = 0, \ldots, (N/2) - 1.$$

Notice that $C(N/2) = C(0)$. Changing k to $(N/2) - k$ and conjugating, we have

$$C^*((N/2) - k) = (2/N)^{1/2} \sum_{j=0}^{(N/2)-1} [g(2j) - ig(2j+1)]\mathbf{1}^{2jk/N}.$$

Then, from Eq. (6.2.3),

$$G(k) = 1/N^{1/2} \sum_{j=0}^{(N/2)-1} [g(2j)\mathbf{1}^{2jk/N} + g(2j+1)\mathbf{1}^{(2j+1)k/N}]$$

$$= (1/2\sqrt{2})[(C(k) + C^*((N/2) - k)) - i\mathbf{1}^{k/N}(C(k) - C^*((N/2) - k))].$$
$$(6.2.33)$$

Program FXRL1 in Appendix D implements Eqs. (6.2.32) and (6.2.33). An alternate, more complicated way of transforming real data is given by Bergland.[4]

6.2.1.6. The Transform of Multidimensional Data. As shown above, a two-dimensional FFT [Eq. (6.2.6)] may be performed as a set of in-place, one-dimensional transforms on the rows and columns [Eqs. (6.2.7) and (6.2.8)]. Programs CFFT2, RFFT2, and HFFT2 in Appendix E implement Eq. (6.2.6) when transforming complex arrays, real arrays, and transforms of real arrays, respectively. An $N_1 \times N_2$ real array is transformed into an $((N_1/2) + 1) \times N_2$ complex array, since

$$G^*(k_1, k_2) = G(-k_1, -k_2), \qquad (6.2.34)$$

taking the subscripts modulo N_1 and N_2.

Faster versions of the FFT can be written by restricting the number of data to a power of two and factoring by fours. This gives a 25% speedup over factoring by twos. Bergland[4] points out that factoring by eights is 10% faster yet, but the program size quadruples. Further speedup can be obtained by computing complex exponentials by Singleton's recursion.[5] The author has written a program named FOUR2 that incorporates factoring by four and use of the above recursion. This program uses in-place mapping and therefore requires no working storage at all. Copies of this program may be obtained from the author upon request.

---

[4] G. D. Bergland, *IEEE Trans. Audio Electroacoust.* **AU17**, 138 (1969).
[5] R. C. Singleton, *IEEE Trans. Audio Electroacoust.* **AU17**, 93 (1969).

6.2.1.7. The Transform of Real Symmetric Data. If the input data $g(j)$ are real and even (or odd) symmetric around the origin, i.e.,

$$g(-j) = g(N-j) = \pm g(j), \qquad (6.2.35)$$

then the transform $G(k)$ will be real even (or imaginary odd). Thus, it is sufficient to retain storage for only $g(j)$, $j = 0, \ldots, N/2$, which is one quarter of the storage needed for N complex data. If N is a multiple of 4, a simple trick (Rabiner et al.[6]) will transform $g(j)$ in this limited space in about one quarter to one third the time of transforming N values of complex data.

First, from the real array $g(j)$, construct the complex array $c(j)$ of length $N/2$ by

$$c(j) = \begin{cases} g(2j) + i[g(2j+1) - g(2j-1)] & \text{for even symmetry,} \\ [g(2j+1) - g(2j-1)] + ig(2j) & \text{for odd symmetry.} \end{cases} \qquad (6.2.36)$$

In either case, $c^*(j) = c((N/2) - j)$. As in the case of Eqs. (6.2.32) and (6.2.33), we take the inverse transform of $c(j)$ to obtain $C(k)$ by means of

$$G(k) = \begin{cases} \left[\dfrac{C(k) + C((N/2) - k)}{2\sqrt{2}} \right] + \left[\dfrac{C(k) - C((N/2) - k)}{4\sqrt{2}\,\sin(2\pi k/N)} \right], \\[4mm] i\left[\dfrac{C(k) + C((N/2) - k)}{4\sqrt{2}\,\sin(2\pi k/N)} \right] + i\left[\dfrac{C((N/2) - k) - C(k)}{2\sqrt{2}} \right]. \end{cases} \qquad (6.2.37)$$

Also,

$$G(0) = 1/N^{1/2} \sum_{j=0}^{N-1} g(j) \quad \text{and} \quad G(N/2) = 1/N^{1/2} \sum_{j=0}^{N-1} (-1)^j g(j).$$

Program FORS1 in Appendix F implements these equations.

6.2.1.8. The Transform of Very Large Arrays. It is sometimes necessary to transform data arrays too large to fit into the internal, core memory of the computer. Either the main memory is very limited, as on most mini-computers, or many data points are necessary for high-frequency resolution. Singleton[2] has given a procedure for the FFT of an array held in an external storage device with linear access, such as tape. If a direct access device, such as a disk or drum, is available, a faster version of the FFT that needs no working storage can be used. Such a program, called FOR2D, is available from the author.

6.2.1.9. Program Accuracy. Gentleman and Sande[7] give an upper bound for the root-mean-square of the relative errors of the transform values:

[6] L. R. Rabiner, R. W. Schafer, and C. M. Rader, *IEEE Trans. Audio Electroacoust.* AU17, 86 (1969).

[7] W. M. Gentleman and G. Sande, *Proc. Fall Joint Comput. Conf., Washington, D.C., 1966* 29, 563 (1966).

$2^{-b}3 \sum_\alpha f_\alpha^{1.5}$, where f_α are the factors of the number of data, and b is the number of bits in the floating point fraction. Since the FFT performs fewer arithmetic operations than conventional methods, it builds up less error.

6.2.2. Program Calling Sequences

Each program included in the Appendixes is prefaced by a description of its use; what follows here is an overview. The calling sequence of the given programs are

CALL FOURG(CDATA1,N1,ISIGN,WORK1)

CALL FOUR1(CDATA1,N1,ISIGN)

CALL FXRL1(RDATA1 or HDATA1,N1,ISIGN,IFORM)

CALL CFFT2(CDATA2,N1,N2,ISIGN)

CALL RFFT2(RDATA2,N1,N2,HDATA2,N1/2+1,ISIGN)

CALL HFFT2(RDATA2,N1,N2,HDATA2,N1/2+1,ISIGN)

CALL FORS1(SDATA1,N1,ISIGN,ISYM)

where

COMPLEX CDATA1(N1),CDATA2(N1,N2),HDATA1(N1/2+1),
HDATA2(N1/2+1,N2),WORK1(N1)

REAL RDATA1(N1),RDATA2(N1,N2),SDATA1(N1/2+2),

ISIGN $= +1$ or -1 represents the transform direction.

In FOUR1, the length of the array must be a power of 2. In FXRL1, the length N1 must be even. In FORS1, the length N1 must be divisible by 4. In the other routines, the lengths of the arrays may be any positive integer, though a highly factorable length runs fastest; cf. Eq. (6.2.17).

In all cases, the transform done is Eq. (6.2.3) or (6.2.4) in each dimension. For ordinary complex (IFORM $= +1$) data, CDATA1$(J+1) = g(J)$. For other kinds of data, only the nonredundant parts are passsed. Real (IFORM $= 0$) data are passed by RDATA1$(J+1) = \text{Re}(g(J))$; transform of real (IFORM $= -1$) data are passed by HDATA1$(J+1) = g(J)$ for $J = 0, \ldots, N/2$. Real symmetric data are passed to FORS1 by SDATA1$(J+1) = \text{Re}(g(J)), J = 0, \ldots, N/2$, though for technical reasons, SDATA1 is of length $(N/2) + 2$; the transform is returned by FORS1 in SDATA1$(K+1) = \text{Re}(G(K))$ for even symmetry, $\text{Im}(G(K))$ for odd. ISYM $= +1$ for even symmetry, -1 for odd.

IFORM $= 0$ arrays transform to IFORM $= -1$, and vice versa; IFORM $= +1$ to IFORM $= +1$.

The data are arranged in column-major order with real and imaginary parts adjacent, as is normal in Fortran. For example, in two dimensions, the order is $(1,1),(2,1),\ldots,(N1,1)$, or $(N1/2+1,1),(1,2),(2,2),\ldots,(N1,N2)$, or $(N1/2+1,N2)$.

6.2.3. Programming Implementation

In the in-place variation of the FFT used in program FOUR1, the data array must be permuted by "bit reversal." That is, DATA(I) must be exchanged with DATA(IREV), where IREV-1 expressed in binary is the mirror image of $I-1$; cf. Eq. (6.2.19). For example, with $N = 16$, the indices $0,\ldots,15$ become $0, 8, 4, 12, 2, 10, 6, 14, 1, 9, 5, 13, 3, 11, 7$, and 15, i.e., $0000, 1000, 0100$, etc.

Program FOUR1 performs the permutation in one pass. The IREV are generated in order, since the reversal of $I+1$ is obtainable easily from the reversal of I by adding 1 to its high-order bit and propagating the carry downward. Then, if IREV is less than I (to prevent reexchanges), DATA(I) is exchanged with DATA(IREV).

6.2.4. Program Testing

In testing the programs with a simple function such as a square wave, do not expect the transform to be that given by the infinite integral Fourier transform. Rather, use the summation definition in Eq. (6.2.3). A good discussion of the difference between the two kinds of Fourier transforms is given by Cooley et al.[1] Most importantly, remember that the FFT considers an input function $g(j)$ to be periodic every N points: $g(j + N) = g(j)$.

Two useful test functions are:

(a) the sawtooth: input $g(j) = j$; output $G(k) = N^{1/2}(1^{k/N} - 1)^{-1}$ for $k \neq 0$, $G(0) = N^{1/2}(N - 1)/2$.

(b) the train of square pulses: input $g(j) = 1$ for $j < M$, 0 for $j \geq M$; output $G(k) = (1^{Mk/N} - 1)/[N^{1/2}(1^{k/N} - 1)]$ for $k \neq 0$, $G(0) = M/N^{1/2}$.

6.3. Data Presentation Techniques

6.3.1. Contour Mapping*

Contours of equal functional value are traced through a given two-dimensional array. The coordinates of points along each contour are passed to a user-supplied subroutine, which plots them on an X–Y plotting device. We include two contour-mapping programs, the first for use with plotting devices having mechanically operated pens, and the second for devices with cathode-ray-tube displays. The first program, Appendix G, produces a series of connected line segments so that a contour may be drawn continuously; the second program, Appendix H, produces contour segments in almost random order. Figure 1 shows an example of a computer-generated contour map.

Let us assume that the given array DATA(I,J) specifies the value of a function of two variables $f(x, y)$ at each point of a rectangular grid (X(I), Y(J)), for I = 1, ..., NUSD1 and J = 1, ..., NUSD2. The arrays X(I) and Y(J) must be monotonic, though not necessarily equispaced. A contour, as traced out, will have no more than one point lying on each segment of the grid; contours may be made more rounded, therefore, by supplying more highly interpolated data in a denser grid. Contour points, if they do not fall on a grid intersection, are computed by linear interpolation.

The contour-mapping program CONTR is written as a subroutine. A single call to CONTR may specify several different contour levels, but only equispaced. Nonequispaced contour levels may be traced by calling CONTR repeatedly.

The algorithm takes into account special cases. Single-point contours and contours parallel to one side or the other of the grid are all found. The logic is straightforward and no scratch storage is required. The contours are found accurately enough to be traced out in the left and right halves of the data grid separately (if the whole is too large to be held in core), with the two sets meshing smoothly.

6.3.1.1. Calling Statements. Both contour-tracing programs are called by CALL   CONTR(DATA,NDIM1,NDIM2,NUSD1,NUSD2,CBEG,CLIM, CSTP), where DATA is the floating point array containing the function values. This array is dimensioned NDIM1 rows by NDIM2 columns, but

---

* Section 6.3.1 is by **Norman Brenner and Stanley H. Zisk**

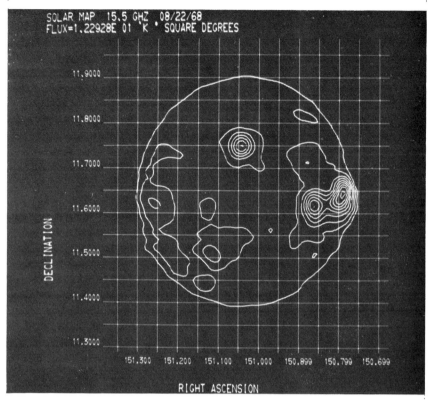

Fig. 1. Contour map of the sun observed with the 120-ft diameter Haystack antenna at a frequency of 15.5 GHz.

only the first NUSD1 rows and the first NUSD2 columns are actually used for data. For example, suppose

DIMENSION DATA(100,100)

were used to anticipate data of unknown amount, but only an array of size 50 by 60 were read in. Then NDIM1 = NDIM2 = 100, NUSD1 = 50, and NUSD2 = 60.

A set of contours will be traced, at heights CBEG, CBEG ± CSTP, CBEG ± 2*CSTP, ..., the sign being chosen to direct values toward CLIM, which may not be exceeded. If CSTP = 0, or CBEG = CLIM, only one contour is plotted.

The user must supply a subroutine CPLOT to act as interface between CONTR and the plotting device. As each new point of a contour is found by CONTR, it calls CPLOT:

CALL CPLOT(IPLOT,X1,X2,CTR)

IPLOT $= 0$, 1, or 2 depending upon whether this is the beginning of a new contour or contour segment (0), a continuation of a contour (1), or the one point on a contour where a label would be appropriate (2). (X1, X2) are the floating point coordinates of the row and column of the current point; $1 \leq X1 \leq NUSD1$ and $1 \leq X2 \leq NUSD2$, though X1 and X2 are not integral. CTR is the value of the current contour. It is CPLOT's duty to convert the current point to plotter coordinates (which may involve conversion to polar or other coordinates), and either draw a line segment from the previously given point to the current one, or write a numerical label giving CTR here.

6.3.1.2. Continuous-Contour Algorithm. Define the "grid segment" to the right of $(X(I),Y(J))$ as the straight line connecting it to $(X(I+1), Y(J))$, and the grid segment above as the line connecting it to $(X(I),Y(J+1))$. A contour is said to cross a grid segment if the contour height falls between the function values of the endpoints; exactly where it crosses is determined by linear interpolation. Crossings are marked as they are plotted to prevent their reuse. If the grid segment to the right of $(X(I),Y(J))$ is crossed, the low bit of the function value there is set to one; if the segment above is crossed, the second lowest bit is set. Initially, all low bits are set to zero.

The method of contour tracing is to search the grid of data values for a first crossing, plot consecutive adjacent crossings away from it in one direction as far as possible, then return to the first crossing and plot away from it in the other direction. Contouring continues until all possible crossings at a given height have been found and plotted. In detail, there are five phases.

Phase 1: *initial crossing.* Every grid segment is searched in order for an unused crossing. If one is found, it is marked and saved for later use as the "first crossing," and the plotting device is positioned here (CALL CPLOT(0,...)). Then phase 2 begins.

Phase 2: *next crossing.* The six grid segments adjacent to the current crossing (or eight, if the crossing lies exactly on a grid point) are searched for an unused crossing. The first segments searched are those lying along the current direction (left/right, up/down) of the contour. If the three segments to one side are all eligible, the nearest is chosen, to prevent an X-type crossing later. If a crossing is found, it is marked and the plotting device draws a straight line here (CALL CPLOT(1,...)); phase 2 continues. If no unused crossing can be found, phase 3 begins.

Phase 3: *termination.* The contour in this direction has come to an end. At most, one more point can be plotted to tie this contour end to a previously drawn contour. Therefore, a marked crossing in the neighborhood of the current crossing is searched for, exactly as in phase 2. (The crossing just previously plotted is ineligible, of course.) If one is found, a line is drawn to it. In any event, phase 4 now begins.

Phase 4: *labeling.* If the last-plotted crossing is the same as the first crossing, phase 5 begins. Else, write a numerical label on top of the current crossing giving the value of the contour height (CALL CPLOT(2,...)). Then position the plotting device at the first crossing again, and return to phase 2 to search for the contour's continuation in the direction opposite to that already drawn. If both directions away from the first crossing have been plotted, return to phase 1, instead.

Phase 5: *labeling a closed contour.* Write a numerical label and return to phase 1. A one-point contour is handled here as a special kind of closed contour.

To prevent the crisscrossing of a plateau with meaningless lines, a grid segment with equal end-point values is automatically ineligible for crossing.

Since changing the low-order bits of a datum changes its value slightly, all comparisons are done within a "fuzz" of a computed, small epsilon. Also, if fixed-point data are used, they should be large enough not to be significantly affected by bit marking.

A listing of this program written in ANS Fortran is contained in Appendix G.

6.3.1.3. Segmented-Contour Algorithm. Nonmechanical plotting devices can make use of a much simpler algorithm. Merely scan through the grid of data values, examining every square formed by four data points. If a contour crossing occurs on any two of the sides of a square, connect them. If contour crossings occur on all four sides, draw two contour segments, connecting appropriate adjacent sides. No bit marking is necessary, as each square of data points is examined only once. Appendix H contains a listing of this program written in ANS Fortran.

Notice that this short algorithm will produce contour segments in an almost random order. A mechanical pen plotter would therefore waste much of its time moving from side to side of the plotting area.

6.3.2. Ruled-Surface Mapping*

The program described here draws a ruled-surface map representing the values of a function of two variables. Pictorially, it appears most like a two-dimensional projection (without perspective) of a mountainous landscape. The independent variables are x and y, and the values of the function are indicated by the height at each point. The view direction may have an azimuth of $0°$, $90°$, $180°$, or $270°$, and any elevation between $0°$ and $90°$.

The drawing is composed of a set of lines across the landscape parallel to the front edge of the picture, each line being a slice through the function

---

* Sections 6.3.2 and 6.3.3 are by Norman Brenner.

Fig. 2. Ruled-surface map of the sun observed with the 120-ft diameter Haystack antenna at a frequency of 15.5 GHz.

values, as could be seen with cardboard cutouts. Nearer peaks occlude farther ones, adding to the three-dimensional effect. Increasing the number of ruled lines increases the effective shading of hollows and hills. Figure 2 shows an example of a ruled-surface presentation.

Ruled surface mapping is, perhaps, superior to the contour drawing when a feel for the relative heights of parts of the data must be had; it is not as good for analytical study, since it is difficult to indicate the precise values of x and y of every datum.

6.3.2.1. Calling Statement. The ruled-surface mapping program, Appendix I, is written as a subroutine with the following calling statement:

CALL RULED(DATA,NROW,NCOL,NRUSD,NCUSD,VMAX,
AZ,EL,NLINE,ACRAT,HT)

The variable DATA is a floating-point, two-dimensional array containing the function values. DATA was dimensioned with NROW rows and NCOL columns, but only the first NRUSD rows are actually used for data and only the first NCUSD columns. For example, suppose

DIMENSION DATA(100,100)

were used to anticipate data of unknown amount, but only an array of size 50 by 60 were read in. Then NROW = 100, NCOL = 100, NRUSD = 50, and NCUSD = 60. Both NRUSD and NCUSD must be less than 300; otherwise, array HIGH in RULED must be dimensioned larger.

VMAX is the value of the largest number in DATA(I,J), the smallest being assumed to approximate zero. VMAX is always floating-point. (VMAX affects only the apparent height of the peaks on the map; its effects are balanced by HT.)

A ruled-surface map is drawn from apparent vantage point at azimuth AZ, elevation EL. AZ is in degrees and is rounded to the nearest multiple of 90° (AZ is taken modulo 360°). Table I indicates the orientation of the picture for each value.

TABLE I.   Display Orientation as Determined by the Parameter AZ

AZ	Left to right	Bottom to top
0°	Increasing column	Increasing row
90°	Increasing row	Decreasing column
180°	Decreasing column	Decreasing row
270°	Decreasing row	Increasing column

When the EL parameter equals 0°, the landscape is being viewed horizontally, giving a silhouette effect; when EL equals 90°, only a set of parallel straight lines appear. A typical value is 30°.

NLINE is the (integer) number of lines drawn. An average map has 20–30 lines. Note that NLINE is independent of either NRUSD or NCUSD, as linear interpolation on the data is performed in the direction along the line of sight. For shading effect, the number of lines should be about 70 to 100.

The ratios of depth to width and height to width are called RCRAT and HT, respectively. RCRAT is literally the ratio of the apparent size of the number of rows to the number of columns. Thus, even though NRUSD = 50 and NCUSD = 60, the landscape may be forced to have a square base by setting RCRAT = 1. If the landscape is to have a rectangular base proportional to the number of rows and columns, set RCRAT = .833. The rectangularity is correctly rotated as the viewer's AZ is changed.

The highest peak visible (normally, that corresponding to the value of VMAX) will have an apparent height, relative to the number-of-columns side, of HT cos EL. A normal value for HT is 1. Larger values will increase the apparent heights of all peaks and deepen all valleys; smaller values will flatten the scene.

6.3.2.2. The Algorithm RULED. For flexibility, RULED does no output to the plotting device directly. Instead, it calls two user-supplied interface subroutines, RLIMS and RPLOT.

CALL RLIMS(IAZ,XMIN,XMAX,YMIN,YMAX) is called at the beginning of each map. Its function is to clear the plotting device and set up the left and right plotting boundaries (XMIN and XMAX) and the bottom and top boundaries (YMIN and YMAX). All four numbers will be computed by RULED and passed to RLIMS. They are generally some permutation of 1, NRUSD, 0, and NCUSD, with factors of cos EL thrown in. RLIMS uses them only to set up the boundaries. IAZ is 0, 1, 2, or 3, as AZ is 0°, 90°, 180°, or 270°. It is useful for titling the display.

CALL RPLOT(IPLOT,XPLOT,YPLOT) is called for each vertex of each line of the map. IPLOT = 0 or 1 depending upon whether this is the beginning of a new line or the continuation of the current line. XPLOT lies between XMIN and XMAX, and YPLOT between YMIN and YMAX. It is RPLOT's duty to draw a line segment from the previously given point to the new one, or just to remember the new one (if IPLOT = 0).

To clear out the XX and YY arrays, RULED calls RPLOT once at the very end:

CALL RPLOT(0,0,0.)

which will automatically plot the last few segments.

RPLOT should not attempt to convert the XPLOT and YPLOT coordinates to any nonrectangular coordinates. If the row and columns of the original data had represented polar coordinates, for example, the picture built up by RULED would be completely distorted, as it assumes rectangular coordinates when deleting hidden lines.

The array of DATA is rotated conceptually, and values are identified by (IWIDE,IDEEP), rather than (I,J) coordinates, as shown in Table II. Com-

TABLE II.   Index Assignment as Determined by Parameter AZ

AZ	IWIDE	IDEEP
0°	J	I
90°	I	NCUSD−J
180°	NCUSD−J	NRUSD−I
270°	NRUSD−I	J

pare this table with Table I. IWIDE runs from 1 to either NRUSD or NCUSD. IDEEP runs from 1 to NLINE; if it falls between two columns or two rows' a linearly interpolated value is used. IWIDE always falls exactly on a row or column. At a particular (IWIDE,IDEEP), RULED/IRULE finds the value DATA(I,J) and computes the apparent position of the datum:

$$XPLOT = IWIDE$$
$$YPLOT = RCRAT'*((NWIDE-1)/(NDEEP-1))*(IDEEP-1)*$$
$$SINEL + HT*(DATA(I,J)/VMAX)*COSEL$$

where RCRAT' is either RCRAT or its reciprocal, NDEEP is NRUSD' or NCUSD (the range of IDEEP), and NWIDE is the range of IWIDE.

The wide range of values possible from this transformation makes it imperative that a subroutine like RLIMS be called with the minimum and maximum values.

In this manner, a set of apparent positions along a slice through the data is computed. Each apparent position is compared with the highest already drawn on the screen in that X position (drawing proceeds from front to back). If the new position is invisible, being hidden by the nearer "peak" from the viewer, no line is drawn to it. If it can be seen, a line segment is drawn to it, and it is the new "highest position" at that X coordinate.

Lines are drawn alternately from left to right and right to left to minimize paper or pen rewinding on the mechanical plotter.

If the width required for the data array exceeds capacity, the following error message is printed: ERROR IN RULED SURFACE MAP. FOR AX = XX, THE MAP IS XXXX POINTS WIDE, WIDER THAN THE MAXIMUM MMM. (MMM is usually 300.) A normal return is made with no lines being drawn.

6.3.3. Gray-Scale Mapping

Radio telescopes cannot directly produce photographs of the sky as do optical telescopes. However, it is possible to generate displays of radio astronomical data in which shades of gray represent intensities of the observed radio signals. We refer to such displays as gray-scale maps, analogous to the contour and ruled-surface maps discussed in the preceding sections. In this section we discuss two programs for generating gray-scale maps. The first program, GRAYM in Appendix J, makes use of a line printer as an output device based on a technique developed by MacLeod,[1] and the second program, GRAYMP in Appendix K, is intended for a plotting device with a cathode-ray-tube display. Figures 3 and 4, respectively, show examples of these display techniques.

[1] I. D. G. MacLeod, *IEEE Trans. Comput.* **EC19**, 160 (1970).

```
GRAY MAP.   WHITE =  -C.869594E 01,  BLACK =   0.754962E 01
ZZZZZ1'1111111111111))))))))))))++++++++++++++++++++++++++++++
ZZZZZZZZ11111111111))))))))))))+++++++++++++++++++++++++++++++
ZZZZZZZZZZ1111111111))))))))))++++++++++++++++++========++++++++++
ZZZZZZZZZZZZZ11111111))))))))++++++++++++++========+++++++
ZZZZZZZZZZZZZZZ111111))))))++++++++++++===============+++++
ZZZZZZZZZZZZZZZ11111))))))++++++++++++==================+++
ZZZZZZXXXXXXZZZZZ1111))))))++++++++++================+++
ZZZXXXXXXXXXXXZZZ1111))))))+++++++++===============+
ZXXXXXAAMMBBBMAXXZZ111))))+++++++=================
XXXXAAMBBBBBBBBMXZZ11)))+++++++==================
XXAAMBBBBBBBBBBBBAXZ11))))++++++===============
XAAMBBBBBBBBBBBBBBMZ11))))+++++===============
AAMBBBBBBBBBBBBBBBBAZ11))))++++=============
AMBBBBBBBBBBBBBBBBBXZ1))))++++=============
ABBBBBBBBBBBBBBBBBB71)))+++=============
MBBBBBBBBBBBBBBBBBBX1)))+++============
BBBBBBBBBBBBBBBBBBBZ1)))+++===========
BBBBBBBBBBBBBBBBBBZ1)))+++==========
BBBBBBBBBBBBBBBBBBBZ1)))+++=======
BBBBBBBBBBBBBBBBBBBZ1)))+++=======
BBBBBBBBBBBBBBBBBBBB71)))+++=======
BBBBBBBBBBBBBBBBBBBBZ1)))+++======
MBBBBBBBBBBBBBBBBBBX1)))+++========
ABBBBBBBBBBBBBBBBBBHZ1)))+++=========
AMBBBBBBBBBBBBBBBBBX/1)))+++==========
AAMBBBBBBBBBBBBBBBBAZ11)))++++==========
XAAMBBBBBBBBBBBBBMZ11)))++++==========
XXAAMBBBBBBBBBBAXZ11))))++++============
XXXXAAMBBBBBBBMXZZ11))))+++++=========
ZXXXXXAAMMBBBMAXXZZ111)))))++++++=========
ZZZXXXXXXXXXXXXZZZ1111))))))++++++===========
ZZZZZZXXXXXZZZZZ1111))))))+++++++==========++
ZZZZZZZZZZZZZZZ11111))))))+++++++++=========+++
ZZZZZZZZZZZZZZ11111111)))))))++++++++++========+++++
ZZZZZZZZZZZ1111111111)))))))++++++++++++=======+++++++
ZZZZZZZZZ11111111111))))))))))+++++++++++++====+++++++++
ZZZZZZZ11111111111111))))))))))+++++++++++++++++++++++++
ZZZZZ111111111111111))))))))))))++++++++++++++++++++++++++
```

```
SAMPLE OF GRAY SHADES
1   2   3   4   5   6   7   8   9 10 11 12 13 14 15 16 17 18 19 20 21 22 23 24 25 26
------===+++))))11ZZZXXXAAAMMMBBBBBBBBBBBBBBBBBBBBBBBBBBBBBBBBBBBBB
```

FIG. 3. Gray-scale map obtained with a computer-output printer. Gray shades are produced by overprinting various characters.

The calling sequence for the printer program is

CALL GRAYM(XIMAG,NDIM1,NDIM2,NUSD1,NUSD2)

The data matrix XIMAG is dimensioned NDIM1 by NDIM2 in the calling program, but only the first NUSD1 by NUSD2 positions are used. NUSD1 must be less than 133. If the function of interest is $f(x, y)$, then XIMAG(I,J) = $f(x_I, y_J)$ for equispaced x_I and y_J. The largest and smallest data values are determined and assigned to white and black, respectively. Intermediate data values are plotted as one of 21 gray shades in MacLeod's gray scale.[1] Shades are produced by overprinting up to eight characters; for example, black is overprinted O, H, X, ', ., B, V, and A. The algorithm for the the line-printer program is straightforward and given by MacLeod.[1]

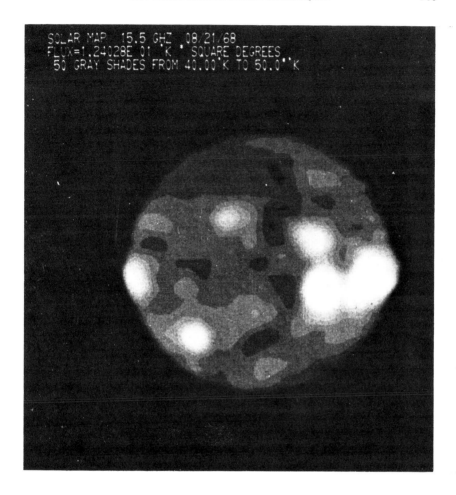

FIG. 4. Gray-scale map of the sun observed with the 120-ft diameter Haystack antenna at a frequency of 15.5 GHz.

The calling sequence of the program for the cathode ray tube is CALL GRAYMP(IDATA,NRDIM,NCDIM,NRUSED,NCUSED,IXO,IYO,IDX, IDY,NHORIZ,NVERT,ITRANS,NGRAY,MINVAL,MAXVAL). Again, the data array IDATA is dimensioned NRDIM by NCDIM, but only NRUSED by NCUSED positions are filled. Again, $IDATA(I,J) = f(x_I, y_J)$.

The map for the cathode ray tube (CRT) is drawn as a set of NRUSED∗ NCUSED∗NHORIZ∗NVERT $\leq 1024^2$ dots, each dot of a particular shade of gray. The dots are drawn on the CRT screen by simple overstriking; i.e., a dot of shade 5 is produced by exposing a dot of standard brightness five times at that position. Normally, the dots are placed close enough to-gether (within three raster units) so that a continuous gradation of shading is visible to the naked eye. Magnification shows the dots easily. The spacing, chosen by the user, is $|IDX|$ raster units in the X (Y) direction, and $|IDY|$ units in the Y (X) direction, as ITRANS = 0 (1). The coordinates of the CRT are shown in the accompanying diagram:

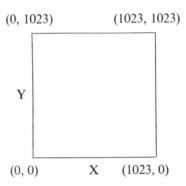

In all references made to (X, Y) coordinates, it is assumed that ITRANS = 0. If ITRANS = 1, transpose to (Y, X). That is, if according to the formulas below, a certain dot is plotted at (IX, IY), it will actually be plotted at (IY,IX) if ITRANS = 1. This parameter, IXO, IYO, and the signs of IDX and IDY permit arbitrary orientation of the array on the screen.

The values in the incoming array, IDATA(I,J), exist only for I = 1(1)NRUSED (i.e., 1, 2, ..., NRUSED), J = 1(1)NCUSED. To obtain enough values for closely packed plotting, GRAYMP interpolates, using double linear interpolation to effectively obtain I = 1(1/NHORIZ)NRUSED, J = 1(1/NVERT)NCUSED. These values are computed as needed, so no extra storage is required.

For each interpolated I and J, a dot is drawn at coordinates

IX = IXO+IDX*(I−1)*NHORIZ
IY = IYO+IDY*(J−1)*NVERT

of shade

ISHADE = (IDATA(I,J)−MINVAL)*NGRAY/(IDMAX−IDMIN)

IDX and IDY may be negative (to enable mirror images).
Certain consistency relations must be satisfied:

NRUSED,NCUSED ≥ 2
1023 ≥ IXO,IYO,IXO+IDX*(NRUSED−1)*NHORIZ,
IYO+IDY*(NCUSED−1)*NVERT ≥ 0

MINVAL and MAXVAL need not be the actual minimum and maximum to be found in IDATA; they may be set to emphasize low level detail, etc. ISHADE is set to 0 for IDATA(I,J) < MINVAL, and is set to NGRAY for IDATA(I,J) > MAXVAL.

The algorithm for the CRT map is as follows. Iterate through the data array, interpolating in two directions linearly. A dot of shade ISHADE is plotted at location (IX,IY) by calling subroutine DPLOT, with parameters (IX,IY), ISHADE times. Subroutine DPLOT(IX,IY) must be supplied by the user to plot a single dot of minimum brightness at the CRT coordinates indicated. For speed, it is strongly recommended that DPLOT save up several hundred plot commands in a buffer before sending them to the CRT. For the same reason, DPLOT might be written in machine language.

Appendixes

Appendix A
Subroutine DOP for Reduction of Radial-Velocity Data

```
      SUBROUTINE DOP (RAHRS,RAMIN,RASEC,DDEG,DMIN,DSEC,NYR,NDAY,
     2 NHUT,NMUT,NSUT,ALAT,OLONG,ELEV,  XLST,VSUN,VMON,VOBS,V1)
C
C     THE NEXT FIVE STATEMENTS DECLARE ALL VARIABLES AND FUNCTIONS TO BE
C     DOUBLE PRECISION.
      IMPLICIT REAL*8 (A-H,O-Z)
      SQRTF(X)=DSQRT(X)
      SIGNF(X,Y)=DSIGN(X,Y)
      SINF(X)=DSIN(X)
      COSF(X)=DCOS(X)
C     FOLLOWING COMPS DEAL WITH SUN MOTION TOWARD RA=AAA, DEC=DD
      Y=2.D0
      ZRO=0.D0
      AAA=18.D0*3.141592653500/12.D0
      DD=30.D0*3.141592653500/180.D0
C     MOVE  PRECESSES THIS DIRECTION TO DATE
      CALL MOVE(1900,1900+NYR,1,NDAY,AAA,DD,DELA,DELDD,DC)
      AAA=AAA+DELA
      DD=DD+DELDD
C     THIS VELOCITY IS CONVERTED TO CARTESIAN COMPONENTS
      XO=20.D0*COSF(AAA)*COSF(DD)
      YO=20.D0*SINF(AAA)*COSF(DD)
      ZO=20.D0*SINF(DD)
C     RA1 IS THE RIGHT ASCENSION (REVS=DAYS)
      RA1=(RAHRS+RAMIN/60.D0+RASEC/3600.D0)/24.D0
C     RA IS THE RIGHT ASCENSION (RADIANS)
      RA=2.D0*3.141592653500*RA1
C     DEC IS THE DECLINATION (RADIANS)
      DEC=3.141592653500*(DDEG+SIGNF(DMIN/60.D0+DSEC/3600.D0,DDEG))/
     2 180.0D0
C     CC, CS, AND S ARE THE DIRECTION COSINES CORRESPONDING TO RA AND DEC
      CC=COSF(DEC)*COSF(RA)
      CS=COSF(DEC)*SINF(RA)
      S=SINF(DEC)
C     VSUN IS THE PROJECTION ONTO THE LINE OF SIGHT TO THE STAR OF THE SUN'S
C     MOTION WITH RESPECT TO THE LOCAL STANDARD OF REST (KM/SEC)
      VSUN=-XO*CC-YO*CS-ZO*S
C
C     COORDINATES OF THE OBSERVER, LATITUDE (RADIANS), AND LONGITUDE (REVS=DAYS)
      CAT=ALAT*3.141592653500/180.D0
      WLONG=OLONG/360.D0
C
C     THE FOLLOWING CALCULATIONS DEAL WITH THE TIME
C     THE EPOCH IS 1900 JANUARY 0.5 UT = JULIAN DAY 2415020.0
C     DU IS THE TIME FROM THE EPOCH TO JAN 0.0 OF THE CURRENT YEAR (DAYS)
      DU=(JULDA(1900+NYR)-2415020)-0.5D0
C     TU IS DU CONVERTED TO JULIAN CENTURIES
      TU=DU/36525.0D0
C     UTDA IS THE GMT FROM JAN 0.0 TO THE PRESENT (DAYS)
      UTDA=DFLOAT(NDAY)+DFLOAT(NHUT)/24.D0+DFLOAT(NMUT)/1440.D0
     ++DFLOAT(NSUT)/86400.D0
C     SMD (SMALL D) IS THE TIME FROM THE EPOCH TO THE PRESENT (DAYS)
      SMD=DU+UTDA
C     T IS SMD CONVERTED TO JULIAN CENTURIES
      T=SMD/36525.D0
C     START IS THE GREENWICH MEAN SIDEREAL TIME ON JAN 0.0 (DAYS)
C     (THE EXTRA 129.1794 SECS CORRESPONDS TO THE 0.7 CENTURY SUBTRACTED FROM T
C     THE PRECISION IS THEREBY IMPROVED.)
```

308

```
      START=(6.D0+38.D0/60.D0+(45.836D0+129.1794D0+8640184.542D0*(TU-0.7)+0.0929
     +D0)+0.0929D0*TU**2)/3600.D0)/24.D0
C   C1 IS THE CONVERSION FACTOR FROM SOLAR TIME TO SIDEREAL TIME
      C1=0.9972695664140D0
C   GST IS THE GREENWICH MEAN SIDEREAL TIME (DAYS)
      GST=START+UTDA/C1
C   XLST IS THE LOCAL MEAN SIDEREAL TIME (FROM JAN 0) (DAYS)
      XLST=GST-WLONG
      XLST=XLST-IDINT(XLST)
C
C   THE FOLLOWING CALCULATIONS DEAL WITH THE OBSERVER'S MOTION WITH
C   RESPECT TO THE EARTH'S CENTER.
C   RHO IS THE RADIUS VECTOR FROM THE EARTH'S CENTER TO THE OBSERVER (METERS)
      RHO=6378160.0D0*(0.998327073D0+0.001676438D0*COSF(2.D0*CAT)-
     1 3.51D-6*COSF(4.D0*CAT)+0.000000008D0*COSF(6.D0*CAT))+ELEV
C   AND VRHO IS THE CORRESPONDING CIRCULAR VELOCITY (METERS/SIDEREAL DAY)
      VRHO=2.D0*3.1415926535D0*RHO
C   CONVERTED TO KILOMETERS/SEC
      VRHO=VRHO/24.0D3/3600.D0/C1
C   REDUCTION OF GEODETIC LATITUDE TO GEOCENTRIC LATITUDE (ARCSECONDS)
      DLAT=-(11.D0*60.D0+32.7430D0)*SINF(2.D0*CAT)+1.1633D0*SINF(4.D0*
     +CAT) -0.0026D0*SINF(6.D0*CAT)
C   CONVERT CAT TO GEOCENTRIC LATITUDE (RADIANS)
      CAT=CAT+DLAT*3.1415926535D0/3600.D0/180.D0
C   VOBS IS THE PROJECTION ONTO THE LINE OF SIGHT TO THE STAR OF THE VELOCITY
C   OF THE OBSERVER WITH RESPECT TO THE EARTH'S CENTER (KM/SEC)
      VOBS=VRHO*COSF(CAT)*COSF(DEC)*SINF(2.D0*3.1415926535D0*(XLST-RA1))
C
C   THE FOLLOWING CALCULATIONS DEAL WITH THE EARTH'S ORBIT ABOUT THE SUN
C   AM IS THE MEAN ANOMALY (OF THE EARTH'S ORBIT) (RADIANS)
      AM=(358.47583D0+0.9856002670D0*SMD-0.000150D0*T**2-
     2 0.000003D0*T**3)*3.1415926535D0/180.0D0
C   PI IS THE MEAN LONGITUDE OF PERIHELION (RADIANS)
      PI=(101.22083D0+0.0000470684D0*SMD+0.000453D0*T**2+
     2 0.000003D0*T**3)*3.1415626535D0/180.0D0
C   E IS THE ECCENTRICITY OF THE ORBIT (DIMENSIONLESS)
      E=0.01675104D0-0.00004180D0*T-0.000000126D0*T**2
C   AI IS THE MEAN OBLIQUITY OF THE ECLIPTIC (RADIANS)
      AI=(23.452294D0-0.0130125D0*T-0.00000164D0*T**2+
     2 0.000000503D0*T**3)*3.1415926535D0/180.0D0
C   VS IS THE TRUE ANOMALY (APPROXIMATE FORMULA) (RADIANS)
C   (EQUATION OF THE CENTER)
      VS=AM+(2.D0*E-0.25D0*E**3)*SINF(AM)+1.25D0*E**2*SINF(2.D0*AM)+
     2 13.D0/12.D0*E**3*SINF(3.D0*AM)
C   XLAM IS THE TRUE LONGITUDE OF THE EARTH AS SEEN FROM THE SUN (RADIANS)
      XLAM=PI+VS
C   ALAM IS THE TRUE LONGITUDE OF THE SUN AS SEEN FROM THE EARTH (RADIANS)
      ALAM=XLAM+3.1415926535D0
C   BETA IS THE LATITUDE OF THE STAR (RADIANS)
C   ALONG IS THE LONGITUDE OF THE STAR (RADIANS)
      CALL COORD (ZRO,ZRO,-3.1415926535D0/2.D0,3.1415926535D0/2.D0-AI,
     2 RA,DEC, ALONG,BETA)
C   AA IS THE SEMI-MAJOR AXIS OF THE EARTH'S ORBIT (KM)
      AA=149598500.0D0
C   AN IS THE MEAN ANGULAR RATE OF THE EARTH ABOUT THE SUN (RADIANS/DAY)
      AN=2.D0*3.1415926535D0/365.2564D0
C   HOP IS H/P FROM SMART = THE COMPONENT OF THE EARTH'S VELOCITY PERPENDICULAR
C   TO THE RADIUS VECTOR (KM/DAY)
      HOP=AN*AA/SQRTF(1.D0-E**2)
C   CONVERTED TO KM/SEC
      HOP=HOP/86400.D0
C   V IS THE PROJECTION ONTO THE LINE OF SIGHT TO THE STAR OF THE VELOCITY
C   OF THE EARTH-MOON BARYCENTER WITH RESPECT TO THE SUN (KM/SEC)
      V=-HOP*COSF(BETA)*(SINF(ALAM-ALONG)-E*SINF(PI-ALONG))
C
```

```
C   THE FOLLOWING CALCULATIONS DEAL WITH THE MOON'S ORBIT AROUND THE
C   EARTH-MOON BARYCENTER
C   OMGA (OMEGA) IS THE LONGITUDE OF THE MEAN ASCENDING NODE OF THE LUNAR ORBIT
C   (DEGREES)
      OMGA =259.183275D0-0.0529539222D0*SMD+0.002078D0*T**2+
    2  0.000002D0*T**3
C   OMGAR IS OMGA IN RADIANS
      OMGAR=OMGA*3.1415926535D0/180.D0
C   AMON IS OMGA PLUS THE MEAN LUNAR LONGITUDE OF THE MOON (DEGREES)
C   (SHOULD BE 13.1763965268)
      AMON=270.434164D0+13.176396527D0*SMD-0.001133D0*T**2+
    2  0.0000019D0*T**3
C   GAMP (GAMMA-PRIME) IS OMGA PLUS THE LUNAR LONGITUDE OF LUNAR PERIGEE (DEGREE
      GAMP=334.329556D0+0.1114040803D0*SMD-0.010325D0*T**2-
    2  0.000012D0*T**3
C   PIM IS THE MEAN LUNAR LONGITUDE OF LUNAR PERIGEE (TO RADIANS)
      PIM=(GAMP-OMGA)*3.1415926535D0/180.D0
C   EM IS THE ECCENTRICITY OF THE LUNAR ORBIT
      EM=0.054900489D0
C   OLAMM IS THE MEAN LUNAR LONGITUDE OF THE MOON (TO RADIANS)
      OLAMM=(AMON-OMGA)*3.1415926535D0/180.D0
C   AIM IS THE INCLINATION OF THE LUNAR ORBIT TO THE ECLIPTIC (RADIANS)
      AIM=5.1453964D0*3.1415926535D0/180.D0
C   AMM IS THE APPROXIMATE MEAN ANOMALY (RADIANS)
C   (IT IS APPROXIMATE BECAUSE PIM SHOULD BE THE TRUE RATHER THAN THE MEAN LUNAR
C   LONGITUDE OF LUNAR PERIGEE)
      AMM=OLAMM-PIM
C   VSM IS THE TRUE ANOMALY (APPROXIMATE FORMULA) (RADIANS)
C   (EQUATION OF THE CENTER)
      VSM=AMM+(2.D0*EM-0.25D0*EM**3)*SINF(AMM)+1.25D0*EM**2*SINF(2.D0*AMM)
     +M) +13.D0/12.D0*EM**3*SINF(3.D0*AMM)
C   ALAMM IS THE TRUE LUNAR LONGITUDE OF THE MOON (RADIANS)
      ALAMM=PIM+VSM
C   ANM IS THE MEAN ANGULAR RATE OF THE LUNAR ROTATION (RADIANS/DAY)
      ANM=2.D0*3.1415926535D0/27.321661D0
C   AAM IS THE SEMI-MAJOR AXIS OF THE LUNAR OBRIT (KILOMETERS)
      AAM=60.2665D0*6378.388D0
C   BETAM IS THE LUNAR LATITUDE OF THE STAR (RADIANS)
C   ALGM IS THE LUNAR LONGITUDE OF THE STAR (RADIANS)
      CALL COORD (OMGAR,ZRO,OMGAR-3.1415926535D0/2.0D0,
    2  3.1415626535D0/2.0D0-AIM,ALONG,BETA, ALGM,BETAM)
C   HOPM IS H/P FROM SMART = THE COMPONENT OF THE LUNAR VELOCITY PERPENDICULAR
C   TO THE RADIUS VECTOR (KM/DAY)
      HOPM=ANM*AAM/SQRTF(1.D0-EM**2)
C   CONVERTED TO KM/SEC
      HOPM=HOPM/86400.D0
C   VMON IS THE PROJECTION ONTO THE LINE OF SIGHT TO THE STAR OF THE VELOCITY
C   OF THE EARTH'S CENTER WITH RESPECT TO THE EARTH-MOON BARYCENTER (KM/SEC)
C   (THE 81.30 IS THE RATIO OF THE EARTH'S MASS TO THE MOON'S MASS)
      VMON=-HOPM/81.30D0*COSF(BETAM)*(SINF(ALAMM-ALGM)-EM*SINF(PIM-ALGM)
     1)
C
      V1=V+VSUN+VMON+VOBS
      RETURN
C
C   THIS PROGRAM OMITS THE PLANETARY PERTURBATIONS ON THE EARTH'S ORBIT. THESE
C   AMOUNT TO ABOUT 0.003 KM/SEC AND ARE THOUGHT TO BE THE LARGEST CONTRIBUTION
C   TO THE ERROR IN THE VELOCITY.
C
      END
```

```
      FUNCTION JULDA(NYR)
C     THIS FUNCTION COMPUTES THE JULIAN DAY NUMBER AT 12 HRS UT ON JANUARY
C     0 OF THE YEAR NYR (GREGORIAN CALENDAR). JULDA IS AN INTEGER BECAUSE
C     OF THIS DEFINITION.  FOR EXAMPLE, JULDA = 2439856 FOR NYR = 1968.
C
      NYRM1=NYR-1
      IC=NYRM1/100
      JULDA=1721425+365*NYRM1+NYRM1/4-IC+IC/4
      RETURN
      END


      SUBROUTINE COORD (AO,BO,AP,BP,A1,B1,  A2,B2)
C
C     THE NEXT STATEMENT DECLARES ALL VARIABLES TO BE DOUBLE PRECISION.
      IMPLICIT REAL*8 (A-H,O-Z)
      SBO=DSIN(BO)
      CBO=DCOS(BO)
      SBP=DSIN(BP)
      CBP=DCOS(BP)
      SB1=DSIN(B1)
      CB1=DCOS(B1)
C
C     THIS SUBROUTINE CONVERTS THE LONGITUDE-LIKE (A1) AND LATITUDE-LIKE (B 1)
C     COORDINATES OF A POINT ON A SPHERE INTO THE CORRESPONDING COORDINATES (A2,
C     B2) IN A DIFFERENT COORDINATE SYSTEM THAT IS SPECIFIED BY THE COORDINATES
C     OF ITS ORIGIN (AO, BO) AND ITS NORTH POLE (AP, BP) IN THE ORIGINAL CO ORDINATE
C     SYSTEM.   THE RANGE OF A2 WILL BE FROM -PI TO PI.
C
C     ALL ARGUMENTS ARE IN RADIANS.
C
C     EXAMPLES OF USE
C        PI = 3.1415926535
C        PIO2 = PI/2.0
C
C     EXAMPLE I--TO CALCULATE AZIMUTH AND ELEVATION FROM HOUR ANGLE AND DEC LINATION
C        CALL COORD (PI,PIO2-LATITUDE,0.0,LATITUDE,HOUR ANGLE,DECLINATION,
C       2 AZIMUTH,ELEVATION)
C     THEN IF AZIMUTH IS DESIRED IN THE RANGE 0 TO PI SET
C        AZIMUTH = AZIMUTH + (PI - SIGNF(PI,AZIMUTH))
C
C     EXAMPLE II--TO CALCULATE HOUR ANGLE AND DECLINATION FROM AZIMUTH AND
C     ELEVATION
C        CALL COORD (PI,PIO2-LATITUDE,0.0,LATITUDE,AZIMUTH,ELEVATION,
C       2 HOUR ANGLE,DECLINATION)
C        YY=1.D0
C
C     EXAMPLE III--TO CALCULATE LI AND BI FROM RIGHT ASCENSION AND DECLINAT ION
C     (EPOCH 1900.0)
C        AP = (12.0+40.0/60.0)*PI/12.0
C     (I.E. 12 HOURS 40 MINUTES CONVERTED TO RADIANS)
C        BP = 28.0*PI/180.0
C        AO = (18.0+40.0/60.0)*PI/12.0
C        BO = 0.0
C     (REFER TO KRAUS, P., RADIO ASTRONOMY, MCGRAW HILL, NEW YORK, 1966.  B UT FOR
C     FURTHER REFINEMENTS, SEE ALSO ALLEN, C. W., ASTROPHYSICAL QUANTITIES,
C     ATHLONE PRESS, LONDON, 1963.)
C        CALL COORD (AO,BO,AP,BP,RIGHT ASCENSION,DECLINATION,LI,BI)
C
C     EXAMPLE IV--TO CALCULATE RIGHT ASCENSION AND DECLINATION (EPOCH 1900. 0) FROM
C     LI AND BI
C     IN GENERAL, WHENEVER WE KNOW THE FORWARD TRANSFORMATION (EXAMPLE III ABOVE)
C     WE MAY DO THE REVERSE TRANSFORMATION WITH AT MOST TWO EXTRA PRELIMINA RY CALLS
C     TO COORD TO CALCULATE THE COORDINATES IN SYSTEM 2 OF THE POLE AND ORI GIN IN
```

```
C  SYSTEM 1.  BUT OFTEN IT IS POSSIBLE TO GET THESE NEEDED COORDINATES BY
C  INSPECTION.  FOR EXAMPLE, BP WILL REMAIN THE SAME FOR THE FORWARD AND REVERS
C  TRANSFORMATIONS.  FOR THIS EXAMPLE WE SEE BY INSPECTION THAT
C      APP = 6.0*PI/12.0
C      BPP = 28.0*PI/180.0
C  (THE SECOND P REPRESENTS PRIME.)  AND WE MAY CALCULATE AOP AND BOP FROM
C      CALL COORD (AO,BO,AP,BP,0.0,0.0,AOP,BOP)
       YY=2.D0
C  WHERE THE AO, ETC. ARE FROM EXAMPLE III.  THEN THE ACTUAL CONVERSION IS
C      CALL COORD (AOP,BOP,APP,BPP,LI,BI,RIGHT ASCENSION,DECLINATION)
C
C  EXAMPLE V--TO CALCULATE LII AND BII FROM RIGHT ASCENSION AND DECLINATION
C  (EPOCH 1950.0)
C      AP = (12.0+49.0/60.0)*PI/12.0
C      BP = 27.4*PI/180.0
C      AO = (17.0+42.4/60.0)*PI/12.0
C      BO = -(28.0+55.0/60.0)*PI/180.0
C      CALL COORD (AO,BO,AP,BP,RIGHT ASCENSION,DECLINATION,LII,BII)
C
C  EXAMPLE VI--TO CALCULATE RIGHT ASCENSION AND DECLINATION (EPOCH 1950.0) FROM
C  LII AND BII
C  FIRST CALCULATE APP AND BPP FROM
C      CALL COORD (AO,BO,AP,BP,0.0,PIO2,APP,BPP)
C  THEN CALCULATE AOP AND BOP FROM
C      CALL COORD (AO,BO,AP,BP,0.0,0.0,AOP,BOP)
C  WHERE THE AO, ETC. ARE FROM EXAMPLE V.  THEN THE ACTUAL CONVERSION IS
C      CALL COORD (AOP,BOP,APP,BPP,LII,BII,RIGHT ASCENSION,DECLINATION)
C
C  EXAMPLE VII--TO CALCULATE (ECLIPTIC) LATITUDE AND LONGITUDE FROM RIGHT
C  ASCENSION AND DECLINATION
C  EPS IS THE OBLIQUITY OF THE ECLIPTIC WHICH IS ABOUT 23.443 DEGREES, BUT IT
C  DEPENDS ON THE EPOCH.  SEE THE AMERICAN EPHEMERIS AND NAUTICAL ALMANAC.
       YY=3.D0
C      EPS=23.443*PI/180.0
C      CALL COORD (0.0,0.0,-PIO2,PIO2-EPS,RIGHT ASCENSION,DECLINATION,
C      2 LATITUDE,LONGITUDE)
C
C  THE NOTATION USES S OR C FOR SINE OR COSINE OF THE CORRESPONDING VARIABLE,
C  FOR EXAMPLE, SBO = SINF(BO), ETC.
C
C  NOTE THAT THE INPUT PARAMETERS ARE PARTIALLY REDUNDANT.  FOR EXAMPLE, IF
C  AP, BP, AND AO ARE SPECIFIED, THEN THERE ARE ONLY TWO DISCRETE VALUES
C  POSSIBLE FOR BO (EXCEPT FOR A FEW DEGENERATE SPECIAL CASES).  SEE BELOW FOR
C  WHAT TO DO IF IT IS NECESSARY TO PRECALCULATE AO AND BO.
C
C  IF, INSTEAD OF AO AND BO, THE LONGITUDE OF THE ASCENDING NODE IS KNOWN IN
C  BOTH THE OLD (AN1) AND NEW (AN2) COORDINATE SYSTEMS, THEN AO AND BO MAY BE
C  CALCULATED BY A PRELIMINARY CALL TO COORD
C      CALL COORD (0.0,0.0,AN1-AP,BP,-AN2,0.0,AO,BO)
C  THEN THIS AO AND BO MAY BE USED FOR A SERIES OF ORDINARY CALLS TO COORD AS
C  DESCRIBED ABOVE.
C
C  IF AP, BP, AND AO ARE KNOWN, THEN THE TWO POSSIBLE VALUES OF BO MAY BE
C  CALCULATED FROM
C      SBO=(SBP $ 2.0*CBP**2*CAPAO*SQRTF(1.0+CAPAO**2))/
C      2 (SBP**2+(CBP*CAPAO)**2)
C  WHERE CAPAO = COSF(AP-AO) AND THE OTHER NOTATION IS EXPLAINED ABOVE, AND
C  WHERE THE $ IS TO BE REPLACED BY + AND -.
C
C  IF AP, BP, AND BO ARE KNOWN, THEN THE TWO POSSIBLE VALUES OF AO MAY BE
C  CALCULATED FROM
       YY=4.D0
C      CAPAO=(1.0-SBO*SBP)/(CBO*CBP)
C  BOTH ANGLES WITH THIS COSINE ARE POSSIBLE.
       SB2=SBP*SB1+CBP*CB1*DCOS(AP-A1)
```

```
      B2=DARSIN(SB2)
C (NOTE BO IS NOT NEEDED TO CALCULATE B2)
      CB2=DCOS(B2)
C
      SAA=DSIN(AP-A1)*CB1/CB2
      CAA=(SB1-SB2*SBP)/(CB2*CBP)
C
      CBB=SBO/CBP
      SBB=DSIN(AP-AO)*CBO
C
      SA2=SAA*CBB-CAA*SBB
      CA2=CAA*CBB+SAA*SBB
C THERE ARE TWO FORMULAE FOR TA202 (TANF(A2/2.0)), ONE USABLE EVERYWHERE
C EXCEPT NEAR CA2=+1, AND THE OTHER EVERYWHERE EXCEPT NEAR CA2=-1.
C PREVIOUS VERSIONS OF THIS PROGRAM WERE INACCURATE NEAR A2=0.
      IF(CA2)10,10,20
   10 TA202=(1.0D0-CA2)/SA2
      GO TO 30
   20 TA202=SA2/(1.0D0+CA2)
   30 A2=2.D0*DATAN(TA202)
C
      RETURN
      END


      SUBROUTINE MOVEA(AYRI,NYRF,MO,NDA,RA,D,DELR,DELD,DC)
C COMPUTES PRECESSION AND SOME TERMS OF NUTATION
C FOR INITIAL EPOCH NOT AN INTEGER YEAR
C THE NEXT STATEMENT DECLARES ALL VARIABLES TO BE DOUBLE PRECISION.
      IMPLICIT REAL*8 (A-H,O-Z)
      DLR=0.D0
      DLD=0.D0
      NYRI=AYRI
      DYRI=AYRI-NYRI
      IF(DYRI.LT.1.D-10)GO TO 10
      NDAY=DYRI*365.2421988D0
      CALL MOVE(NYRI,NYRI,1,NDAY,RA,D,DLR,DLD,DC)
   10 CALL MOVE(NYRI,NYRF,MO,NDA,RA,D,DELR,DELD,DC)
      DELR=DELR-DLR
      DELD=DELD-DLD
      RETURN
      END


      SUBROUTINE MOVE  (NYRI,NYRF,MO,NDA,RA,D,DELR,DELD,DC)
C
C THE NEXT STATEMENT DECLARES ALL VARIABLES TO BE DOUBLE PRECISION.
      IMPLICIT REAL*8 (A-H,O-Z)
C
C MOVE  CALCULATES THE CORRECTION (DELR) IN RIGHT ASCENSION (RA) AND THE
C CORRECTION (DELD) IN DECLINATION (D) (ALL IN RADIANS) TO BE ADDED TO THE
C MEAN COORDINATES FOR EPOCH NYRI (EG 1960).  TO GIVE THE APPARENT POSITIONS
C OF A DATE SPECIFIED BY THE YEAR (NYRF, E.G. 1968), MONTH (MO, 1 TO 12), AND
C DAY (NDA).  IF THE DAY-NUMBER IS KNOWN, USE IT FOR NDA AND SET MO=1.
C MOVE  ALSO CALCULATES THE EQUATION OF THE EQUINOXES (DC, IN MINUTES OF TIME)
C WHICH MAY BE ADDED TO THE MEAN SIDEREAL TIME TO GIVE THE APPARENT SIDEREAL
C TIME (AENA-469).  DELR AND DELD CONTAIN CORRECTIONS FOR PRECESSION, ANNUAL
C ABERRATION, AND SOME TERMS OF NUTATION.  IF RA AND D ARE FOR THE MEAN EPOCH
C (I.E. HALFWAY BETWEEN NYRI AND NYRF) THEN THE PRECISION OF DELR AND DELD IS
C ABOUT 2 ARCSECONDS (SEE NEGLECTED TERMS IN ESE-44).  IF RA AND D ARE EITHER
C OF THE END POINTS OF THE INTERVAL, THEN THE PRECISION MAY BE SOMEWHAT WORSE.
C AENA   THE AMERICAN EPHEMERIS AND NAUTICAL ALMANAC (THE BLUE BOOK).
C ESE    THE EXPLANATORY SUPPLEMENT TO ABOVE (THE GREEN BOOK).
```

```
C
      SND=DSIN(D)
      CSD=DCOS(D)
      TND=SND/CSD
C
      CSR=DCOS(RA)
      SNR=DSIN(RA)
C
C  AL IS AN APPROXIMATE DAY NUMBER (I.E. THE NUMBER OF DAYS SINCE JANUARY 0
C  OF THE YEAR NYRF).
      AL=30*(MO-1)+NDA
C
C  TO IS THE TIME FROM 1900 TO AYRI (CENTURIES)
C  T IS THE TIME FROM AYRI TO DATE (NYRF, MO, NDA) (CENTURIES)
C  (365.2421988 IS THE NUMBER OF EPHEMERIS DAYS IN A TROPICAL YEAR)
      TO=DFLOAT(NYRI-1900)/100.DO
      T=(DFLOAT(NYRF-NYRI)+AL/365.2421988DO)/100.DO
C  ZETAO IS A PRECESSIONAL ANGLE FROM ESE-29 (ARCSECONDS)
      ZETAO=(2304.250DO+1.396DO*TO)*T+0.302DO*T**2+0.018DO*T**3
C  DITTO FOR Z
      Z=ZETAO+0.791DO*T**2
C  AND THETA
      THETA=(2004.682DO-0.853DO*TO)*T-0.426DO*T**2-0.042DO*T**3
C  AM AND AN ARE THE M AND N PRECESSIONAL NUMBERS (SEE AENA-50, 474) (RADIANS)
      AM=(ZETAO+Z)*4.848136811D-6
      AN=THETA*4.848136811D-6
C
C  ALAM IS AN APPROXIMATE MEAN LONGITUDE FOR THE SUN (AENA-50) (RADIANS)
      ALAM=(0.985647DO*AL+278.5DO)*0.0174532925DO
      SNL=DSIN(ALAM)
      CSL=DCOS(ALAM)
C  DELR IS THE ANNUAL ABERRATION TERM IN RA (RADIANS) (ESE-47,48)
C  (0.91745051   COS(OBLIQUITY OF ECLIPTIC))
C  (-9.92413605E-5   K   20.47 ARCSECONDS   CONSTANT OF ABERRATION) (ESE-48))
      DELR=-9.92413605D-5*(SNL*SNR+0.91745051DO*CSL*CSR)/CSD
C  PLUS PRECESSION TERMS (SEE AENA-50 AND ESE-38)
     2 +AM+AN*SNR*TND
C  DELD IS DITTO ABOVE IN DECLINATION
      DELD=-9.92413605D-5*(SNL*CSR*SND-0.91745051DO*CSL*SNR*SND
C  (0.39784993   SIN(OBLIQUITY OF ECLIPTIC))
     2 +0.39784993DO*CSL*CSD) +AN*CSR
C
C  THE FOLLOWING CALCULATES THE NUTATION (APPROXIMATELY) (ESE-41,45)
C  OMEGA IS THE ANGLE OF THE FIRST TERM OF NUTATION (ESE-44) (APPROXIMATE
C  FORMULA) (DEGREES)
      OMEGA=259.183275DO-1934.142DO*(TO+T)
C  ARG IS OMEGA CONVERTED TO RADIANS
      ARG=OMEGA*0.0174532925DO
C  DLONG IS THE NUTATION IN LONGITUDE (DELTA-PSI) (RADIANS)
      DLONG=-8.3597D-5*DSIN(ARG)
C  DOBLQ IS THE NUTATION IN OBLIQUITY (DELTA-EPSILON) (RADIANS)
      DOBLQ= 4.4678D-5*DCOS(ARG)
C
C  ADD NUTATION IN RA INTO DELR (ESE-43)
      DELR=DELR+DLONG*(0.91745051DO+0.39784993DO*SNR*TND)-CSR*TND*DOBLQ
C  AND DEC.
      DELD=DELD+0.39784993DO*CSR*DLONG+SNR*DOBLQ
C  DC IS THE EQUATION OF THE EQUINOXES (MINUTES OF TIME) (ESE-43)
      DC=DLONG*210.264169DO
      RETURN
      END
```

Appendix B

Subroutine FOURG, Short Version of the Fast Fourier Transform for One-Dimensional, Complex Data Stored in Core

```
      SUBROUTINE FOURG(DATA,N,ISIGN,WORK)
C     ONE-DIMENSIONAL COMPLEX FAST FOURIER TRANSFORM FOR ANY NUMBER OF
C     POINTS. DATA IS COMPLEX OF LENGTH N, AN ARBITRARY INTEGER.
C     THE FOLLOWING TRANSFORM REPLACES DATA IN STORAGE:
C     TRAN(K) = 1/SQRT(N)  *  SUM(DATA(J)*EXP(2*PI*I*ISIGN*(J-1)*(K-1)/N))
C     SUMMED OVER ALL J FROM 1 TO N, FOR ALL K FROM 1 TO N.
C     ISIGN = +1 OR -1, THE DIRECTION OF THE TRANSFORM.
C     WORK IS A COMPLEX WORKING STORAGE ARRAY OF LENGTH N.
C     RUNNING TIME IS PROPORTIONAL TO N*(SUM OF PRIME FACTORS OF N).
C     NORMAN BRENNER, MIT, NOVEMBER 1971.
      DIMENSION DATA(1),WORK(1)
      IP0=2
      IF(N)20,20,10
10    IF(IABS(ISIGN)-1)20,40,20
20    WRITE(6,30)N,ISIGN
30    FORMAT(21H0ERROR IN FOURG.   N =,I10,28H IS NON-POSITIVE, OR ISIGN
     1=,I10,17H IS NOT +1 OR -1.)
      RETURN
40    NREM=N
      IN=-1
      IDIV=2
      IGO=-1
      GO TO 80
50    IDIV=1
      IGO=0
60    IDIV=IDIV+2
      IF(IDIV**2-NREM)80,80,70
70    IDIV=NREM
      IGO=1
      IF(IDIV-1)160,160,80
80    IF(NREM-IDIV*(NREM/IDIV))150,90,150
90    NDONE=N/NREM
      NREM=NREM/IDIV
      DO 140 I3=1,NDONE
      DO 140 I2=1,IDIV
      I23=((I2-1)*NDONE+I3-1)*NREM
      THETA=6.283185*FLOAT(I23)/FLOAT(ISIGN*N)
      WR=COS(THETA)
      WI=SIN(THETA)
      ISTEP=NREM*IP0
      DO 140 I1=1,NREM
      I231=(I23+I1-1)*IP0+1
      I321=((I3*IDIV-1)*NREM+I1-1)*IP0+1
      IF(IN)100,100,120
100   WORK(I231)=DATA(I321)
      WORK(I231+1)=DATA(I321+1)
      DO 110 J2=2,IDIV
      I321=I321-ISTEP
      TEMP=WORK(I231)
      WORK(I231)=WR*TEMP-WI*WORK(I231+1)+DATA(I321)
110   WORK(I231+1)=WR*WORK(I231+1)+WI*TEMP+DATA(I321+1)
      GO TO 140
120   DATA(I231)=WORK(I321)
      DATA(I231+1)=WORK(I321+1)
      DO 130 J2=2,IDIV
      I321=I321-ISTEP
      TEMP=DATA(I231)
      DATA(I231)=WR*TEMP-WI*DATA(I231+1)+WORK(I321)
```

```
130    DATA(I231+1)=WR*DATA(I231+1)+WI*TEMP+WORK(I321+1)
140    CONTINUE
       IN=-IN
       GO TO 80
150    IF(IGO)50,60,160
160    S=SQRT(FLOAT(N))
       IMAX=IPO*N
       IF(IN)170,170,190
170    DO 180 I=1,IMAX,IPO
       DATA(I)=DATA(I)/S
180    DATA(I+1)=DATA(I+1)/S
       RETURN
190    DO 200 I=1,IMAX,IPO
       DATA(I)=WORK(I)/S
200    DATA(I+1)=WORK(I+1)/S
       RETURN
       END
```

Appendix C

Subroutine FOUR1, Short Version of the Fast Fourier Transform for
One-Dimensional, Complex Data, 2^n Points in Core

```
       SUBROUTINE FOUR1(DATA,N,ISIGN)
C      ONE-DIMENSIONAL COMPLEX FAST FOURIER TRANSFORM OF ARRAY DATA,
C      OF LENGTH N = 2**LOG2N POINTS.
C      THE FOLLOWING TRANSFORM REPLACES DATA IN STORAGE:
C      TRAN(K) = 1/SQRT(N) * SUM(DATA(J)*EXP(2*PI*I*ISIGN*(J-1)*(K-1)/N))
C      SUMMED OVER ALL J FROM 1 TO N, FOR ALL K FROM 1 TO N.
C      ISIGN = +1 OR -1, THE DIRECTION OF THE TRANSFORM.
C      RUNNING TIME IS PROPORTIONAL TO N*LOG2(N).
C      NORMAN BRENNER, MIT, OCTOBER 1971.
       DIMENSION DATA(1),ITWO(20)
       IPO=2
       IF(N-1)90,110,10
10     LOG2N=ALOG10(FLOAT(N))/.30103+.5
       IF(IABS(ISIGN)-1+N-2**LOG2N)90,20,90
20     ITWO(1)=IPO
       DO 30 L=2,LOG2N
30     ITWO(L)=2*ITWO(L-1)
       SQRTN=SQRT(FLOAT(N))
       IMAX=N*IPO
       IREV=1
       DO 70 I=1,IMAX,IPO
C      IN BINARY, I-1 AND IREV-1 ARE MIRROR IMAGES, AS 1101 AND 1011.
C      FOR EACH I, EXCHANGE DATA(I) AND DATA(IREV). TO PREVENT
C      RE-EXCHANGING, I MUST NOT BE GREATER THAN IREV.
       IF(I-IREV)40,40,50
40     TEMPR=DATA(I)
       TEMPI=DATA(I+1)
       DATA(I)=DATA(IREV)/SQRTN
       DATA(I+1)=DATA(IREV+1)/SQRTN
       DATA(IREV)=TEMPR/SQRTN
       DATA(IREV+1)=TEMPI/SQRTN
C      COMPUTE THE REVERSAL OF I BY ADDING 1 TO THE HIGH ORDER BIT OF
C      THE REVERSAL OF I-1 AND PROPAGATING THE CARRY DOWNWARDS.
50     DO 60 L=1,LOG2N
       ML=LOG2N-L+1
       IF(IREV-ITWO(ML))70,70,60
60     IREV=IREV-ITWO(ML)
```

```
70      IREV=IREV+ITWO(ML)
C       SEPARATE THE DATA INTO EVEN AND ODD SUBSEQUENCES, TRANSFORM EACH,
C       PHASE SHIFT THE ODD TRANSFORM AND ADD THEM. THIS RECURSIVE
C       FORMULATION OF THE FFT IS HERE WRITTEN ITERATIVELY, FROM THE
C       INNERMOST SUBTRANSFORM OUTWARD.
        DO 80 L=1,LOG2N
        ITWOL=ITWO(L)
        DO 80 I1=1,ITWOL,IPO
        THETA=3.1415926535*FLOAT(ISIGN*(I1-1))/FLOAT(ITWOL)
        WR=COS(THETA)
        WI=SIN(THETA)
        ISTEP=2*ITWOL
        DO 80 IA=I1,IMAX,ISTEP
        IB=IA+ITWOL
        TEMPR=WR*DATA(IB)-WI*DATA(IB+1)
        TEMPI=WR*DATA(IB+1)+WI*DATA(IB)
        DATA(IB)=DATA(IA)-TEMPR
        DATA(IB+1)=DATA(IA+1)-TEMPI
        DATA(IA)=DATA(IA)+TEMPR
80      DATA(IA+1)=DATA(IA+1)+TEMPI
        RETURN
90      WRITE(6,100)N,ISIGN
100     FORMAT(28H0ERROR IN FOUR1. EITHER N =,I10,50H IS NON-POSITIVE OR
       1NOT A POWER OF TWO, OR ISIGN =,I10,17H IS NOT +1 OR -1.)
110     RETURN
        END
```

Appendix D

Subroutine FXRL1, Module for Fast Fourier Transform of Real Data when Combined with FOURG or FOUR1

```
        SUBROUTINE FXRL1(DATA,N,ISIGN,IFORM)
C       FIX UP A REAL OR HALF-SIZE COMPLEX ARRAY FOR FAST FOURIER TRANS-
C       FORMING. FOR IFORM=0: DATA IS A REAL ARRAY OF LENGTH N WHICH
C       HAS JUST BEEN TRANSFORMED BY FOUR1 OR FOURG WITH AN ASSUMED
C       COMPLEX LENGTH OF N/2. AFTER FOUR1 OR FOURG, CALL FXRL1,
C       AND THEN DATA WILL BE A COMPLEX ARRAY OF LENGTH N/2+1, HOLDING
C       THE FIRST N/2+1 TRANSFORM VALUES. (THE OTHER N/2-1 VALUES ARE
C       THE REVERSE ORDER CONJUGATE OF THOSE SUPPLIED.)
C       FOR IFORM=-1: DATA IS COMPLEX N/2+1 LONG, THE TRANSFORM OF A
C       REAL ARRAY N LONG. FIRST CALL FXRL1, THEN CALL FOUR1 OR FOURG.
C       DATA NOW HOLDS N REAL VALUES, THE TRANSFORM.
C       ISIGN = +1 OR -1, AND MUST BE THE SAME AS IN FOUR1 OR FOURG.
C       N MUST BE EVEN. FOR EXAMPLE, TO TRANSFORM A REAL ARRAY:
C       REAL DATA(200)
C       COMPLEX CDATA(101),WORK(100)
C       EQUIVALENCE (DATA,CDATA)
C       READ(5,10)(DATA(I),I=1,200)        THE REAL INPUT VALUES
C       CALL FOURG(DATA,100,+1,WORK)
C       CALL FXRL1(DATA,200,+1,0)
C       WRITE(6,20)(CDATA(I),I=1,101)       THE COMPLEX TRANSFORM
C       NORMAN BRENNER, MIT, NOVEMBER 1971.
        DIMENSION DATA(1)
        IF(IABS(N-1)-N+IABS(N)-2*(IABS(N)/2)+IABS(IABS(ISIGN)-1)+
       1 IABS(2*IFORM+1)/2)30,30,10
10      WRITE(6,20)N,ISIGN,IFORM
20      FORMAT(22H0ERROR IN FXRL1. N = ,I10,35H IS ODD OR NON-POSITIVE, O
       1R ISIGN =,I10,11H OR IFORM =,I10,14H IS INCORRECT.)
        RETURN
```

```
30    S=(1.-FLOAT(IFORM))/SQRT(2.)
      IP0=2
      NHALF=IP0*N/2
      IF(IFORM)50,40,10
40    DATA(NHALF+1)=DATA(1)
      DATA(NHALF+2)=DATA(2)
50    IMAX=NHALF/2+1
      DO 70 I=1,IMAX,IP0
      ICONJ=NHALF+2-I
      THETA=6.283185307*(.25+FLOAT(I-1)/FLOAT(2*ISIGN*N))
      ZR=.5*(1.+FLOAT(2*IFORM+1)*COS(THETA))
      ZI=(FLOAT(IFORM)+.5)*SIN(THETA)
      DIFR=DATA(I)-DATA(ICONJ)
      DIFI=DATA(I+1)+DATA(ICONJ+1)
      TEMPR=ZR*DIFR-ZI*DIFI
      TEMPI=ZR*DIFI+ZI*DIFR
      DATA(I)=(DATA(I)-TEMPR)*S
      DATA(I+1)=(DATA(I+1)-TEMPI)*S
      IF(I-ICONJ)60,80,80
60    DATA(ICONJ)=(DATA(ICONJ)+TEMPR)*S
70    DATA(ICONJ+1)=(DATA(ICONJ+1)-TEMPI)*S
80    RETURN
      END
```

Appendix E

Subroutines CFFT2, RFFT2, and HFFT2, Programs for Transform of Two-Dimensional Complex or Real Data

```
      SUBROUTINE CFFT2(DATA,N1,N2,ISIGN)
C     FFT A COMPLEX TWO-DIMENSIONAL ARRAY BY TRANSFORMING THE COLUMNS
C     AND ROWS SEPARATELY.
      COMPLEX DATA(N1,N2),WORK(1000),WORK2(1000)
C     COMPLEX WORK(MAX(N1,N2)),WORK2(MAX(N1,N2))
      DO 20 I2=1,N2
      DO 10 I1=1,N1
10    WORK(I1)=DATA(I1,I2)
      CALL FOURG(WORK,N1,ISIGN,WORK2)
C OR  CALL FOUR1(WORK,N1,ISIGN)
      DO 20 I1=1,N1
20    DATA(I1,I2)=WORK(I1)
      DO 40 I1=1,N1
      DO 30 I2=1,N2
30    WORK(I2)=DATA(I1,I2)
      CALL FOURG(WORK,N2,ISIGN,WORK2)
C OR  CALL FOUR1(WORK,N2,ISIGN)
      DO 40 I2=1,N2
40    DATA(I1,I2)=WORK(I2)
      RETURN
      END


      SUBROUTINE RFFT2(RDATA,N1,N2,CDATA,N1HP1,ISIGN)
C     FFT A REAL TWO-DIMENSIONAL ARRAY INTO A HALF-SIZE COMPLEX ARRAY
C     BY FIRST TRANSFORMING THE REAL COLUMNS, FIXING THEM UP, AND THEN
C     TRANSFORMING THE COMPLEX COLUMNS, ALL IN PLACE.
      REAL RDATA(N1,N2),RWORK(2002)
      COMPLEX CDATA(N1HP1,N2),CWORK(1001),CWORK2(1000)
      EQUIVALENCE(RWORK,CWORK)
C     REAL RDATA(N1,N2),RWORK(MAX(N1,N2)+2)
```

```
C       COMPLEX CDATA(N1/2+1,N2),CWORK(MAX(N1/2+1,N2)),CWORK2(MAX(N1/2,N2)
C       N1HP1=N1/2+1
        DO 20 I2=1,N2
        I2R=N2+1-I2
        DO 10 I1=1,N1
10      RWORK(I1)=RDATA(I1,I2R)
        CALL FOURG(RWORK,N1/2,ISIGN,CWORK2)
C OR    CALL FOUR1(RWORK,N1/2,ISIGN)
        CALL FXRL1(RWORK,N1,ISIGN,0)
        DO 20 I1=1,N1HP1
20      CDATA(I1,I2R)=CWORK(I1)
C       ITERATE COLUMNS IN REVERSE ORDER SINCE FORTRAN ARRAYS ARE
C       ORDERED COLUMNWISE AND WE REPLACE WITH MORE WE TOOK OUT.
        DO 40 I1=1,N1HP1
        DO 30 I2=1,N2
30      CWORK(I2)=CDATA(I1,I2)
        CALL FOURG(CWORK,N2,ISIGN,CWORK2)
C OR    CALL FOUR1(CWORK,N2,ISIGN)
        DO 40 I2=1,N2
40      CDATA(I1,I2)=CWORK(I2)
        RETURN
        END


        SUBROUTINE HFFT2(RDATA,N1,N2,CDATA,N1HP1,ISIGN)
C       FFT A HALF-SIZE COMPLEX TWO-DIMENSIONAL ARRAY BY FIRST TRANS-
C       FORMING THE COMPLEX ROWS IN PLACE, THEN FIXING UP THE COMPLEX
C       COLUMNS AND TRANSFORMING THEM.
        REAL RDATA(N1,N2),RWORK(2002)
        COMPLEX CDATA(N1HP1,N2),CWORK(1001),CWORK2(1000)
        EQUIVALENCE(RWORK,CWORK)
C       REAL RDATA(N1,N2),RWORK(MAX(N1,N2)+2)
C       COMPLEX CDATA(N1/2+1,N2),CWORK(MAX(N1/2+1,N2)),CWORK2(MAX(N1/2,N2)
C       N1HP1=N1/2+1
        DO 20 I1=1,N1HP1
        DO 10 I2=1,N2
10      CWORK(I2)=CDATA(I1,I2)
        CALL FOURG(CWORK,N2,ISIGN,CWORK2)
C OR    CALL FOUR1(CWORK,N2,ISIGN)
        DO 20 I2=1,N2
20      CDATA(I1,I2)=CWORK(I2)
        DO 40 I2=1,N2
C       ITERATE COLUMNS IN FORWARD ORDER SINCE FORTRAN ARRAYS ARE
C       ORDERED COLUMNWISE AND WE REPLACE WITH LESS THAN WE TOOK OUT.
        DO 30 I1=1,N1HP1
30      CWORK(I1)=CDATA(I1,I2)
        CALL FXRL1(CWORK,N1,ISIGN,-1)
        CALL FOURG(CWORK,N1/2,ISIGN,CWORK2)
C OR    CALL FOUR1(CWORK,N1/2,ISIGN)
        DO 40 I1=1,N1
40      RDATA(I1,I2)=RWORK(I1)
        RETURN
        END
```

Appendix F

Subroutine FORS1, Module for Fast Fourier Transform of Real, Symmetric Data in Core

```
      SUBROUTINE FORS1 (DATA,N,ISIGN,ISYM)
C     FAST FOURIER TRANSFORM OF ONE-DIMENSIONAL, REAL, SYMMETRIC DATA.
C     TRAN(K) = 1/SQRT(N) * SUM(DATA(J)*EXP(2*PI*I*ISIGN*(J-1)*(K-1)/N))
C     SUMMED OVER ALL J FROM 1 TO N, FOR ALL K FROM 1 TO N.
C     INPUT ARRAY DATA IS REAL, LENGTH N/2+2, REPRESENTING A REAL ARRAY
C     OF LENGTH N.  N IS AN ARBITRARY MULTIPLE OF FOUR, THO IF IT IS NOT
C     A POWER OF TWO, REPLACE FOUR2 BY FOURT (OR FOUR1 BY FOURG).
C     THE FULL ARRAY OF LENGTH N HAS EITHER EVEN OR ODD SYMMETRY (ISYM =
C     +1 OR -1) ACCORDING TO WHETHER DATA(N+2-I) = + DATA(I) OR
C     - DATA(I), FOR I = 2,3,...N.  DATA(1) = + OR - DATA(1).   ODD
C     SYMMETRY IMPLIES THAT DATA(1)=DATA(N/2+1)=0.  ISIGN = +1 OR -1,
C     THE DIRECTION OF THE TRANSFORM.  THE FIRST N/2+1 POINTS OF THE
C     TRANSFORM, REAL AND SAME SYMMETRY AS THE INPUT, WILL BE RETURNED
C     IN DATA, REPLACING THE INPUT.  DATA(N/2+2) HOLDS NO DATA VALUE,
C     BUT IS USED FOR WORKING STORAGE.  RUNNING TIME AND ARRAY
C     STORAGE ARE EQUIVALENT TO THE TRANSFORM OF N/4 COMPLEX, ASYM-
C     METRIC DATA.  NOTE THAT THE COMPLEX TRANSFORM OF EVEN OR ODD
C     SYMMETRIC DATA IS EQUIVALENT TO THE COSINE OR SINE TRANSFORM,
C     RESPECTIVELY, OF ASYMMETRIC DATA.  NORMAN BRENNER, MIT, OCT. 1968.
      DIMENSION DATA(1)
      IF (IABS(N-1)-N+1+IABS(N)-4*(IABS(N)/4)+IABS(IABS(ISIGN)-1)
     1  +IABS(IABS(ISYM)-1))10,10,140
C     REARRANGE DATA INTO A CONJUGATE-SYMMETRIC COMPLEX ARRAY.
10    NHALF=N/2
      ODDSM=0.
      PREV=DATA(2)*FLOAT(ISYM)
      DO 30 I=2,NHALF,2
      ODDSM=ODDSM+DATA(I)
      TEMP=DATA(I)
      DATA(I)=DATA(I)-PREV
      PREV=TEMP
      IF (ISYM)20,30,30
20    TEMP=DATA(I-1)
      DATA(I-1)=DATA(I)
      DATA(I)=TEMP
30    CONTINUE
      IF (ISYM)40,50,50
40    ODDSM=0.
      DATA(NHALF+1)=-2.*PREV
50    DATA(NHALF+2)=0.
C     FOURIER TRANSFORM THE CONJUGATE-SYMMETRIC COMPLEX ARRAY INTO A
C     REAL ARRAY, BOTH OF OSTENSIBLE LENGTH N/2.
      CALL FXRL1(DATA,N/2,ISIGN,-1)
      CALL FOUR1(DATA,N/4,ISIGN)
C OR JUST    CALL FOUR2(DATA,N/2,1,ISIGN,-1)    IN PLACE OF BOTH CALLS
C     FIX UP THE REAL ARRAY.
      SQRT8=SQRT(8.)
      S=SQRT(FLOAT(N)/8.)
      THETA=6.2831853071795865D0/FLOAT(ISIGN*N)
      SPREV=-2.*SIN(THETA)
      TEMP=0.
      SINTH=SIN(THETA/2.)
      F=-4.*SINTH*SINTH
      NRCUR=8
      DATA(NHALF+1)=DATA(1)-ODDSM
      DATA(1)=DATA(1)+ODDSM
      DO 120 I=1,NHALF
```

```
        ICONJ=NHALF+2-I
        IF (I-ICONJ)60,60,130
60      SUM=(DATA(ICONJ)+DATA(I))/SQRT8
        DIF=(DATA(ICONJ)-DATA(I))/SQRT8
C       S=2.*SIN(ISIGN*2*PI*(I-1)/N)
        IF (ISYM)70,80,80
70      SUM=SUM/S
        GO TO 90
80      DIF=-DIF/S
90      DATA(I)=SUM+DIF
        DATA(ICONJ)=SUM-DIF
        IF(I-NRCUR*(I/NRCUR))110,110,100
100     S=F*TEMP+2.*TEMP-SPREV
        SPREV=TEMP
        GO TO 120
110     S=2.*SIN(THETA*FLOAT(I))
        SPREV=2.*SIN(THETA*FLOAT(I-1))
120     TEMP=S
130     RETURN
140     WRITE (6,150) N,ISIGN,ISYM
150     FORMAT (28HOERROR IN FORS1.   EITHER N =,I10,11H OR ISIGN =,I10,
       1 10H OR ISYM =,I10,12H IS ILLEGAL.)
        RETURN
        END
```

Appendix G
Subroutine CONTR for Tracing Continuous Contours

```
      SUBROUTINE CONTR(DATA,NDIM1,NDIM2,NUSD1,NUSD2,CBEG,CLIM,CSTP)
C     FIND CONTOURS OF SPECIFIED HEIGHTS THRU GIVEN TWO-DIMENSIONAL
C     ARRAY DATA.  PROGRAM IN ANSI BASIC FORTRAN BY NORMAN BRENNER, MIT,
C     BASIC ALGORISM BY STANLEY ZISK.  SEE THE LONG WRITEUP.  OCT. 1972.
C     CONTOUR HEIGHTS PLOTTED ARE CTR = CBEG, CBEG+STEP, CBEG+2*STEP,
C     ... SO LONG AS CTR DOES NOT EXCEED CLIM EITHER WHILE INCREASING OR
C     DECREASING.  STEP IS CSTP WITH THE SIGN OF CLIM-CBEG.  ONLY
C     CTR = CBEG IS PLOTTED IF CSTP = 0.
C     DATA WAS DIMENSIONED NDIM1 BY NDIM2 IN A DIMENSION STATEMENT IN
C     THE CALLING PROGRAM, BUT ONLY THE FIRST NUSD1 BY NUSD2 ARE
C     ACTUALLY USED.  (FIRST SUBSCRIPT MUST INCREASE FASTEST IN STORAGE
C     ORDER, AS USUAL).  ALL ACTUAL PLOTTING IS DONE IN THE USER-
C     SUPPLIED SUBROUTINE CPLOT, WHICH MUST BE HEADED AS FOLLOWS--
C     SUBROUTINE CPLOT(IPLOT,X1,X2,CTR)
C     CONTR EXTRUDES ONE POINT AT A TIME TO CPLOT.  (X1,X2) IS THE
C     COORDINATE PAIR, X1 FROM 1.0 TO FLOAT(NUSD1), X2 FROM
C     1.0 TO FLOAT(NUSD2).  IF DESIRED FOR EFFICIENCY, CPLOT MAY
C     SAVE UP POINTS IN A WORK ARRAY.
C     FOR IPLOT=0, CPLOT SHOULD JUST REMEMBER THIS POINT (MOVE TO
C     (X1,X2) WITH PEN UP FOR A CALCOMP PLOTTER).
C     FOR IPLOT=1, CPLOT SHOULD DRAW A LINE SEGMENT FROM THE LAST POINT
C     TO THE CURRENT POINT AND REMEMBER THE CURRENT POINT (MOVE TO
C     (X1,X2) WITH PEN DOWN).
C     FOR IPLOT=2, CPLOT SHOULD PLOT A NUMERIC LABEL GIVING THE VALUE
C     OF CTR AT THIS POINT.  EACH PHYSICALLY SEPARATE CONTOUR DRAWN
C     WILL RECEIVE ONE LABEL.
C     EACH DATUM IS INCREASED BY 1.E-20 AND ALTERED IN ITS TWO LOW ORDER
C     BITS.  NON-ZERO FLOATING POINT VALUES ARE HARDLY AFFECTED, THOUGH
C     FIXED POINT VALUES MIGHT BE.  (CONVERT TO FIXED POINT BY SAYING--
C     INTEGER DATA,CTRM,CTRZ,CTRD
C     AND CHANGING ABS(DATA...) TO IABS(DATA...).)
      DIMENSION DATA(1),NUSD(2),IX(2),IA(2),IB(2),IDELT(2),IFORB(2)
     1     ,X(2),XTENT(2),XPREV(2),XFRST(2),XSEC(2),XDIF(2)
      NUSD(1)=NUSD1
      NUSD(2)=NUSD2
      CTR=CBEG
C     ZERO THE MARK BITS OF EVERY DATUM
10    I2MAX=NDIM1*NUSD2
      DO 20 I1=1,NUSD1
      DO 20 I2=I1,I2MAX,NDIM1
C     ADDING 1.E-20 PREVENTS THE UNDERFLOW CAUSED BY BIT-MARKING ZERO
      DATA(I2)=DATA(I2)+1.E-20
20    CALL ZERLO(DATA(I2))
      CTRZ=CTR+1.E-20
      CALL ZERLO(CTRZ)
      CTRM=CTRZ
      CALL MRKLO(CTRM,1,0.)
C     EPS IS CHOSEN LARGER THAN THE EFFECT OF ALTERING THE LOW ORDER BIT
      EPS=.1/SQRT((.01*CTRM)/(CTRM-CTRZ)+1.)
C
C     SEARCH FOR STARTING POINT FOR EACH CONTOUR AT HEIGHT CTR
      DO 390 I1=1,NUSD1
      DO 390 I2=1,NUSD2
      IPHAS=1
      CTRD=CTRZ
      IDIR=1
      XPREV(2)=-1.
```

```
          X(2)=0.
          IDELT(1)=1
          IDELT(2)=1
          IB(1)=I1
          IB(2)=I2
C         CHECK FOR EITHER A ROW CROSSING OR A COLUMN CROSSING
          DO 40 IVARY=1,2
          IF(IB(IVARY)-1)40,40,30
30        IA(IVARY)=IB(IVARY)-1
          IUNVY=3-IVARY
          IA(IUNVY)=IB(IUNVY)
          GO TO 50
40        CONTINUE
          GO TO 390
C
C         INTERNAL SUBROUTINE TO DETERMINE WHETHER THE SEGMENT FROM
C         (IA(1),IA(2)) TO (IB(1),IB(2)) IS CROSSED AND HAS NOT BEEN
C         PLOTTED ALREADY.  A CROSSING IF FOUND IS AT (XTENT(1),XTENT(2)).
50        LA=IA(1)+NDIM1*(IA(2)-1)
          LB=IB(1)+NDIM1*(IB(2)-1)
          IF(DATA(LA)-CTRD)60,80,70
60        IF(DATA(LB)-CTRD)120,80,80
70        IF(DATA(LB)-CTRD)80,80,120
80        IF(ABS(DATA(LA)-DATA(LB))-(CTRM-CTRZ))120,120,90
90        XTENT(IVARY)=FLOAT(IA(IVARY))+(FLOAT(IDELT(IVARY))*
      1     (CTRD-DATA(LA)))/(DATA(LB)-DATA(LA))
          XTENT(IUNVY)=IA(IUNVY)
C         HAS THIS TENTATIVELY FOUND CROSSING BEEN PLOTTED BEFORE
          LX=IFIX(XTENT(1)+EPS)+NDIM1*IFIX(XTENT(2)+EPS-1.)
          IF(IFMLO(DATA(LX),IVARY))130,130,100
C         YES, BUT PERHAPS IT BE CAN BE PLOTTED AGAIN TO END A CONTOUR
100       IF(IPHAS-3)120,110,120
110       IF(ABS(XTENT(1)-XPREV(1))+ABS(XTENT(2)-XPREV(2))-EPS)120,120,130
120       IF(IPHAS-1)40,40,300
C
C         PLOT AND MARK THE CURRENT CROSSING OR LABEL
130       IF(XPREV(2))140,150,160
C         SAVE THE FIRST AND SECOND CROSSINGS FOR REVERSE DIRECTION SEARCH
140       XFRST(1)=XTENT(1)
          XFRST(2)=XTENT(2)
          GO TO 170
150       XSEC(1)=XTENT(1)
          XSEC(2)=XTENT(2)
160       XPREV(1)=X(1)
170       XPREV(2)=X(2)
          X(1)=XTENT(1)
          X(2)=XTENT(2)
          IF(IDIR)180,190,190
180       CALL CPLOT(IPHAS/4,XFRST(1),XFRST(2),CTR)
          IDIR=0
190       CALL CPLOT(IPHAS/2,X(1),X(2),CTR)
          GO TO (200,210,350,380,390),IPHAS
200       IPHAS=2
210       DO 220 I=1,2
          IX(I)=X(I)+EPS
          XDIF(I)=X(I)-FLOAT(IX(I))-EPS
220       IFORB(I)=(IX(I)+NUSD(I)-4)/(NUSD(I)-2)-1
          IF(IDIR)240,230,230
230       LX=IX(1)+NDIM1*(IX(2)-1)
C         BOTH DIRECTIONS ARE MARKED IF THE CONTOUR PASSES THRU THE CORNER
          CALL MRKLO(DATA(LX),IVARY,XDIF(IVARY))
C
C         LOOK IN ALL EIGHT POSSIBLE DIRECTIONS FROM THE CURRENT CROSSING
C         FOR ANOTHER CROSSING.  LOOK IN THE CURRENT DIRECTION FIRST, AT
C         ADJACENT SEGMENTS TO PREVENT FIGURE EIGHTS.
```

```
240     IVARY=3-IVARY
        IVPRV=IVARY
        DO 330 LOOK=1,8
        DO 290 IVORY=1,2
        IF(XDIF(IVORY))250,250,270
C       IGNORE SEGMENTS OUTSIDE DATA OR WHICH INCLUDE THE CURRENT CROSSING
250     IF(IDELT(IVORY)-IFORB(IVORY))260,300,260
260     IB(IVORY)=IX(IVORY)+IDELT(IVORY)
        GO TO 290
270     IF(IDELT(IVORY)+1-IVARY+IVORY)280,300,280
280     IB(IVORY)=IX(IVORY)+(IDELT(IVORY)+1)/2
290     IA(IVORY)=IB(IVORY)
        IA(IVARY)=IA(IVARY)-IDELT(IVARY)
        IUNVY=3-IVARY
        GO TO 50
300     IDELT(IVPRV)=-IDELT(IVPRV)
        GO TO (330,310,330,320,330,310,330,320),LOOK
310     IVARY=3-IVARY
        GO TO 330
320     IUNVP=3-IVPRV
        IDELT(IUNVP)=-IDELT(IUNVP)
330     CONTINUE
C
C       END THIS CONTOUR SEGMENT IF ALL DIRECTIONS FROM HERE HAVE BEEN
C       SEARCHED.  AT MOST ONE CONNECTING POINT OR A LABEL IS NOW PLOTTED.
        IF(IPHAS-2)340,340,350
C       TRY CLOSING THE CONTOUR TO A NEARBY (EXCEPT THE PREVIOUS) CROSSING
340     IPHAS=3
        CTRD=CTRM
        GO TO 240
C       WRITE A CONTOUR LABEL AT THE END OF THIS CONTOUR
350     IPHAS=4
        IF(IDIR)390,390,360
C       DOES THE CONTOUR CLOSE ON ITSELF (E.G., A ONE-POINT CONTOUR)
360     IF(ABS(X(1)-XFRST(1))+ABS(X(2)-XFRST(2))-EPS)370,370,190
370     IPHAS=5
        IF(XPREV(2))180,180,190
C       IF THE CONTOUR HAS NOT CLOSED ON ITSELF, RETURN TO THE STARTING
C       POINT AND SEARCH FOR CONTINUATION IN ANOTHER DIRECTION
380     IPHAS=2
        CTRD=CTRZ
        IDIR=-1
        X(1)=XFRST(1)
        X(2)=XFRST(2)
        XPREV(1)=XSEC(1)
        XPREV(2)=XSEC(2)
        GO TO 210
390     CONTINUE
C
        IF(CSTP)400,410,400
400     CTR=CTR+SIGN(CSTP,CLIM-CBEG)
        IF((CTR-CLIM)*SIGN(1.,CLIM-CBEG)-.001*ABS(CSTP))10,10,410
C       IN CASE SUBROUTINE CPLOT IS SAVING UP POINTS, CLEAR IT OUT
410     CALL CPLOT(0,0.,0.,0.)
        RETURN
        END
        SUBROUTINE ZERLO(IDATA)
C       SET THE TWO LOW ORDER BITS OF IDATA TO ZERO.
C       IBM 360 REAL*4 OR INTEGER*4.
        IF(IDATA)10,20,30
10      IDATA=-4*((3-IDATA)/4)
20      RETURN
30      IDATA=4*(IDATA/4)
        RETURN
        END
```

```
      FUNCTION IFMLO(IDATA,IVARY)
C     RETURN 0 IF BIT IVARY (= 1 OR 2) OF IDATA IS 0, 1 IF NOT.
C     IBM 360 REAL*4 OR INTEGER*4.
      IF(IDATA)10,20,20
10    IFMLO=MOD((IVARY-1-IDATA)/IVARY,2)
      RETURN
20    IFMLO=MOD(IDATA/IVARY,2)
      RETURN
      END
      SUBROUTINE MRKLO(IDATA,IVARY,DIF)
C     SET BIT IVARY (= 1 OR 2) OF IDATA TO 1.  (TO BE CALLED ONLY ONCE
C     PER POINT).  SET BOTH LOW ORDER BITS TO 1 IF DIF = 0.
C     IBM 360 REAL*4 OR INTEGER*4.
      IDATA=IDATA+IVARY
      IF(DIF)10,10,20
10    IDATA=IDATA+3-IVARY
20    RETURN
      END
```

Appendix H

Subroutine CONTR for Tracing Contours in Segments

```
      SUBROUTINE CONTR(DATA,NDIM1,NDIM2,NUSD1,NUSD2,CBEG,CLIM,CSTP)
C     FIND CONTOURS OF SPECIFIED HEIGHTS THRU GIVEN TWO-DIMENSIONAL
C     ARRAY DATA.  ALL PARAMETERS HAVE THE SAME MEANING AS IN THE
C     LONGER SUBROUTINE CONTR.  THIS SUBROUTINE SHOULD NOT BE USED WITH
C     A CALCOMP OR MECHANICAL PEN PLOTTER, AS THE CONTOUR SEGMENTS ARE
C     PLOTTED IN A NEAR-RANDOM ORDER.  ONLY A CATHODE RAY TUBE OR OTHER
C     RANDOM ACCESS TYPE PLOTTER SHOULD BE USED.
      DIMENSION DATA(1),X(2),Y(2),IDELT(6),DVAL(5)
      DATA IDELT/0,-1,-1,0,0,-1/
      NCTR=1
      IF(CSTP)10,20,10
10    NCTR=ABS((CLIM-CBEG)/CSTP)+1.001
20    CTR=CBEG
      DO 140 ICTR=1,NCTR
      LABEL=0
      DO 130 I=2,NUSD1
      DO 130 J=2,NUSD2
      NPT=0
      IJ=I-1+NDIM1*(J-2)
      IJ1=IJ+NDIM1
      DVAL(1)=DATA(IJ1)
      DVAL(2)=DATA(IJ)
      DVAL(3)=DATA(IJ+1)
      DVAL(4)=DATA(IJ1+1)
      DVAL(5)=DVAL(1)
      DO 120 ICORN=1,4
      IF(DVAL(ICORN)-CTR)30,40,40
30    IF(DVAL(ICORN+1)-CTR)120,50,50
40    IF(DVAL(ICORN+1)-CTR)50,120,120
50    NPT=NPT+1
      DELTA=(CTR-DVAL(ICORN))/(DVAL(ICORN+1)-DVAL(ICORN))
      GO TO (60,70,60,70),ICORN
60    X(NPT)=I+IDELT(ICORN+1)
      Y(NPT)=FLOAT(J+IDELT(ICORN))+DELTA*FLOAT(IDELT(ICORN+1)-
     *  IDELT(ICORN))
      GO TO 80
```

```
70    X(NPT)=FLOAT(I+IDELT(ICORN+1))+DELTA*FLOAT(IDELT(ICORN+2)-
     *  IDELT(ICORN+1))
      Y(NPT)=J+IDELT(ICORN)
80    IF(NPT-2)120,90,90
90    CALL CPLOT(0,X(1),Y(1),CTR)
      CALL CPLOT(1,X(2),Y(2),CTR)
      IF(LABEL)100,100,110
100   CALL CPLOT(2,X(1),Y(1),CTR)
      LABEL=1
110   NPT=0
120   CONTINUE
130   CONTINUE
140   CTR=CTR+SIGN(CSTP,CLIM-CBEG)
      RETURN
      END
```

Appendix I

Subroutine RULED, for Tracing a Ruled–Surface Map

```
      SUBROUTINE RULED(DATA,NROW,NCOL,NRUSD,NCUSD,VMAX,AZ,EL,NLINE,
     1 RCRAT,HT)
C     TO CONVERT THIS SUBROUTINE TO FIXED POINT, MERELY DECLARE DATA
C     TO BE INTEGER.
C
C     DRAW A RULED SURFACE MAP OF THREE DIMENSIONAL DATA.
C     EFFECTIVELY, TAKE A MATRIX (DATA) OF NROW ROWS AND NCOL COLUMNS
C     (AS DIMENSIONED) OF WHICH ONLY THE FIRST NRUSD ROWS AND THE
C     FIRST NCUSD COLUMNS ARE USED, AND CONSTRUCT A WOODEN MODEL OF IT.
C     THE HEIGHT AT ANY POINT IS GIVEN BY THE THE VALUE DATA(I,J) FOR
C     I THE ROW (FROM 1 TO NRUSD) AND J THE COLUMN (FROM 1 TO NCUSD).
C     THE DIMENSIONS OF THE MODEL ARE 1. (COLUMNS) BY RCRAT (ROWS)
C     BY HT (VALUE), AS SEEN FROM THE SOUTH, OR USUAL, VIEW
C     (AZ = 0).  THE WOODEN MODEL IS PLACED ON THE X-Y PLANE
C     ITS EDGES PARALLEL TO THE X-Y AXES, TURNED TO ONE OF FOUR AZI-
C     MUTHAL POSITIONS (AZ = 0, 90, 180 OR 270, FOR FRONT EDGE FACING
C     SOUTH, WEST, NORTH OR EAST).  THE LOWER LEFT
C     CORNER OF THE MODEL IS PLACED AT COORDINATES (0.,0.).
C     AN OBSERVER'S EYE IS PLACED DUE SOUTH ABOVE THE NEGATIVE Y-AXIS
C     IN THE Y-Z PLANE AT INFINITE DISTANCE AND ELEVATION EL FROM THE
C     ORIGIN OF THE X-Y-Z SYSTEM.  A SCREEN IS PLACED WITH ITS BOTTOM
C     EDGE PARALLEL TO THE X-AXIS AT A DISTANCE FROM THE ORI-
C     GIN.  IT IS TILTED SO AS TO BE NORMAL TO THE RAY CONNECTING THE
C     EYE TO THE ORIGIN.    THE PROJECTION OF THE WOODEN MODEL AS SEEN
C     ON THE SCREEN BY THE EYE IS WHAT IS DISPLAYED ON THE PLOTTING
C     DEVICE.   THERE IS NO PERSPECTIVE.  VMAX = MAX(DATA(I,J)).
C
C     THE MODEL IS CONSTRUCTED AS A SERIES OF CUTS PARALLEL TO THE EDGE
C     CLOSEST TO THE EYE.   AS SEEN ON THE SCREEN THESE CUTS OCCLUDE
C     FARTHER ONES, SO THAT A NEAR PEAK WILL BLOCK FARTHER PEAKS.
C     NLINE CUTS ARE MADE THRU THE DATA.  OBVIOUSLY, MORE CUTS GIVE
C     FINER DETAIL.  LINEAR INTERPOLATION IS PERFORMED IF A CUT DOES
C     NOT COINCIDE WITH A ROW OR COLUMN.  EACH CUT IS SET UPRIGHT
C     (I.E., NORMAL TO THE X-Y PLANE).  WHEN EL IS 90 DEGREES AND THE
C     EYE IS VERY FAR AWAY, THEREFORE, ALL THAT WILL BE SEEN ARE A SET
C     OF PARALLEL LINES, SINCE ONE IS LOOKING DOWN FROM ABOVE.  USUAL
C     VIEWING PARAMETERS ARE FOR EL ABOUT 45 DEGREES, AZ 0,90,180 OR 270
C     NOTE THAT THE HIGHEST VALUE IN THE DATA MATRIX IS AUTOMATICALLY
C     SET TO HT, SO THAT INCREASING HT WILL INCREASE THE APPARENT
C     HEIGHTS OF ALL PEAKS.
C
C     SUBROUTINE RULED IS INDEPENDENT OF ANY PARTICULAR PLOTTING
C     DEVICE.   THE USER SUPPLIES TWO SUBROUTINES TO DO THE ACTUAL
C     PLOTTING.   RLIMS IS CALLED BY RULED AS FOLLOWS--
C     CALL RLIMS(IAZ,XMIN,XMAX,YMIN,YMAX)      (IAZ = 0, 1, 2 OR 3)
C     WHERE IAZ = AZ/90.  RLIMS IS TO SET UP THE PLOTTING DEVICE
C     SO THAT THE EDGES OF A PLOTTING SQUARE ARE XMIN, XMAX, ETC.
C
C     SUBROUTINE RPLOT IS CALLED BY RULED AS FOLLOWS--
C     CALL RPLOT(IPLOT,X,Y)
C     RPLOT IS TO DRAW A LINE SEGMENT FROM THE LAST PLOTTED (X,Y)
C     POINT TO THIS NEW ONE (IF IPLOT = 1) OR JUST TO REMEMBER THAT
C     THIS (X,Y) POINT IS THE START OF A NEW CURVE AND PLOT NOTHING
C     (FOR IPLOT = 0).
C
C     IF EITHER NRUSD OR NCUSD IS LARGER THAN MAXW, THE LENGTH OF ARRAY
C     HIGH (CURRENTLY 300), THE PROGRAM CANNOT RUN.
```

```
C
      DIMENSION DATA(1), HIGH(300)
      MAXW=300
C     THE LENGTH OF ARRAY HIGH
C
C     FOR THE APPROPRIATE ORIENTATION, FIND WIDTH, DEPTH AND ANCILLARY
C     CONSTANTS. (WIDTH AND DEPTH ARE 1. AND RCRAT APPROPRIATELY
C     ORIENTED.)
      IAZ=(AZ+765.)/90.
      IAZ=IAZ-4*(IAZ/4)
      IGO=1+IAZ
      GO TO (10,20,30,40), IGO
10    ICWID=NROW
      ICDEP=1
      ICON=-NROW
      GO TO 50
20    ICWID=1
      ICDEP=-NROW
      ICON=NROW*NCUSD
      GO TO 60
30    ICWID=-NROW
      ICDEP=-1
      ICON=NRUSD-1+NROW*NCUSD
      GO TO 50
40    ICWID=-1
      ICDEP=NROW
      ICON=NRUSD-1-NROW
      GO TO 60
50    WIDTH=1.
      DEPTH=RCRAT
      NWIDE=NCUSD
      NDEEP=NRUSD
      GO TO 70
60    WIDTH=RCRAT
      DEPTH=1.
      NWIDE=NRUSD
      NDEEP=NCUSD
70    THETA=EL*3.1415926535/180.
      SINEL=SIN(THETA)
      COSEL=COS(THETA)
      DEPTH=(DEPTH/WIDTH)*FLOAT(NWIDE-1)
      WIDTH=NWIDE-1
      IF (NWIDE-MAXW)80,80,190
80    DIMAX=AMAX1(WIDTH,DEPTH)
      XPROJ=1.+WIDTH/2.
      YPROJ=0.
C     DISPLAY ON THE PLOTTING DEVICE THAT PART OF THE PROJECTION SCREEN
C     ONTO WHICH THE WOODEN MODEL IS PROJECTED.  SPECIFICALLY THE
C     BOTTOM CENTER POINT OF THE PLOTTING AREA IS THE PROJECTION
C     OF THE FRONT CENTER POINT OF THE MODEL, AND THE PLOTTING AREA IS
C     SQUARE, OF SIDE MAX(WIDTH,DEPTH).
C     NOTE THAT THE PLOTTING COORDINATES ARE ESSENTIALLY ARBITRARY.
      CALL RLIMS (IAZ,XPROJ-DIMAX/2.,XPROJ+DIMAX/2.,YPROJ,YPROJ+DIMAX)
C     FAR PEAKS ARE BLOCKED BY REMEMBERING WHAT THE HIGHEST POINT
C     ACTUALLY PLOTTED AT THAT WIDTH COORDINATE IS.  INITIALLY, THIS IS
C     THE LOWER EDGE OF THE PLOTTING AREA.
      DO 90 IWIDE=1,NWIDE
90    HIGH(IWIDE)=YPROJ
C     MAKE NLINE CUTS THRU THE DATA, EQUISPACED FROM THE NEAREST ROW
C     TO THE FARTHEST.  IF A CUT IS BETWEEN TWO ROWS (OR COLUMNS) DO
C     LINEAR INTERPOLATION.   (THIS LOOP ITERATES DEPTHWISE).
      DO 180 ILINE=1,NLINE
      IPAR=ILINE-2*(ILINE/2)
C     IPAR=0 IF ILINE IS EVEN, 1 IF ODD.
```

```
      YDEEP=FLOAT((ILINE-1)*(NDEEP-1))/FLOAT(NLINE-1)+1
      Y=DEPTH*(YDEEP-1.)/FLOAT(NDEEP-1)
      IDEEP=YDEEP
      FRAC=YDEEP-FLOAT(IDEEP)
      JWIDE=1
C     THE FIRST POINT OF THE LINE
      IWIDE=JWIDE*(1-IPAR)+(NWIDE+1-JWIDE)*IPAR
C     ITERATE FROM LEFT TO RIGHT ON EVEN-NUMBERED LINES, AND RIGHT
C     TO LEFT ON ODD-NUMBERED LINES.  THIS WILL REDUCE REWINDING OF
C     THE PAPER ON A CALCOMP.  CRT PLOTS WILL BE UNAFFECTED.
      I1=ICWID*IWIDE+ICDEP*IDEEP+ICON
      I2=I1+ICDEP
      VDIF=DATA(I2)-DATA(I1)
      VALUE=DATA(I1)
      VALUE=VALUE+FRAC*VDIF
      Z=HT*VALUE/VMAX
      XPROJ=IWIDE
      YPROJ=Y*SINEL+Z*COSEL
C     IS THE FIRST POINT OF THE CUT HIDDEN BEHIND A PREVIOUS PEAK Q.
C     IHOLD=1 IF YES, 0 IF NO
      IHOLD=1
      IF (YPROJ-HIGH(IWIDE))110,100,100
C     IF NOT, REMEMBER THIS POINT AS THE START OF A CUT TO BE PLOTTED.
100   IHOLD=0
      OLDH=HIGH(IWIDE)
      HIGH(IWIDE)=YPROJ
      YPOLD=YPROJ
      CALL RPLOT (0,XPROJ,YPROJ)
C     ITERATE ACROSS THE CUT WIDTHWISE, SUPPRESSING PLOTTING WHEN
C     SEGMENTS ARE HIDDEN BEHIND PREVIOUS PEAKS.
110   DO 180 JWIDE=2,NWIDE
      IWIDE=JWIDE*(1-IPAR)+(NWIDE+1-JWIDE)*IPAR
      I1=ICWID*IWIDE+ICDEP*IDEEP+ICON
      I2=I1+ICDEP
      VDIF=DATA(I2)-DATA(I1)
      VALUE=DATA(I1)
      VALUE=VALUE+FRAC*VDIF
      Z=HT*VALUE/VMAX
      XPROJ=IWIDE
      YPROJ=Y*SINEL+Z*COSEL
C     IS THIS POINT HIDDEN BEHIND A PREVIOUS PEAK Q.
      IHNEW=1
      IF (YPROJ-HIGH(IWIDE))130,120,120
120   IHNEW=0
      OLDH=HIGH(IWIDE)
      HIGH(IWIDE)=YPROJ
C     SEPARATE THE FOUR CASES.
130   IGO=1+IHOLD+2*IHNEW
      GO TO (140,150,160,170), IGO
C     NON-HIDDEN POINT TO NON-HIDDEN POINT.  ADD THIS SEGMENT TO THE
C     CURVE BEING PLOTTED.
140   CALL RPLOT (1,XPROJ,YPROJ)
      GO TO 170
C     HIDDEN POINT TO NON-HIDDEN POINT.  START A CURVE HERE, USING AS
C     THE FIRST POINT THE INTERSECTION OF THE PARTIALLY HIDDEN SEGMENT
C     WITH THE SEGMENT PREVIOUSLY PLOTTED THAT PARTIALLY HIDES IT.
150   IWPRV=IWIDE+2*IPAR-1
C     THE PREVIOUS POINT IN THE CORRECT DIRECTION
      HIGH1=HIGH(IWPRV)
      HIGH2=OLDH
      DENOM=HIGH1+YPROJ-HIGH2-YPOLD
      XPLOT=XPROJ+FLOAT(2*IPAR-1)*(YPROJ-HIGH2)/DENOM
      YPLOT=(HIGH1*YPROJ-HIGH2*YPOLD)/DENOM
      CALL RPLOT (0,XPLOT,YPLOT)
```

```
      GO TO 170
C     NON-HIDDEN POINT TO HIDDEN POINT.  END A CURVE HERE.
160   HIGH1=OLDH
      HIGH2=HIGH(IWIDE)
      DENOM=HIGH1+YPROJ-HIGH2-YPOLD
      XPLOT=XPROJ+FLOAT(2*IPAR-1)*(YPROJ-HIGH2)/DENOM
      YPLOT=(HIGH1*YPROJ-HIGH2*YPOLD)/DENOM
      CALL RPLOT (1,XPLOT,YPLOT)
C     HIDDEN POINT TO HIDDEN POINT.  IGNORE BOTH POINTS.
C     SAVE THE OLD POINT FOR POSSIBLE USE IF THE NEXT POINT ISNT HIDDEN.
170   IHOLD=IHNEW
180   YPOLD=YPROJ
      CALL RPLOT (0,0.,0.)
      RETURN
190   WRITE (6,200) AZ,NWIDE,MAXW
200   FORMAT (39HOERROR IN RULED SURFACE MAP.  FOR AZ = ,F5.1,13H, THE M
     1AP IS ,I4,37H POINTS WIDE, WIDER THAN THE MAXIMUM ,I4,1H.)
      RETURN
      END
```

Appendix J

Subroutine GRAYM for Generating a Gray-Scale Map on a Line Printer

```
      SUBROUTINE GRAYM(DATA,NDIM1,NDIM2,NUSD1,NUSD2)
C     PRODUCE A GRAY SCALE MAP OF ARRAY DATA ON THE HIGH SPEED PRINTER.
C     THE ARRAY IS DISPLAYED USING A 21-LEVEL GRAY SCALE.  THE FIRST
C     SUBSCRIPT GOES ALONG PAGE ROWS, THE SECOND SUBSCRIPT DOWN COLUMNS.
C     THE GRAY LEVEL CHARACTER SET WAS DESIGNED USING A PHOTOMETER.
C     N1 MUST BE LESS THAN OR EQUAL TO 132.  ALSO, SINCE PRINTER CHAR-
C     ACTERS ARE 5/3 AS HIGH AS THEY ARE BROAD, THERE WILL BE
C     VERTICAL DISTENSION OF THE IMAGE.
C     NORMAN BRENNER, MIT AND IBM RESEARCH, JULY 1969.
C     UNDER OS/ASP OPERATING SYSTEM, INSERT THE FOLLOWING CARD AFTER THE
C     JOB CARD TO PREVENT LINE SKIP AT PAGE BOTTOM:
C     /*FORMAT PR,DDNAME=FT06F001,OVFL=OFF
      DIMENSION DATA(NDIM1,NDIM2)
C     DATA IS DIMENSIONED NDIM1 BY NDIM2, BUT WE USE ONLY NUSD1 BY NUSD2
      DIMENSION LINE(132),NCHAR(26),CHAR(8,26),ICHAR(132)
      INTEGER CHAR
      DATA NCHAR/13*1,3*2,3,2*4,3*5,2*6,7,8/
      DATA CHAR                    /1H ,1H ,1H ,1H ,1H ,1H ,1H ,1H ,
     1 1H ,1H ,1H ,1H ,1H ,1H ,1H ,1H , 1H ,1H ,1H ,1H ,1H ,1H ,1H ,
     2 1H-,1H ,1H ,1H ,1H ,1H ,1H , 1H-,1H ,1H ,1H ,1H ,1H ,1H ,
     3 1H=,1H ,1H ,1H ,1H ,1H ,1H , 1H+,1H ,1H ,1H ,1H ,1H ,1H ,
     4 1H),1H ,1H ,1H ,1H ,1H ,1H , 1H1,1H ,1H ,1H ,1H ,1H ,1H ,
     5 1HZ,1H ,1H ,1H ,1H ,1H ,1H , 1HX,1H ,1H ,1H ,1H ,1H ,1H ,
     6 1HA,1H ,1H ,1H ,1H ,1H ,1H , 1HM,1H ,1H ,1H ,1H ,1H ,1H ,
     7 1HO,1H-,1H ,1H ,1H ,1H ,1H , 1HO,1H=,1H ,1H ,1H ,1H ,1H ,
     8 1HO,1H+,1H ,1H ,1H ,1H ,1H , 1HO,1H+,1H',1H ,1H ,1H ,1H ,
     9 1HO,1H+,1H',1H.,1H ,1H ,1H , 1HO,1H+,1H',1H.,1H ,1H ,1H ,
     A 1HO,1H+,1H',1H.,1H=,1H ,1H , 1HO,1H+,1H',1H.,1H=,1H ,1H ,
     B 1HO,1HX,1H',1H.,1H-,1H ,1H , 1HO,1HX,1H',1H.,1HH,1HC,1H ,
     C 1HO,1HX,1H',1H.,1HH,1HB,1H , 1HO,1HX,1H',1H.,1HH,1HB,1HV,1H ,
     D 1HO,1HX,1H',1H.,1HH,1HB,1HV,1HA/
C     USE MCLEOD'S LINEAR GRAY SCALE CHARACTER SET (IEEE COMP. TRANS.,
C     FEB 1970).  THE TRUE DENSITIES OF THE 26 CHARACTERS ARE:
C     .00 .00 .00 .15 .15 .22 .25 .29 .33 .37 .40 .42 .45 .53 .56 .60
C     .64 .67 .79 .79 .85 .89 .93 .97 1.00
      FUN(X)=SIGN(ALOG(AMAX1(ABS(X),1.)),X)
C     OR ANY OTHER FUNCTION FOR MAPPING THE FUNCTION VALUES
      IF(NUSD1.GT.132)WRITE(6,10)NUSD1
   10 FORMAT(28H0WARNING.  IN GRAYM, NUSD1 =,I10,25H IS TAKEN TO BE ONLY
     1 132.)
      IMAX=MIN0(NUSD1,132)
      DMAX=-1.E25
      DMIN=-DMAX
      DO 20 J=1,NUSD2
      DO 20 I=1,IMAX
      D=FUN(DATA(I,J))
      DMAX=AMAX1(D,DMAX)
   20 DMIN=AMIN1(D,DMIN)
C     ZMIN IS SHOWN AS WHITE, ZMAX AS BLACK, PROPORTIONAL IN BETWEEN.
      WRITE (6,30) DMIN,DMAX
   30 FORMAT(19H1GRAY MAP.  WHITE =,E15.6,9H, BLACK =,E15.6)
      NGRAY=26
      ENGRY=FLOAT(NGRAY)-.01
      JMAX=NUSD2+1
      DO 110 J=1,JMAX
      IF(J.EQ.JMAX)GO TO 50
      MAXCHR=0
```

```
        DO 40 I=1,IMAX
        D=FUN(DATA(I,J))
        ICHAR(I)=1+IFIX(ENGRY*(D-DMIN)/(DMAX-DMIN))
40      MAXCHR=MAX0(MAXCHR,ICHAR(I))
        NLINE=NCHAR(MAXCHR)
        GO TO 80
C       PRINT A SAMPLE BAR
50      NLINE=NCHAR(NGRAY)
        IMAX=3*NGRAY
        DO 60 I=1,IMAX
60      ICHAR(I)=(I+2)/3
        WRITE(6,70)(I,I=1,NGRAY)
70      FORMAT(22HOSAMPLE OF GRAY SHADES/26I3)
C       GO TO THE NEXT LINE
80      WRITE(6,90)
90      FORMAT(1X)
C       OVERPRINT UP TO EIGHT LINES.  '+' IS THE OVERPRINT CHARACTER.
        DO 110 L=1,NLINE
        DO 100 I=1,IMAX
        IC=ICHAR(I)
100     LINE(I)=CHAR(L,IC)
110     WRITE(6,120)(LINE(I),I=1,IMAX)
120     FORMAT(1H+,132A1)
        RETURN
        END
```

Appendix K

Subroutine GRAYMP for Generating a Gray–Scale Map on a Cathode Ray Tube

```
        SUBROUTINE GRAYMP(IDATA,NRDIM,NCDIM,NRUSED,NCUSED,IX0,IY0,IDX,IDY,
     *  NHORIZ,NVERT,ITRANS,NGRAY,MINVAL,MAXVAL)
C       PLOT A GRAY-SCALE MAP ON A CATHODE RAY TUBE OR ASSOCIATED MICRO-
C       FILM RECORDER.  THE USER MUST SUPPLY SUBROUTINE DPLOT(IX,IY),
C       WHOSE PURPOSE IS PLOT A SINGLE DOT OF MINIMUM BRIGHTNESS AT CRT
C       COORDINATES (IX,IY).  IT IS STRONGLY SUGGESTED THAT FOR SPEED,
C       SUBROUTINE DPLOT SAVE UP SEVERAL HUNDRED SUCH COMMANDS AND
C       PLOT THEM ALL AT ONCE.  FOR THIS REASON, IT MIGHT BE NECESSARY
C       TO WRITE SUBROUTINE DPLOT IN MACHINE LANGUAGE.  NOTE THAT DPLOT
C       IS CALLED ONCE AT THE END WITH PARAMETERS (-1,-1) AS A SIGNAL TO
C       END PLOTTING.
        DIMENSION IDATA(1)
        NRDIF=NRDIM-NRUSED
        ISCALE=NVERT*NHORIZ
        IDIF=MAXVAL-MINVAL
        MAXSHA=IDIF*ISCALE
        IDENOM=MAXSHA/NGRAY
        IX=IX0
        DO 100 J=2,NCUSED
        IY=IY0
        DO 90 I=2,NRUSED
        IJ=I+NRDIM*(J-1)
        IJM1=IJ-NRDIM
        ISTART=ISCALE*(IDATA(IJM1-1)-MINVAL)
        ISTRST=NHORIZ*(IDATA(IJM1)-IDATA(IJM1-1))
        ISTEP=NVERT*(IDATA(IJ-1)-IDATA(IJM1-1))
        ISTPST=IDATA(IJ)-IDATA(IJM1)-IDATA(IJ-1)+IDATA(IJM1-1)
        IYY=IY
        DO 80 II=1,NVERT
```

```
        IVAL=ISTART
        IXX=IX
        DO 70 JJ=1,NHORIZ
        ISHADE=MAX0(0,MINO(MAXSHA,IVAL))/IDENOM
        IF(ISHADE)60,60,10
10      IF(ITRANS)40,20,40
20      DO 30 ITIME=1,ISHADE
30      CALL DPLOT(IXX,IYY)
        GO TO 60
40      DO 50 ITIME=1,ISHADE
50      CALL DPLOT(IYY,IXX)
60      IXX=IXX+IDX
70      IVAL=IVAL+ISTEP
        IYY=IYY+IDY
        ISTART=ISTART+ISTRST
80      ISTEP=ISTEP+ISTPST
90      IY=IYY
100     IX=IXX
        CALL DPLOT(-1,-1)
        RETURN
        END
```

ACKNOWLEDGMENT

The Editor wishes to thank Robert N. Davis of the Lincoln Laboratory, Massachusetts Institute of Technology, for his help in obtaining the program listings reproduced in these Appendixes. The listings were printed automatically from decks of punched cards provided by the authors of Part 6.

INDEX FOR VOLUME I2, PART C

INDEX FOR VOLUME 12, PART B